Exploring Science

Junior Certificate Science

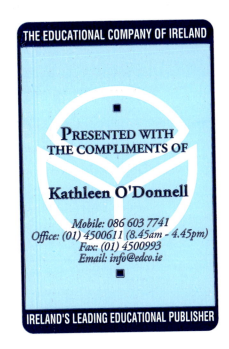
Michael O'Callaghan
Seamus Reilly
Aidan Seery

THE EDUCATIONAL COMPANY

First published 2007
The Educational Company of Ireland
Ballymount Road
Walkinstown
Dublin 12

A member of the Smurfit Kappa Group

© *Michael O'Callaghan, Seamus Reilly, Aidan Seery*

The paper used in this book comes from Managed Forests in Northern Europe For every tree felled, at least one new tree is planted

Editor: Kate Duffy
Book design: Brendan O'Connell
Typesetting and layout: Brendan O'Connell
Picture research: Kate Duffy
Artwork: Daghda
Cartoons: Maria Murray
Cover design: Design Image

Printed by: ColourBooks

Acknowledgements
Acknowledgement is made to the following for supplying photographs and for permission to reproduce copyright photographs: Alamy, Corbis, Getty Images, Irish Image Collection, Science Photo Library, Woodfall Wild Images

CONTENTS

PHYSICS

INTRODUCTION

This textbook is written for the new Junior Certificate Science course. It has been revised in line with user feedback, the most recent examination papers and in accordance with clarifications provided at in-service courses.

The emphasis that the course places on practical work is reflected in the book. All the mandatory activities are included in the text (highlighted in purple–coloured panels). These activities must be carried out and written up by the students to qualify for the marks (10 per cent) for coursework A. Guidelines as to how this work should be presented are included on this page. In order to avoid confusion, the inclusive term 'activity' is used in this book to refer to both experiments and investigations.

The other practical work presented in the book (in yellow-coloured panels) does not have to be written up by the students. However, it may be examined in the written paper.

The book is designed for use by both Ordinary and Higher Level students. The Higher Level material is indicated by a purple bar on the side of the page.

The sequence of the text follows the pattern of the examination papers, i.e. biology, chemistry and physics. The sequence of the chapters generally follows the layout of the syllabus. The material and the depth of treatment is very tightly aligned with both the syllabus and the recent exam papers.

Many definitions are included in coloured panels. These definitions are also included in a substantial glossary at the end of the book.

Each chapter contains a large number of clearly drawn and fully labelled diagrams. A comprehensive range of photographs is included. It is hoped that the high quality of all this artwork will complement and enhance the written text and enthuse the students in their study of the course.

Each chapter is followed by an extensive summary in key points. These key points are ideal for last minute exam revision.

A large number of questions is included at the end of each chapter. These questions mirror the sequence of topics in the chapter and are graded so that the more difficult questions are at the end. The questions include samples of the type that are specified for coursework B. These questions are experiment-based and encourage the students to plan and analyse how practical problems should be attempted. Questions from past examination papers are included at the end of any relevant chapters.

While it is not compulsory, teachers are encouraged to use data logging when and where it is appropriate.

It is not realistic to include details of how data logging could be used due to the wide disparity in the amount and nature of the equipment available in different classrooms. Those activities for which data logging is suitable are indicated in the list of mandatory activities on page vi.

General guidelines for completion of the student reports

Students must record details of 30 activities. These are referred to in the text as mandatory activities and are listed on pages vi–vii. The following represents a suitable format in which to present your work.

1 **Introduction**: Include the date or dates of the mandatory activity. Listing the tasks ensures the student thinks about and breaks down the investigation into simpler steps.

2 **Preparation and planning**: The student should list all the apparatus (to include glassware, measuring instruments, chemicals, materials and other equipment) needed for the activity.

3 **Procedure**: The student should explain the methods used in the activity.

   Diagrams: Diagrams should be drawn in pencil, as large as possible, and should be fully labelled.

   Graphs: Where necessary, graphs should be drawn in pencil. The axes should be labelled, the units used should be named and the points should be clearly visible.

   Risks and safety precautions: The dangers and the steps taken to reduce these dangers should be listed.

   Results: The results, measurements or observations from the activity should be clearly stated.

4 **Conclusion and evaluation of results**: The overall conclusion is given and an attempt should be made to explain if the result makes sense.

5 **Comments**: The student should explain any difficulties they found, any steps that might have given rise to errors, any ways of improving the activity or how it relates to everyday examples.

6 **Name and date of completion**: By signing the report the students have their own personal records of the activity for the assessment of marks in third year.

MANDATORY ACTIVITIES

Indicates mandatory activities suitable for data logging

PHYSICS

Laboratory Equipment

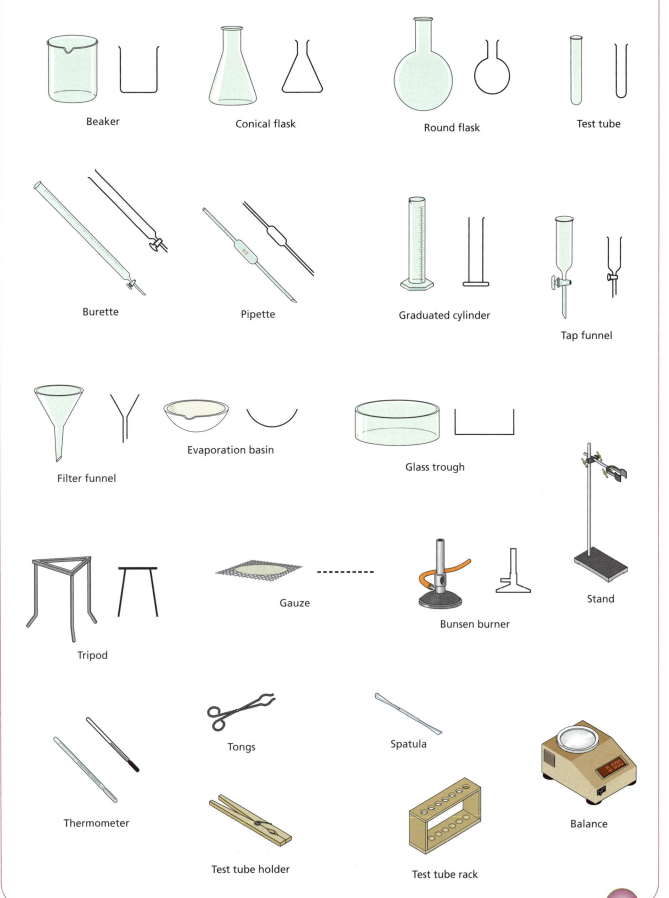

Beaker

Conical flask

Round flask

Test tube

Burette

Pipette

Graduated cylinder

Tap funnel

Filter funnel

Evaporation basin

Glass trough

Tripod

Gauze

Bunsen burner

Stand

Thermometer

Tongs

Spatula

Balance

Test tube holder

Test tube rack

Biology

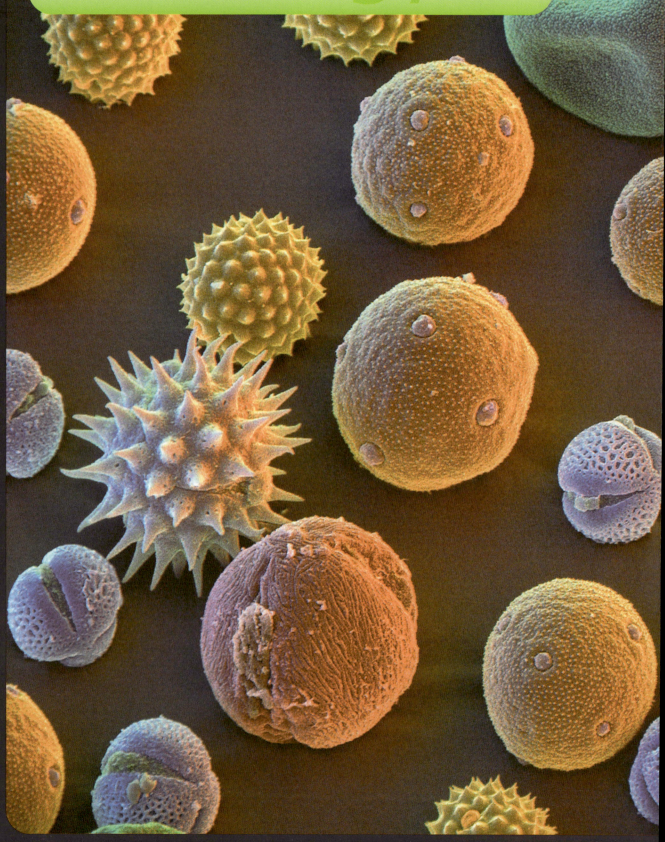

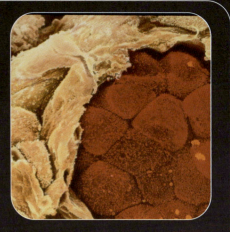

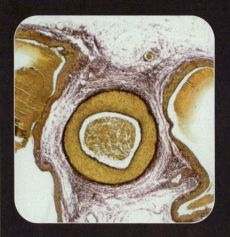

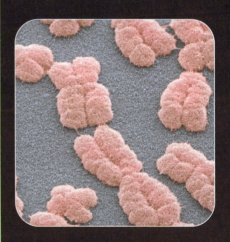

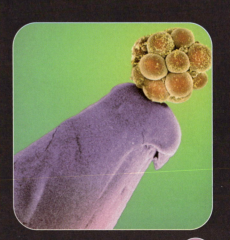

Chapter 1

LIVING THINGS

Characteristics of living things

Biology is the study of living things. Living things are also called organisms.

Trying to tell the difference between living and non-living things can sometimes be a difficult task. There is no simple definition of life.

In general, living things are said to have seven characteristics or features. It is not enough to have **some** of these characteristics. Living things must have **all** of the following seven characteristics:

1 nutrition
2 respiration
3 excretion
4 growth
5 reproduction
6 movement
7 response.

To help remember these characteristics use the first letter of each word in the phrase: **N**ature **R**arely **E**njoys **G**iving **R**ubbish **M**uch **R**espect.

A greenfly feeding on a leaf

Nutrition

Nutrition is the way in which an organism gets its food.

Animals take in food by eating plants or other animals. Plants make their own food in photosynthesis.

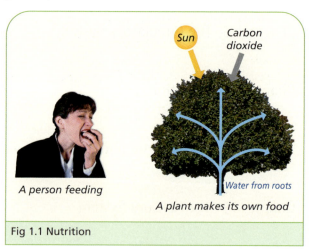

Sun

Carbon dioxide

Water from roots

A person feeding

A plant makes its own food

Fig 1.1 Nutrition

Respiration

Respiration is the way in which living things break down glucose to release energy. This happens in all living cells to supply them with energy.

If oxygen is used in this process it is called **aerobic respiration**.

The equation for aerobic respiration is:

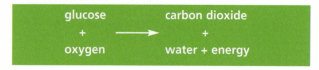

glucose	carbon dioxide
+	+
oxygen →	water + energy

Excretion

Excretion is the removal of the waste products of chemical reactions from the body.

Animals excrete wastes such as carbon dioxide, salts and water. Plants do not excrete as much as animals. However, carbon dioxide passes out from leaves (especially at night).

Fig 1.2 Excretion

Fig 1.4 Reproduction

Growth

All living things grow. In some cases this involves the cell or cells increasing in size. More often growth happens when cells divide to form extra new cells.

For example, humans grow from a single cell so that the adult has billions of cells. A tree grows from a tiny seed.

Movement

Movement in animals usually occurs more quickly than in plants. It is easy to see animals walking, running, flying or swimming.

It is harder to notice movement in plants. However, roots move (grow) through the soil, branches become longer from year to year and flowers open out from buds.

Fig 1.3 Growth

A leopard running

A bud opening to form a flower

Fig 1.5 Movement

Reproduction

Reproduction is the formation of new individuals. If living things did not reproduce then that type of organism would die out (become extinct).

Response

Living things are aware of changes in their surroundings. They are sensitive to different features (or stimuli) and react or respond to these features.

The sensory system in animals allows them to respond very rapidly. This is seen when we jump at hearing a loud sound or run indoors from the rain.

Plants respond more slowly to features such as light or gravity. For example, seedlings will grow towards a source of light. The roots of a plant grow down towards gravity.

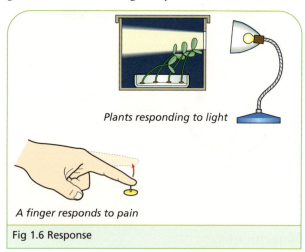

Plants responding to light

A finger responds to pain

Fig 1.6 Response

The variety of living things

Living things are arranged or classified into groups in order to simplify their study.

The two basic groups into which living things are arranged are:

- the animal kingdom
- the plant kingdom.

Classifying animals and plants

Animals and plants can be distinguished under the headings of movement, method of getting food and whether they have cell walls.

Movement

Animals move from place to place.

Plants do not move from place to place. However, plants can move their parts. For example, the daisy opens and closes its flowers by day or by night.

Food

Animals cannot make their own food. They get their food by eating plants or other animals.

Plants can make their own food. They do this in a process called **photosynthesis**. In this process plants use the energy in sunlight to convert carbon dioxide gas and water into a sugar called glucose and a gas called oxygen.

The process of photosynthesis requires the presence of a green pigment or dye called **chlorophyll**. Plants have chlorophyll, while animals do not.

Cell walls

Animal cells do not have a cell wall around them. They are only surrounded by a membrane.

Plant cells also have a membrane. However, outside the membrane plant cells have a strong cell wall. The cell wall means that plant cells are stronger than animal cells.

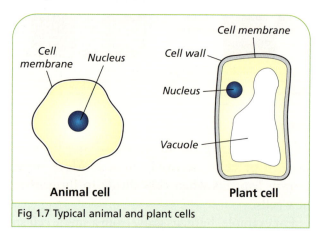

Fig 1.7 Typical animal and plant cells

Table 1.1 Differences between animals and plants	
Animals	**Plants**
Move from place to place	Do not move from place to place
Do not make their own food	Make their own food
Do not carry out photosynthesis	Carry out photosynthesis
Do not have chlorophyll	Have chlorophyll
Are non-green	Are green
Do not have cell walls	Have cell walls

A leaf showing a network of veins

Of course animals and plants can each be classified in greater detail. For example, plants can be classified as being flowering or non-flowering. Animals can be classified as invertebrates or vertebrates.

Invertebrates and vertebrates

> **Invertebrates** are animals that do not have a backbone.

The vast majority (97%) of animals are invertebrates. Examples include jellyfish, earthworms, slugs, crabs, insects and beetles.

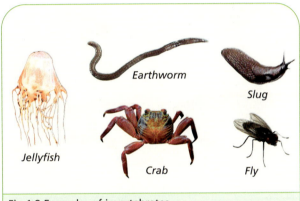

Fig 1.8 Examples of invertebrates

Jellyfish, Earthworm, Slug, Crab, Fly

> **Vertebrates** are animals that have a backbone.

A vertebra is an individual bone in the backbone. Vertebrates are bigger and more complex animals than invertebrates.

Examples of vertebrates include fish, frogs, snakes, birds, mice, dogs, horses and humans.

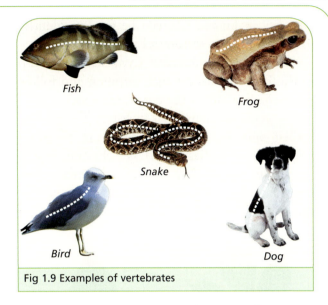

Fish, Frog, Snake, Bird, Dog

Fig 1.9 Examples of vertebrates

Identifying plants and animals

There are so many different types (or species) of plants and animals that it is very difficult to identify and name each one. A useful method of identifying living things is to use a key.

A key contains questions. Each question relates to just one characteristic. Depending on the answers to the questions the key leads to identifying and naming a particular organism.

Different keys are used to identify plants and animals. Some plant keys are based on the shape and size of the plant, others feature the leaves, while some are based on the flowers.

Different animal keys are available to name different types of animals such as insects, birds or fish.

Two sample keys, one for plants and one for animals, are given in the following section.

A simple plant key

The key shown in Figure 1.10 is called a spider key because its shape resembles a spider's legs.

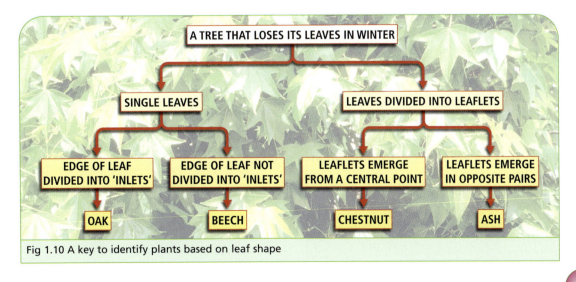

A TREE THAT LOSES ITS LEAVES IN WINTER

SINGLE LEAVES — LEAVES DIVIDED INTO LEAFLETS

EDGE OF LEAF DIVIDED INTO 'INLETS' — EDGE OF LEAF NOT DIVIDED INTO 'INLETS' — LEAFLETS EMERGE FROM A CENTRAL POINT — LEAFLETS EMERGE IN OPPOSITE PAIRS

OAK — BEECH — CHESTNUT — ASH

Fig 1.10 A key to identify plants based on leaf shape

BIOLOGY

5

The key in Figure 1.10 can be used to identify each of the leaves shown in Figure 1.11.

To identify plant number 1 you start at the top of the key. You answer each question and follow down the 'leg' under the correct answer.

For example, plant 1 has its leaves divided into leaflets and the leaflets emerge from a central point. This identifies plant 1 as a chestnut tree.

Identify plants 2, 3 and 4 using the same procedure (the answers are given at the end of this chapter, just before Key Points).

Plant 1   *Plant 2*

Plant 3   *Plant 4*

Fig 1.11 Leaves to be identified using the key in Figure 1.10

A simple animal key

The key shown below can be used to put vertebrates into their correct category.

1	If the animal has hair or feathers	Go to 2
	If the animal has no hair or feathers	Go to 3
2	If there is hair or whiskers on the face or body	Mammal
	If feathers cover the body	Bird
3	If the body is covered in scales	Go to 4
	If the body has no scales and is wet and smooth	Amphibian
4	If the animal has nostrils and lays eggs on land	Reptile
	If the animal has gills and fins and lives in water	Fish

This is a numbered key. It can be used to classify the organisms in Figure 1.12.

To identify animal number 1 read question 1. Decide if the animal has hair or feathers. As it does not have either of these go to question 3.

Decide if the animal has scales or has no scales and is wet and smooth. As it has scales go to question 4.

The animal has nostrils and lays eggs on land. It is a reptile.

Animal 1   *Animal 2*

Animal 3   *Animal 4*

Fig 1.12 A key to identify vertebrates

Use the same method to allocate animals 2, 3 and 4 to their correct group (the answers are given at the end of this chapter, just before Key Points).

★ MANDATORY ACTIVITY 1

To investigate living things

Aim

To investigate the variety of living things by direct observation of animals and plants in their environment, to classify living organisms as plants or animals and to classify animals as vertebrates or invertebrates.

Method

1 Visit an area (such as a nearby field, hedge, seashore or woodland) and look for as many living things as you can see.

2 Do not pick, collect or damage any living things in the area.

3 Use the information in the previous sections of this chapter to classify the organisms as plants or animals and to further classify the animals as vertebrates or invertebrates.

BIOLOGY

Results

● The results should be presented in the form of a chart as shown. (Note that this chart refers to living things found in and near a hedge.)

Plants	Animals	
	Vertebrates	Invertebrates
Blackberry	Robin	Greenfly
Hawthorn	Thrush	Butterfly
Ferns	Fox	Bee
Nettles	Mouse	Earthworm
Primroses	Frog	Woodlouse
Daisy	Rabbit	Spider
Dandelion	Sparrowhawk	Snail
Grass	Blackbird	Slug

BIOLOGY

Answers

The names of the plants in Figure 1.11:
Plant 2 is a beech, plant 3 is an ash and plant 4 is an oak.

The groups for each of the animals in Figure 1.12:
Animal 2 is a mammal, animal 3 is an amphibian and animal 4 is a fish.

Key Points

● Biology is the study of living things (organisms).

● Living things have the following seven characteristics:
 – nutrition
 – respiration
 – excretion
 – growth
 – reproduction
 – movement
 – response.

● The main differences between animals and plants are:
 – Animals move from place to place but plants do not.
 – Animals take in food but plants make their own food.
 – Animal cells do not have a cell wall but plant cells do have a cell wall.

● Invertebrates are animals that do not have backbones.

● Vertebrates are animals that have a backbone.

● Living things can be identified and named using keys.

? Questions

1 What is meant by (a) biology? (b) organism?

2 Rewrite the table to match up the following characteristics with the correct explanation.

Characteristics	Description
Nutrition	Getting energy from food
Reproduction	Getting rid of waste
Response	Getting food
Movement	Producing new cells which then enlarge
Excretion	Reacting to a stimulus
Respiration	Producing offspring
Growth	Walking, running, flying or swimming

3 Write out and complete the following:

(a) Living things or _____ can be _____ into plants and animals. Animals can _____ from place to place, while plants can only _____ their parts. Animals feed on _____ or other _____ while plants can make their own _____. Animal cells are not enclosed by a _____ _____ while plant cells are.

(b) The process by which plants make their own food is called _____. This process requires the green chemical called _____. This process converts _____ energy, carbon _____ and water into food and the gas _____ is produced.

(c) The characteristic of life that cannot take place in very young humans is _____.

4 (a) Give three features that allow a human to be classified as an animal.

(b) Give three features that allow a tree to be classified as a plant.

5 Suggest one reason for each of the following:

(a) Plant cells are stronger than animal cells.

(b) Plant cells are often green but animal cells are not.

(c) A snail is called an invertebrate.

(d) A cat is called a vertebrate.

6 Use the following key to name the four animals shown in Figure 1.13.

1	Body not divided into sections Body divided into sections	2 4
2	Body looks like a worm, no shell Feelers on the head	Roundworm 3
3	Shell present Shell absent	Snail Slug
4	More than eight legs Eight legs or less (including no legs)	5 7
5	Over 20 sections 20 sections or less	6 Woodlouse
6	Two pairs of legs per section One pair of legs per section	Millipede Centipede
7	No legs present Legs present	Insect larva 8
8	Eight legs Six legs	Spider Insect

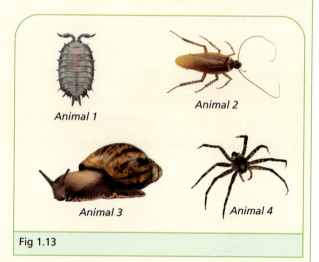

Animal 1

Animal 2

Animal 3

Animal 4

Fig 1.13

7 Use the key in question 6 to answer the following:

(a) Name one feature that snails and slugs have in common.

(b) Give one difference between a snail and a slug.

(c) Give three features of roundworms.

Questions

8 Draw a branched key (similar to Figure 1.10) for the following students using the information below:

Name	Eye colour	Hair colour	Ear lobes
Mary	Green	Brown	Fixed
Fiona	Blue	Blonde	Free
Shane	Brown	Black	Fixed
James	Green	Blonde	Fixed
Sinéad	Blue	Brown	Free

9 The following is a numbered key to identify invertebrates.

1	The animal has jointed legs	Yes – go to 2 No – go to 3
2	The animal has at least four pairs of jointed legs	Yes – go to 4 No – beetle
3	The animal has a shell	Yes – snail No – go to 5
4	The animal lives in the sea	Yes – crab No – spider
5	The animal has stinging tentacles	Yes – jellyfish No – go to 6
6	The animal's body is divided into many segments	Yes – lugworm No – go to 7
7	The animal has five arms	Yes – starfish No – flatworm

(a) Use the key above to identify the animals 1 to 8 shown in Figure 1.14.
(b) Which labelled animal could not be identified using the key?
(c) Give one reason why this animal could not be identified using the key.

Fig 1.14

10 Suggest a reason for each of the following:
(a) A rock is not considered to be living.
(b) Plants are green.
(c) An oak is neither a vertebrate nor an invertebrate.
(d) A piece of chalk is not an organism.
(e) Plant cells are stronger than animal cells.
(f) All organisms need to respire.
(g) Plants have chlorophyll.

BIOLOGY

Chapter 2

CELLS AND THE MICROSCOPE

Cells

Along with the seven characteristics of life all living things are also made up of cells. For example, there are about 100 000 000 000 000 (one hundred million million) cells in an adult human body. Cells are said to be the building blocks of living things.

Most cells are too small to be seen with the naked eye (although eggs are single cells that are very large). To see cells it is necessary to use a microscope.

The structure of a cell

Animal cells have three main parts:

- a cell membrane
- a nucleus
- cytoplasm.

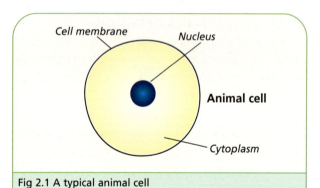

Fig 2.1 A typical animal cell

The **cell membrane** surrounds the cell. Cell membranes help to keep the contents of the cell in place. In addition, cell membranes have tiny pores which allow some molecules to pass in or out of the cell. Other molecules are prevented from passing through the cell membrane.

The **nucleus** is the control centre of the cell. It contains the genes that control inherited characteristics. In addition, the nucleus can divide into two identical nuclei. When a cell divides it forms two smaller cells. Each of these two cells will contain one of the two identical nuclei.

Cell division is the basis for the growth of organisms. This means that at each cell division two similar cells are formed.

Cytoplasm is a watery fluid found between the nucleus and the cell membrane. Many of the reactions that take place in a cell happen in the cytoplasm. The cytoplasm contains many very tiny structures which carry out different functions.

Plant cells

Plant cells have three main features not found in animal cells:

- cell wall
- vacuole
- chloroplasts.

A **cell wall** is found outside the cell membrane. Cell walls are made of a strong substance called **cellulose**. This helps to give strength to a plant cell.

Cellulose is the substance that provides fibre in the human diet. It is also the material of which paper is made.

Vacuoles are mostly found in plant cells and are used to support the cell and store substances such as food, wastes and water. Most plant cells have a large central vacuole.

Plant cells may also have tiny green structures called chloroplasts. **Chloroplasts** contain the green pigment called chlorophyll. Photosynthesis takes place in the chloroplasts.

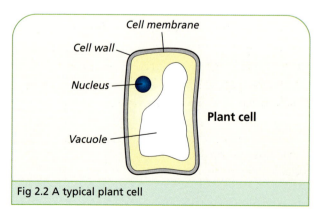

Fig 2.2 A typical plant cell

Table 2.1 The differences between animal cells and plant cells	
Animal cell	**Plant cell**
No cell wall	Cell wall
Rarely have vacuoles	Large vacuole
Never have chloroplasts	May contain chloroplasts

The way in which cells are organised

Cells do not work alone in living things. They co-operate with other similar cells. Groups of similar cells then combine together to co-operate with each other, as outlined below.

Tissues

Plants and animals contain many different types of cells. For example, the cells on the inside of your mouth are very different from red blood cells. The cells on the outside of a root of a plant are very different from the green cells in a leaf.

Each of these four types of cells contains large numbers of cells that are similar in size, colour, structure and shape. Each of them also carries out a different function. They are said to form different tissues.

> A **tissue** is a group of similar cells which carry out the same function.

Examples of animal tissues and their functions are given below.

Table 2.2 Animal tissues		
Organism	**Tissue**	**Function**
Animal	Red blood	Carries oxygen
	Muscle	Contracts to allow movement

Organs

Very often a number of tissues are found working together in a particular structure in a plant or animal. These structures are called organs.

For example, the heart is an organ. It contains muscle tissue and has a good blood supply.

> An **organ** is a structure that contains two or more tissues working together.

Examples of animal organs include the heart, lungs, brain and kidneys. Plant organs include the leaves, roots and flowers.

Systems

Organs which work together form a system. For example, the eyes, ears, skin, nose and tongue are organs that join together to form the sensory system.

> A **system** consists of a number of organs working together.

Organisms

An organism is a living thing, e.g. a human or a tree. Each organism is composed of a number of systems. These systems work together to allow the organism to work properly and stay alive.

This means that organisms are organised in the following way.

Cells → Tissues → Organs → Systems → Organisms

The microscope

A microscope is used to view objects which are too small to be seen using eyesight alone.

The word magnification means how many times bigger the object appears to be when seen through a microscope. Most school microscopes magnify an object up to about 400 times.

Microscopes work by passing light through a thin slice of an object. The microscope is focused to give a clear view of the object. Controlling the amount of light shining through the object also helps to form a clear view.

A highly magnified image of a plant cell dividing to show chromosomes

BIOLOGY

11

The parts of a microscope

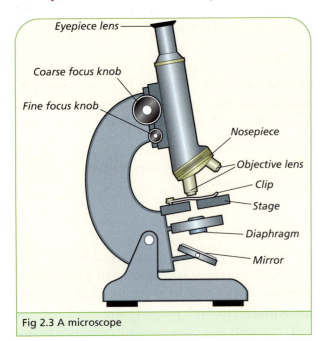

Fig 2.3 A microscope

Eyepiece lens

The eyepiece is the lens that is nearest the eye. It usually has the magnification marked on it. For example if the eyepiece is marked as ×10 this means it magnifies an object ten times.

Focus knobs

The focus knobs allow an image of the object to be seen clearly.

● The coarse focus knob brings the object roughly into focus, using the low power lens.
● The fine focus knob (which may not be present on all microscopes) allows for precise focusing, using the high power lens.

Nosepiece

The nosepiece can revolve. This allows the use of the different lenses located on the nosepiece.

Objective lens

The objective lenses are located on the nosepiece. Each objective lens gives a different magnification.

The total magnification of the image is found by multiplying the power of the eyepiece lens and the objective lens being used. For example, an eyepiece marked ×5 and an objective marked ×20 combine to give a magnification of ×100.

When using a microscope the lowest magnification lenses are used at the start. Later the magnification can be increased.

Stage

To view an object under a microscope it must be put on a glass slide. The slide is placed on the centre of the stage and held in place by clips.

Diaphragm

The diaphragm is used to control the amount of light shining through the object. This helps to give a clearer view of the object.

Mirror

The mirror is used to reflect light up through the object. Normally a lamp shines onto the mirror. The angles of the mirror can be altered to see the object more clearly.

Some microscopes have a light bulb instead of a mirror.

►►► ACTIVITY 2.1

To examine animal cells under a microscope

Method

1 Get a prepared slide of human cheek cells (or any sort of animal cells) from your teacher.

2 Adjust the mirror so that the light is shining through the opening in the centre of the stage (or switch on the microscope light, if one is present).

3 Turn the nosepiece so that the lowest power objective lens is in place.

4 Place the slide under the clips on the stage.

5 Look at the stage from the side and rotate the coarse focus knob so that the objective lens moves as close as possible to the slide. (Be careful not to bring the objective lens down onto the slide or it will be broken.)

6 Look through the eyepiece and rotate the coarse focus knob so that the objective lens moves slowly up, away from the slide.

7 When the cells are visible move the slide on the stage, if necessary, so that the cell or cells of interest are in the centre of the view.

8 Adjust the amount of light coming through the slide.

9 Switch to a higher power objective lens and use the fine focus knob to form a clear image. (Only use the coarse focus knob along with the low power objective lens, otherwise you risk breaking the slide.) ➡

Result

A clear image is seen of the material on the slide at different magnifications.

10 Draw diagrams of the cells as they appear under the different magnifications. Note the magnification under each diagram.

Label the cell membrane, cytoplasm and nucleus where possible.

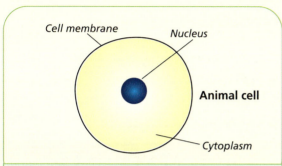

Fig 2.4 A cheek cell as viewed under the microscope

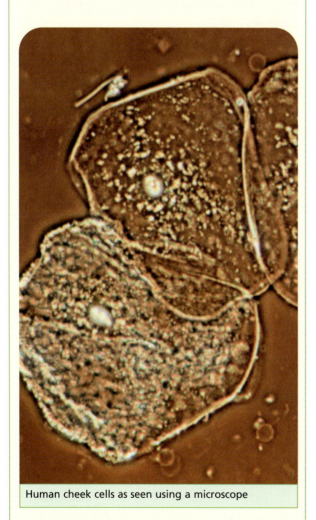

Human cheek cells as seen using a microscope

▶▶▶ **ACTIVITY 2.2**

To examine plant cells under a microscope

Method

1 Repeat the previous activity using a prepared slide of plant cells obtained from your teacher.

Result

A clear image is seen of the material on the slide at different magnifications.

2 Draw diagrams of the cells as they appear under the different magnifications. Note the magnification under each diagram.

Label the cell wall, the cell membrane, cytoplasm and nucleus where possible. (**Note** that it is difficult to see vacuoles.)

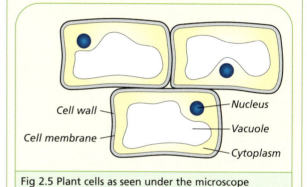

Fig 2.5 Plant cells as seen under the microscope

★ **MANDATORY ACTIVITY 2**

To prepare a slide from plant tissue and draw it as seen under a microscope

Preparing the slide

1 Use forceps, if necessary, to pull a small thin piece of tissue from the inside of a cut-up onion.

2 Place a few drops of water on a slide.

3 Place the onion tissue (which is made of a thin layer of cells) into the water. Make sure the layer of cells does not fold over.

4 Gently lower a cover slip onto the slide, as shown in Figure 2.6.

5 Add a few drops of iodine solution to the microscope slide alongside the cover slip. (Iodine will stain the nucleus a yellow/orange colour.)

6 Place some tissue or filter paper on the other side of the cover slip to draw the iodine across the cells.

BIOLOGY

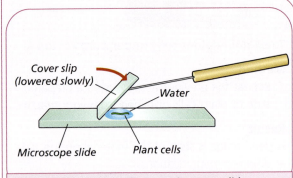

Fig 2.6 Adding a cover slip to a microscope slide

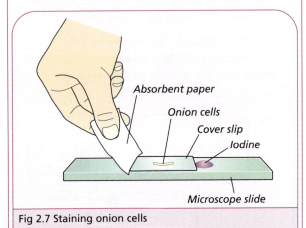

Fig 2.7 Staining onion cells

Viewing the slide under the microscope

1 View the slide in the same way you viewed the prepared slides in the last two activities.

2 Draw diagrams of the cells as they appear under the different magnifications. Note the magnification under each diagram.

Label the cell wall, the cell membrane, cytoplasm and nucleus where possible.

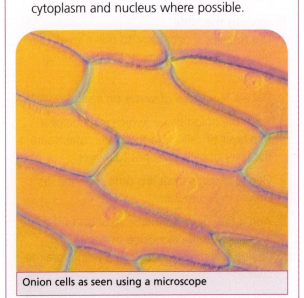

Onion cells as seen using a microscope

 ## Key Points

- The main parts in plant and animal cells are:
  - The cell membrane which controls what passes in or out of the cell
  - The nucleus which controls the cell
  - Cytoplasm which carries out many cell reactions.

- Plant cells have three features not present in animal cells:
  - A cell wall for strength
  - A large vacuole for storage
  - Chloroplasts for photosynthesis.

- Cell division is the basis for the growth of an organism.

- Organisms are organised into:

Cells ➤ Tissues ➤ Organs ➤ Systems ➤ Organisms

- A microscope is used to view small objects in detail.

- The parts of a microscope and their functions are as follows:
  - Eyepiece: this lens enlarges the image of the object.
  - Focus knobs: these allow a clear sharp image to be seen.
  - Nosepiece: this can rotate so that different lenses can be used.
  - Objective: this lens on the nosepiece enlarges the image of the object.
  - Stage: the slide is placed on the stage for viewing.
  - Diaphragm: this changes the amount of light passing through the slide.
  - Mirror: this is used to shine a light source through the slide.

- The total magnification is calculated by multiplying the power of the eyepiece by the power of the objective lens being used.

- To examine animal or plant cells under a microscope:
  - Make sure there is light passing through the opening in the stage.
  - Use the lowest power lens on the nosepiece.
  - Clip the slide onto the stage.
  - Turn the coarse focus knob to bring the stage close to the slide.
  - Turn the coarse focus knob to bring the lens away from the slide.
  - Move the slide so that the cells of interest are in the centre.
  - Adjust the amount of light passing through the slide.
  - Change the objective lens to a higher power and re-focus using the fine focus knob.
  - Draw labelled diagrams of the cells at each magnification.

- To prepare a slide from plant tissue:
  - Cut an onion and remove a thin layer of cells.
  - Place the onion cells in water on a microscope slide.
  - Place a cover slip over the onion cells.
  - Draw a few drops of iodine solution across the cells using absorbent paper.

❓ Questions

1 (a) Give one function for each of the following structures found in cells:
    (i)   nucleus
    (ii) chloroplast
    (iii) cell membrane
    (iv) vacuole
    (v)   cytoplasm
    (vi) cell wall.
  (b) Name the structures from part (a) which are only found in plant cells.

2 Figure 2.8 shows a plant cell and an animal cell as seen under a microscope.
  (a) State which cell, X or Y, represents the plant cell.
  (b) Name the structures labelled A, B, C, D and E.
  (c) Which of these structures (i) controls the cell, (ii) gives strength to the cell, (iii) allows substances in or out of the cell, (iv) is a watery fluid?

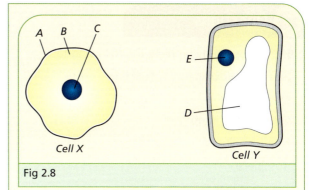

Fig 2.8

3 (a) Arrange the following in the correct order, starting with the simplest: organ, cell, system, tissue, organism.
  (b) Give a named example of each of the following in a human: organ, system and tissue.

4 Match up the following words with their explanations.

Word	Explanation
Tissue	A group of organs working together
Cell	A number of tissues working together
Organ	A group of cells carrying out the same function
System	The basic building block of living things

5 (a) Name the parts of the microscope labelled A, B, C and D.
  (b) Give one use for each of the parts labelled B and D.
  (c) If part A is labelled ×10 and part C is labelled ×20, what is the total magnification?

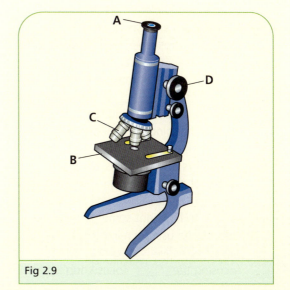

Fig 2.9

? Questions

6 Copy out and complete the following table.

Part of microscope	Function
Eyepiece	
Coarse focus knob	
Fine focus knob	
Nosepiece	
Objective lens	
Stage	
Diaphragm	
Mirror or light	

7 (a) Why are some cells stained before being examined under a microscope?

    (b) Name a stain you used for this purpose.

8 (a) Give two reasons why the Figure 2.10 represents a plant cell rather than an animal cell.

    (b) Name the parts labelled A, B and C in the diagram.

    (c) State two ways in which the diagram would be different for an animal cell.

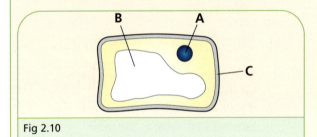

Fig 2.10

9 Write out and complete the following:

    (a) Microscopes are used to look at _____ objects. Microscopes work by passing _____ through a section of material. Lenses are used to make the image of the object appear _____. A clear image can be obtained by focusing the _____ lens and by changing the amount of _____ passing through the slide.

    (b) Thin layers of _____ cells are a good source of plant material for viewing under a _____. Plant cells can be stained with _____ to highlight some parts of the cells. When using a microscope the _____ power objective is used first. When focusing the _____ focus knob is used first.

10 Give a reason for each of the following:

    (a) Using a microscope in biology.

    (b) Using an onion to view plant cells.

    (c) Moving the position of the slide on the microscope stage.

    (d) Having clips on a microscope stage.

    (e) Staining material before looking at it under a microscope.

11 Draw a diagram of an animal cell as seen under a microscope. Label the following parts: cell membrane, cytoplasm and nucleus.

12 Draw a diagram of a plant cell as seen under a microscope. Label the following parts: cell wall, cell membrane, vacuole, cytoplasm and nucleus.

Examination Question

13 (a) Name the piece of equipment shown in the diagram.

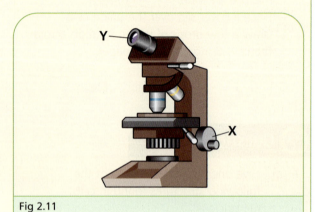

Fig 2.11

    (b) Give one use for this piece of equipment.

    (c) What are the functions of the parts labelled X and Y?

    (d) Onion epidermis is a tissue only one cell thick. It is used in school laboratories on microscope slides to investigate plant cell structure. Describe how to prepare a microscope slide from a plant tissue. (Adapted from JC, OL, Sample Paper; JC, OL, 2006; JC, HL, 2006)

BIOLOGY

Chapter **3** ooo

FOOD

Introduction

The need for food is one of the characteristics or features of living things.

The way in which living things get their food is called **nutrition.**

Humans need food to give them energy, to allow them to grow and repair damaged and worn parts of their bodies and to prevent them getting diseases.

Balanced diet

> A balanced diet contains the right amounts of each of the six different types (or constituents) of food.

The six constituents of a balanced diet are:

1 carbohydrates (which include fibre)
2 fats
3 proteins
4 vitamins
5 minerals
6 water.

The functions of these constituents and foods which contain them are given in Table 3.1.

Failure to eat a balanced diet results in a person being malnourished and unhealthy. For example:

- If people eat too many high-energy foods they may become overweight.
- If they do not get enough energy in their diet they may be tired and slow.
- Lack of iron may result in a person being pale and tired, a condition called **anaemia**.

Food pyramid

The food we eat can be classified into the following five food groups:

1 cereals, bread and potatoes
2 fruit and vegetables
3 dairy products
4 meat and fish
5 other types of food.

In order to maintain a balanced diet it is necessary to eat different amounts of each food group.

The number of servings of each food group that we should eat each day is given in a **food pyramid**. A food pyramid is a guideline to the types and amounts of food that should be eaten.

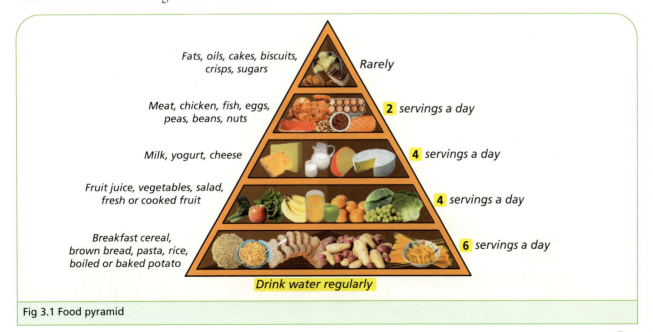

Fig 3.1 Food pyramid

Foods at the top of the food pyramid should be eaten rarely as they cause people to be overweight (they are high in fats). Overeating them may lead to disorders such as diabetes and they do not contain useful minerals or vitamins.

The amount and type of food needed by each person varies according to factors such as:

- age (young people need more food than older people)
- gender (males need more food than females)
- activity levels
- health.

Table 3.1 A balanced diet		
Constituent	Function	Common sources
Carbohydrates		
Sugars	Fast supply of energy	Fruits, honey, soft drinks, chocolate
Starch	Slower supply of energy	Bread, potatoes, rice, pasta
Fibre (or roughage)	Helps to move food through the intestines by a process called peristalsis (i.e. prevents constipation)	Cereals, brown bread, fruit, vegetables
Fats	May supply energy	Butter, cream, milk, oils, margarine
	Store energy in the body	
	Insulate the body	
Proteins	Make muscle, hair, hormones, enzymes	Meat, fish, eggs, milk, cheese, nuts, peas, beans
*Vitamins		
Vitamin C	Keeps skin and gums healthy	Fruit, green vegetables, potatoes
Vitamin D	Forms strong bones	Milk, butter, cheese, eggs, fish liver oil
*Minerals		
Iron (Fe)	Part of red blood (forms haemoglobin which carries oxygen)	Green vegetables, red meat, egg yolk
Calcium (Ca)	Forms stong bones	Milk, cheese, yoghurt
Water	Allows cells to function	Drinks, vegetables

*Many other different vitamins and minerals are needed in a balanced diet. Only two sample vitamins and two sample minerals are given in this table.

A variety of foods forming part of a balanced diet

BIOLOGY

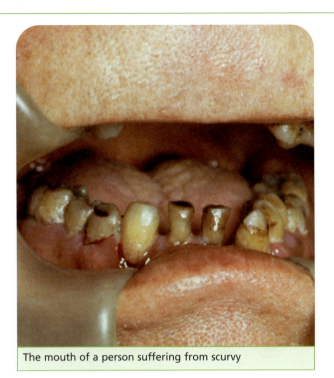

The mouth of a person suffering from scurvy

JAM
Nutritional information
Typical values per 100g

Energy	1 127 kJ
Protein	0.1 g
Carbohydrate	65.5 g
(of which sugars)	*65.4 g*
Fat	0.0 g
(of which saturates)	*0.0 g*
Fibre	1.0 g

SOUP
Nutritional information
Typical values per 100g

Energy	205 kJ
Protein	1.0 g
Carbohydrate	4.0 g
Fat	3.3 g
Fibre	0.0 g

BEANS
Nutritional information
Typical values per 100g

Energy	326 kJ
Protein	4.3 g
Carbohydrate	13.8 g
Fat	0.5 g
Fibre	2.9 g

BISCUITS
Nutritional information
Typical values per 100g

Energy	1616 kJ
Protein	2.8 g
Carbohydrate	66.2 g
Fat	13.6 g
Fibre	0.4 g

Fig 3.2 Examples of nutrition information panels

Energy values

Different foods contain different amounts of energy. The amount of energy in a food is known as its energy value.

Energy is measured in units called **joules** (J). However, the normal units used for the energy value of a food are kilojoules per gram (kJ/g).

In general, carbohydrates and proteins have the same energy values, while fats have over twice the energy value of the other two.

Although carbohydrates and proteins have the same energy values, the human body only uses protein for energy when it is close to starvation. This prevents the body from digesting body parts such as muscles and the heart.

Typical daily energy requirements (kJ) for 12 to 15 year olds are given in Table 3.2.

Table 3.2 Energy needed (kJ)		
	Inactive	Active
Girl	10 000	11 000
Boy	12 000	15 000

The energy content of a food is often given on a nutrition information panel on the side of the food container, as shown in Figure 3.2.

From the information given in Figure 3.2 it can be seen that jam and biscuits have high energy values. Soup and beans have much lower energy values.

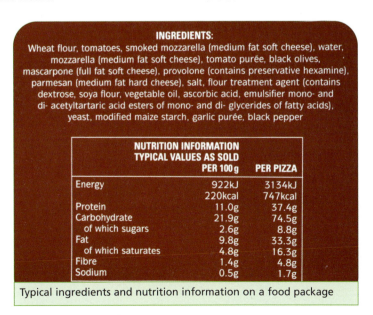

INGREDIENTS:
Wheat flour, tomatoes, smoked mozzarella (medium fat soft cheese), water, mozzarella (medium fat soft cheese), tomato purée, black olives, mascarpone (full fat soft cheese), provolone (contains preservative hexamine), parmesan (medium fat hard cheese), salt, flour treatment agent (contains dextrose, soya flour, vegetable oil, ascorbic acid, emulsifier mono- and di- acetyltartaric acid esters of mono- and di- glycerides of fatty acids), yeast, modified maize starch, garlic purée, black pepper

NUTRITION INFORMATION TYPICAL VALUES AS SOLD		
	PER 100 g	PER PIZZA
Energy	922kJ	3134kJ
	220kcal	747kcal
Protein	11.0g	37.4g
Carbohydrate	21.9g	74.5g
of which sugars	2.6g	8.8g
Fat	9.8g	33.3g
of which saturates	4.8g	16.3g
Fibre	1.4g	4.8g
Sodium	0.5g	1.7g

Typical ingredients and nutrition information on a food package

MANDATORY ACTIVITY 3A

To test a food for starch

Method

1 Add a few drops of **iodine** (which is a yellow/brown colour) to a piece of food such as white bread or a potato. (If necessary, the food sample can be ground up in a mortar and pestle. The ground-up food sample is then dissolved in a test tube of water. A few drops of iodine can be added to the solution in the test tube.)

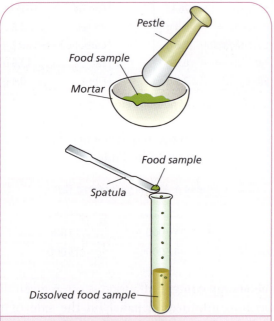

Pestle
Food sample
Mortar

Food sample
Spatula

Dissolved food sample

Fig 3.3 Dissolving a food sample in water

Result

● If the sample remains a yellow/brown colour, starch is not present.

● If the sample turns a blue-black colour, starch is present.

Conclusion

Iodine and starch give a blue-black colour.

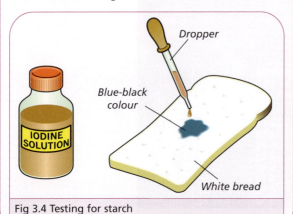

Dropper
Blue-black colour

IODINE SOLUTION

White bread

Fig 3.4 Testing for starch

MANDATORY ACTIVITY 3B

To test a food for reducing (or simple) sugar

Reducing sugars are simple sugars such as glucose. However, table sugar (sucrose) is a simple sugar but not a reducing sugar.

Method

1 Dissolve a sample of the food in water.

2 Add a few drops of **Benedict's solution** (also called **Benedict's reagent**), which is a blue colour.

3 Heat the mixture in a water bath for a few minutes, as shown in the diagram.

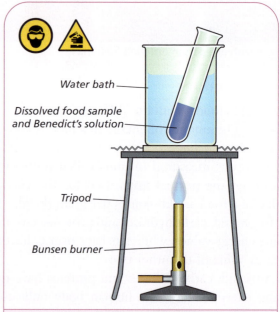

Water bath
Dissolved food sample and Benedict's solution

Tripod

Bunsen burner

Fig 3.5 Testing for sugar

Result

● If the solution remains a blue colour, reducing sugar is absent.

● If the solution turns a deep (or brick) red colour, glucose (or any simple sugar) is present.

Conclusion

Benedict's solution and simple sugar give a red colour.

BIOLOGY

MANDATORY ACTIVITY 3C

To test a food for fat

Method

1 Place a piece of the food on a sheet of brown paper or greaseproof paper.

2 Fold the paper over the piece of food.

3 Rub the piece of food in the brown paper.

Result

If a permanent stain appears in the brown paper, fat is present. (Water will also stain the paper, but a water stain will dry out if left in a warm place. A fat stain will not dry out.)

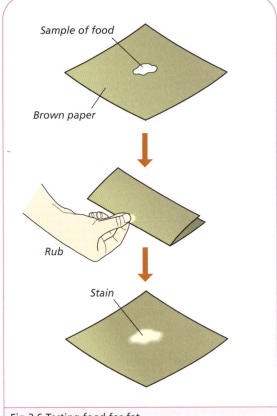

Fig 3.6 Testing food for fat

Conclusion

Fat stains brown paper.

 MANDATORY ACTIVITY 3D

To test a food for protein (this is called the Biuret test)

Method

1 Dissolve a sample of the food in water.

2 Add a few drops of **sodium hydroxide** (which is colourless).

3 Add a few drops of **copper sulfate** (which is blue).

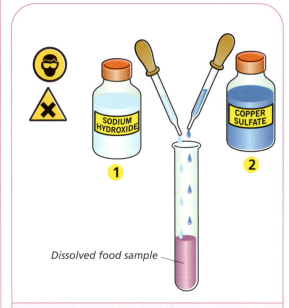

Fig 3.7 Testing food for protein

Result

● If the colour remains blue, protein is not present.
● If the colour changes from blue to purple, protein is present.

Conclusion

Protein turns purple in the presence of sodium hydroxide and copper sulfate.

BIOLOGY

MANDATORY ACTIVITY 4

To investigate the conversion of the chemical energy in food into heat energy

The energy in a food is located in the chemical bonds of the food. We learn in physics (Chapter 38) that energy cannot be created or destroyed, it can only be converted from one form to another. When food is burned, the chemical energy in the food is converted to heat energy.

Method

1 Half fill a test tube with water.

2 Place the test tube of water in a clamp attached to a stand.

3 Record the temperature of the water using a thermometer.

4 Stick a long needle into a piece of cream cracker.

5 Light the cream cracker by holding it in the flame of a Bunsen burner.

6 Hold the burning cream cracker under the test tube of water, as shown in the diagram.

7 When the cream cracker has finished burning, record the temperature of the water again.

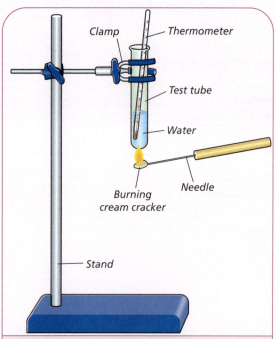

Fig 3.8 The conversion of chemical energy to heat energy

Result

The temperature of the water rises.

Conclusion

The chemical energy in the cream cracker is converted to heat energy by burning.

Key Points

- Food is needed for energy, growth, repair and to prevent diseases.

- A balanced diet consists of the right amounts of carbohydrates (including fibre), fats, proteins, vitamins, minerals and water.

- The functions of the main food constituents are:
  - Carbohydrates give energy.
  - Fibre helps to move food through the intestines.
  - Fat stores energy and insulates.
  - Proteins make and repair body parts.
  - Vitamin C keeps skin and gums healthy; vitamin D helps to form strong bones.
  - Minerals such as iron form red blood; calcium forms healthy bones.
  - Water allows cells to function properly.

- A food pyramid suggests the daily number of servings of each of the five food groups.

- Fats have twice the energy value of carbohydrates or proteins.

Food tested	Chemicals used	Positive result
Starch	Iodine	Blue-black
Glucose	Benedict's solution	Red
Fat	Brown paper	Permanent stain
Protein	Sodium hydroxide and copper sulfate	Purple

- To investigate the conversion of chemical energy in food to heat energy:
  - A burning cream cracker is placed under a test tube of water.
  - The temperature of the water rises.

? Questions

1 What is meant by nutrition?

2 (a) What is meant by a balanced diet?
   (b) List the six constituents of a balanced diet.
   (c) Give two possible results that might occur if a person failed to eat a balanced diet.

3 Write out the table to match up the following foods and their functions.

Food	Function
Fat	Forms healthy skin
Vitamin D	Supplies energy
Glucose	Retains heat in the body
Protein	Is the main material found in cells
Vitamin C	Forms strong bones
Iron	Forms red blood cells
Water	Makes muscles

4 (a) Name two good sources of fibre in the diet.
   (b) What is the value of eating fibre?
   (c) Suggest one result that might arise from a failure to eat enough fibre.

5 (a) Name the main food constituent in each of the following:
   fish, eggs, rice, bread, cream, chicken, butter, chips, chocolate, potato, pasta, nuts.
   (b) Name one food in each case from the above list that should be eaten (i) rarely, (ii) six times a day.

6 Write out and complete the following:
   (a) The units of energy in a food are _____. The food type with the highest energy is _____. The food type from which we never get energy is _____. Active people need _____ energy than _____ people. Girls need _____ energy than boys.
   (b) A good source of iron in the diet is _____. Iron is needed to make _____ which forms part of _____ blood cells. Oranges provide a good source of vitamin _____, which is necessary to maintain healthy _____.

7 Give a reason for each of the following:
   (a) Weightlifters eat lots of protein.
   (b) Lack of iron causes paleness.
   (c) Sportspeople need large amounts of carbohydrates.
   (d) Athletes should eat very little fried food.
   (e) Teenagers tend to be hungrier than their parents.
   (f) Young people require lots of calcium.
   (g) A person suffering from constipation is advised to eat lots of fruit and vegetables.
   (h) Pregnant women need extra minerals in their diet.
   (i) Seals have thick layers of fatty blubber.
   (j) Grilled food is healthier than fried food.

8 The table shows the energy contents of the foods in two meals.

Meal 1		Meal 2	
Fish	800 kJ	Grilled chicken	500 kJ
Chips	1000 kJ	Salad	50 kJ
Soft drink	600 kJ	Brown bread	300 kJ
		Glass of milk	150 kJ

   (a) What is the total energy value of each meal?
   (b) Which meal is the healthier? Give three reasons for your answer.
   (c) Which meal contains more calcium?
   (d) Name one food from each meal that is high in fibre.

9 A box of cereal has the following nutritional information:

Typical value per 100 g	
Energy	1550 kJ
Protein	7 g
Carbohydrates	83 g
of which sugars	7 g
Starch	76 g
Fat	1 g
of which saturates	0.2 g
Fibre	2.5 g

   (a) Suggest one benefit of eating this type of cereal.
   (b) How much energy is in a 50 g serving of this cereal?

BIOLOGY

10 Copy out and complete the following chart.

Food tested	Chemicals used	Original colour	Colour if food present
Glucose			
		No stain	
	Iodine		
			Purple

11 You are asked to investigate the energy values of carbohydrates, fats and proteins. Basing your investigation on one of the mandatory activities in this chapter, answer the following questions:
  (a) Name a suitable carbohydrate, fat and protein food you could use.
  (b) Draw a labelled diagram of the apparatus you would use.
  (c) Why would it be important to:
    (i) Use the same amount of water in the test tube all the time?
    (ii) Start with a fresh supply of water for each investigation?
    (iii) Use the same mass (or weight) of food sample each time?
    (iv) Keep the burning food at the same distance from the water for each food?
  (d) How would you judge which food had the highest energy value?

Examination Question

12 (a) This nutritional information was given on a packet of wheat bran. Wheat bran is used with breakfast cereals and is added to brown bread.

Nutritional information per 100 g	
Energy	872 kJ/206 kcal
Protein	15 g
Carbohydrate of which sugars	26.8 g 3.8 g
Fat of which saturates	2.5 g 0.5 g
Fibre	36.5 g
Sodium	0.028 g

Select any two nutrients from the list given and say what role each one has in maintaining health.

  (b) The diagram shows a food pyramid.

Fig 3.9

  (i) Name one item of food that could be found at X in the pyramid.
  (ii) Why should only a small amount of the food at the top of the pyramid be eaten?

  (c) Tests were carried out on three foods by pupils in a school laboratory.

The results of these tests are given in the table.

Food test				
Food tested	Starch	Reducing sugar	Protein	Fat
Food A	+	−	−	+
Food B	−	−	+	+
Food C	+	−	+	+

A plus (+) sign means a positive result to a test.

A minus (−) sign means a negative result to a test.

  (i) Which one of the foods, A, B or C would most likely be cheese, meat or fish?
  (ii) Which one of the foods, A, B or C would most likely be crisps, or chips?
(Adapted from JC, HL, 2006; JC, OL, 2006)

Chapter 4 ●●●

DIGESTION AND ENZYMES

Introduction

It is essential that nutrients from the food we eat enter the cells of our body. To allow this to happen it is necessary to break down our food into smaller molecules.

> **Digestion** is the breakdown of food.

Digestion is carried out by our digestive system (also called the alimentary canal or gut).

The broken-down molecules can dissolve in water and pass from our digestive system into the bloodstream. The blood carries the soluble nutrients to all the cells of our body.

The process of nutrition or feeding involves five main steps:

1 Food is taken into the mouth.
2 The food is broken down or digested. This makes it small enough to pass from the digestive system into the blood.
3 The dissolved particles of food are carried by the bloodstream to all the cells of the body.
4 The food is used by the body for different purposes (as outlined in Chapter 3).
5 The waste material not taken into the blood is released from the digestive system.

The digestive system

The functions of the main parts of the digestive system

Mouth

The function of the mouth is to take in and digest food. Two types of digestion take place in the mouth: physical and chemical digestion.

Physical digestion

Physical (also called mechanical) digestion occurs when the teeth cut and chew the food into smaller pieces, so that the second type of digestion (chemical digestion) can work better.

Chemical digestion

Chemical digestion takes place when the food is broken down by enzymes.

> **An enzyme** is a chemical (protein) which speeds up chemical reactions without the enzyme being used up.

Enzymes are similar to catalysts used in chemistry, except that enzymes are made of protein while catalysts are not made of protein.

Salivary glands in the cheeks and under the tongue produce a liquid called saliva.

Saliva helps to soften food. It also contains an enzyme called (salivary) amylase. This enzyme chemically digests starch, breaking it down to a simpler sugar called maltose.

$$\text{starch} \xrightarrow{\text{amylase}} \text{maltase}$$

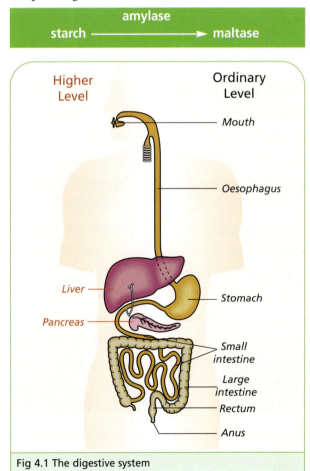

Higher Level / Ordinary Level

- Mouth
- Oesophagus
- Liver
- Pancreas
- Stomach
- Small intestine
- Large intestine
- Rectum
- Anus

Fig 4.1 The digestive system

Teeth

There are four types of teeth in the adult human jaw. The sequence of the teeth from the front to the back of the jaw is: incisors, canines, premolars and molars.

The functions of each type of tooth are as follows:

- **Incisors** have sharp edges like a chisel. They are used to cut, slice and nibble food.
- **Canines** are long and pointed. They are used to grip and tear food.
- **Premolars** are large rounded teeth. They are used for chewing, crushing and grinding food.
- **Molars** are larger teeth and are also used for chewing, crushing and grinding food.

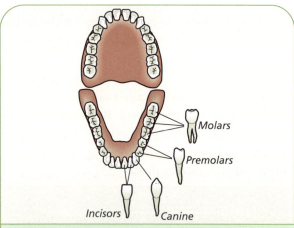

Fig 4.2 The location and types of teeth in the jaw

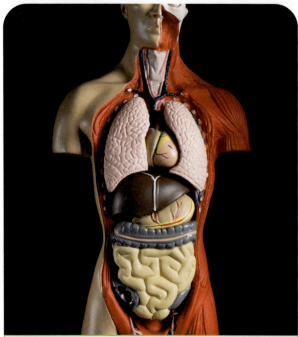

Model of the internal structure of a human. Note the location of the liver (brown just below the lungs and heart), stomach, large intestine (grey) and the much-folded small intestine (yellow)

Oesophagus

The oesophagus or foodpipe is a muscular tube. It forces food down to the stomach using a wave of muscular action called peristalsis.

Stomach

The stomach is a muscular bag which traps food for a few hours.

Hydrochloric acid in the stomach kills bacteria and other germs and also softens food. If this acid gets into the oesophagus it causes a stinging sensation called heartburn.

The stomach produces a number of enzymes which chemically digest the food. The stomach also churns and physically mixes the food.

A cavity (hole) in a tooth

Small intestine

The small intestine produces many more enzymes which complete the breakdown of food.

In the lower part of the small intestine the digested food passes into the bloodstream. The food is said to be absorbed from the small intestine.

Liver

The liver is a complex organ which carries out a range of functions.

One of its functions is to produce a liquid called bile. Bile passes from the liver into the small intestine. Bile helps to digest fats in the small intestine.

Pancreas

The pancreas produces a range of enzymes. These enzymes pass from the pancreas into the small intestine where they help to digest food.

BIOLOGY

Large intestine

The material entering the large intestine contains a lot of liquid. If all this liquid were allowed to pass out of the digestive system, we would suffer from diarrhoea.

The large intestine takes water back into the bloodstream. The semi-solid waste material left in the large intestine is called faeces. If too much water is taken back the waste becomes too solid, a condition called constipation.

Faeces are stored in the rectum and pass out of the intestine through the anus.

Enzyme action

A huge number of enzymes are needed to allow the human body (and all living things) to work properly.

Along with their role in the digestive system enzymes are also involved in speeding up reactions such as the growth of muscles, the production of hair, the formation of blood cells and the breakdown of poisonous substances such as drugs and alcohol (in the liver).

Each enzyme must have a particular shape so that it can attach to the molecule it is going to act on. The molecule produced as a result of the action of an enzyme is called the **product**.

For example, in the mouth the enzyme amylase acts on starch. The product of this reaction is called maltose. Amylase only acts on starch. It does not digest other foods such as proteins or fats.

Enzyme	Acts on	Product
Amylase	Starch	Maltose

The substance on which an enzyme acts is called the **substrate**. In the mouth the substrate is starch, the enzyme is amylase and the product of the reaction is maltose.

Enzyme	Substrate	Product
Amylase	Starch	Maltose

★ MANDATORY ACTIVITY 5

To investigate the action of amylase on starch

Method

1 Dissolve some starch in water to form a starch solution.

2 Add equal amounts of starch solution to each of two test tubes.

3 Add saliva (which contains the enzyme amylase) to one of the tubes, called tube A. Shake the test tube to mix the contents. Do not add saliva to tube B.

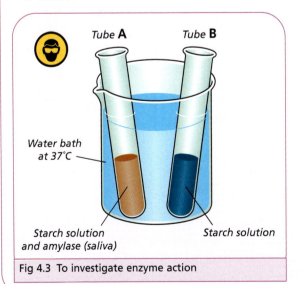

Fig 4.3 To investigate enzyme action

4 Leave both test tubes for 10 minutes in a water bath at 37°C (this is body temperature).

5 Add a few drops of the solution from each tube to a few drops of iodine solution on a dropping tile.

Result

- The contents of tube A stay red/yellow in the presence of iodine. This shows there is no starch present in tube A.
- The contents of tube B turn blue-black. This shows that tube B still contains starch.

Conclusion

The enzyme amylase in tube A breaks down starch. The starch in tube B does not break down because there is no enzyme present.

	Contents	Colour when tested with iodine	Result
Tube A	Starch and amylase	Red/yellow	No starch present
Tube B	Starch	Blue-black	Starch present

BIOLOGY

Note

- At the end of the investigation Benedict's solution can be added to tubes A and B. Both tubes can then be placed in a boiling water bath for a few minutes.

- The results in this case are that the contents of tube A turn red, showing that simple (or reducing) sugar is present.

  The contents of tube B remain blue, showing that there is no simple (or reducing) sugar present.

- This shows that starch is broken down to simple (or reducing) sugar in the presence of amylase.

	Contents	Colour when tested with Benedict's solution	Result
Tube A	Starch and amylase	Red	Simple sugar present
Tube B	Starch	Blue	No simple sugar present

 Key Points

- Digestion is the breaking down of food.
- The functions of the main parts of the digestive system or alimentary canal are as follows:
  - The mouth carries out physical digestion (using teeth) and chemical digestion (using the enzyme amylase).
  - The oesophagus moves food to the stomach by peristalsis.
  - The stomach has acid which softens food and kills bacteria. Enzymes in the stomach break down food. The stomach churns food to cause physical digestion.
  - The liver makes bile which breaks down fats.
  - The pancreas makes enzymes which digest food.
  - The first part of the small intestine completes the breakdown of food.
  - The second part of the small intestine allows the food to be absorbed from the intestine into the bloodstream.
  - The large intestine takes water back into the bloodstream.
  - The rectum stores faeces.

- The functions of the four types of teeth are as follows:
  - Incisors cut and slice food.
  - Canines grip and tear food.
  - Premolars chew, crush and grind food.
  - Molars also chew, crush and grind food.

- Enzymes are proteins which speed up chemical reactions without the enzymes being used up.

- The substance produced by an enzyme is called the product.

- The substance the enzyme acts on is called the substrate.

- The enzyme amylase acts on starch to produce a reducing sugar called maltose.

- To investigate the action of amylase on starch:
  - If saliva (amylase) is added to a starch solution, the solution will turn red/yellow when iodine is later added.
  - If saliva is not added to a starch solution, the solution will turn blue-black when iodine is later added.

? Questions

1 (a) What is meant by digestion?
   (b) Name the type of digestion carried out by (i) the teeth, (ii) enzymes.

2 Rewrite the following parts of the alimentary canal in the order in which food passes through them:
oesophagus, large intestine, mouth, anus, small intestine, stomach.

3 Figure 4.4 shows the human digestive system.

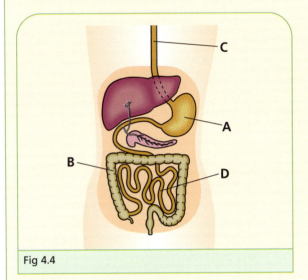

Fig 4.4

   (a) Name the parts labelled A, B, C and D.
   (b) What is the main function of part B?
   (c) Name the labelled part that is acidic.
   (d) Name the labelled part that produces faeces.
   (e) Name the labelled part from which food is most likely to enter the bloodstream.
   (f) In which part are bacteria killed?

4 Write out and complete the following:
   (a) Digestion is the _____ of food.
      Digestion is necessary to allow food nutrients to pass into the _____.
      Digested food is carried by the _____ to all parts of the body.
   (b) The four types of teeth are incisors, _____, _____ and _____. The function of the _____ teeth is to grip and tear food. The teeth used to chew and crush food are the _____ and _____. Teeth carry out _____ digestion.
   (c) The liquid produced in the mouth is called _____. The enzyme in this liquid is called _____. This enzyme acts on _____ and changes it to _____.

5 Copy out and complete the following table.

Structure	Function
Mouth	
Oesophagus	
Stomach	
Small intestine	
Large intestine	

6 Give a reason for each of the following:
   (a) The stomach contains acid.
   (b) Dogs have large canine teeth.
   (c) Food must be broken down into soluble pieces.
   (d) Food can pass from the mouth to the stomach even when we are lying down.
   (e) The large intestine must not take back too much water from our waste products.

7 (a) What are enzymes?
   (b) What is the main advantage of enzymes?

8 Name
   (a) an enzyme
   (b) the place in the body where it is produced
   (c) the substance the enzyme acts on
   (d) the product of the action of the enzyme.

9 Apart from their use in digestion give two other uses for enzymes in the human body.

10 (a) Name the chemical used to test for starch.
   (b) State the colour of this chemical (i) in the absence of starch, (ii) in the presence of starch.

11 (a) Name the chemical used to test for reducing sugars.
   (b) State the colour of this chemical (i) in the absence of reducing sugar, (ii) in the presence of reducing sugar.

BIOLOGY

29

❓ Questions

12 The following apparatus was set up for 15 minutes.

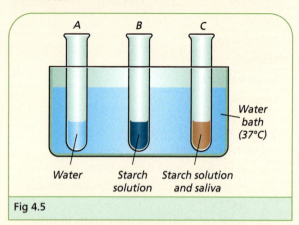

Fig 4.5

(a) Why was the water bath kept at 37°C?
(b) State the colour expected when testing each of the test tubes for starch at the beginning of this period of time. Give a reason for your answer in each case.
(c) State the colour expected when testing each of the test tubes for starch at the end of this experiment. Give a reason for your answer in each case.

13 In an experiment the following apparatus was set up. Each test tube was tested for both starch and reducing sugar at the start of the investigation and again after 15 minutes.

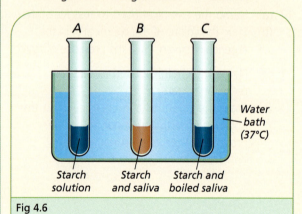

Fig 4.6

(a) Name the enzyme present in saliva.
(b) Name one location in the body where this enzyme is normally active.
(c) Copy out the following table and fill in the words 'present' or 'absent' in each box.

	Starch test (present/absent)	Test for simple sugar (present/absent)
Tube A at start		
Tube A after 15 minutes		
Tube B at start		
Tube B after 15 minutes		
Tube C at start		
Tube C after 15 minutes		

14 (a) State one function in each case for (i) the liver, (ii) the pancreas, (iii) bile.
(b) Explain each of the following words: (i) enzyme, (ii) substrate, (iii) product.
(c) Name any enzyme and state its substrate and product.

15 You are asked to design an experiment to test the effect of changing temperature on the efficiency of enzyme action.

(a) Name an enzyme you could use. State a source of this enzyme.
(b) What factor or variable would you change in your investigation? Suggest how you would change this factor.
(c) Name two factors or variables that you would keep constant. Suggest how you would keep them constant.
(d) At what temperature would you expect the enzyme to be most efficient?

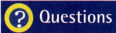

Questions

Examination Question

16 Digestion of food is important so that we can obtain energy from our food.

  (a) Name the parts of the digestive system labelled A, B and C in the diagram.

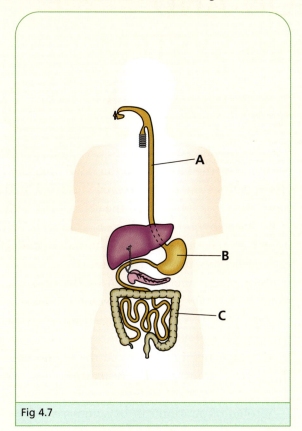

Fig 4.7

  (b) Give one function of the part of the digestive system labelled B.

  (c) Salivary amylase found in the mouth acts on starch in the food we eat. This action can be investigated in the laboratory.

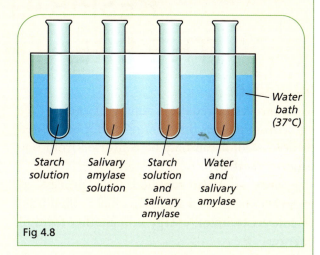

Starch solution   Salivary amylase solution   Starch solution and salivary amylase   Water and salivary amylase   Water bath (37°C)

Fig 4.8

  (i) Name the chemical used to test for the presence of starch at the beginning of the experiment.

  (ii) When the salivary amylase is added to starch solution and the mixture is placed in a water bath at 37°C for 5 minutes, a new product is formed. Name the product formed.

  (iii) Another chemical is used to test for the presence of this new product. This chemical reacts with the new product to produce a brick-red colour when they are heated together in a hot water bath for 5 minutes. Name this chemical.

  (JC, OL, 2006)

BIOLOGY

Chapter 5

RESPIRATION AND THE BREATHING SYSTEM

Introduction

Many processes require a supply of energy. We know from physics that energy cannot be created or destroyed, it can only be converted from one form to another.

Energy is often supplied to a process as chemical energy in a fuel. This fuel is then used to convert chemical energy into another form of energy.

For example, in a Bunsen burner the energy is supplied as chemical energy in the gas. The gas reacts with oxygen from the air and produces heat (and light) energy. The waste products formed in this reaction are carbon dioxide and water vapour.

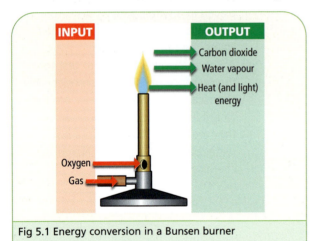

Fig 5.1 Energy conversion in a Bunsen burner

Aerobic respiration

Living things need energy to allow them to move, grow, stay warm and repair damaged parts. They get their energy from food in a process called respiration.

> **Respiration** is the release of energy from food.

The majority of living things need oxygen for respiration. This type of respiration is called aerobic respiration.

> **Aerobic respiration** needs oxygen to release energy from food.

To allow aerobic respiration to take place, glucose is carried by the blood from the small intestine to all the cells of the body. Oxygen is carried by the blood from the lungs to all the cells of the body.

In each living cell glucose combines with oxygen to release energy and the waste products carbon dioxide and water vapour.

Some of the energy is used by the cells while some is lost as heat. The waste products are carried by the blood to the lungs from where they pass out of the body.

As well as taking place in all the living cells in the human body, aerobic respiration also takes place in most animal and plant cells.

Aerobic respiration can be summarised by the word equation:

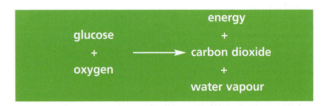

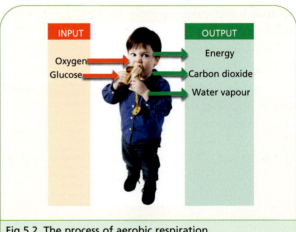

Fig 5.2 The process of aerobic respiration

▶▶▶ ACTIVITY 5.1

To demonstrate carbon dioxide as a product of respiration

Method

1 Place some clear limewater in a test tube.

2 Breathe out through the limewater a number of times, as shown in Figure 5.3.

3 As a control, do not breathe out through a second test tube of clear limewater.

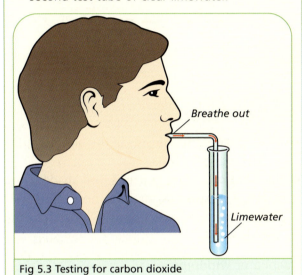

Fig 5.3 Testing for carbon dioxide

Result

● The result is that the limewater, which was breathed through, turns cloudy or milky.
● The limewater that is not breathed through remains clear.

Conclusion

This shows that carbon dioxide was present in the air breathed out.

Anaerobic respiration

Some living things can get energy from food without using oxygen. This type of respiration is called anaerobic respiration.

Anaerobic respiration takes place when a fungus called yeast converts sugars to alcohol.

Anaerobic respiration also occurs when bacteria cause milk to turn sour.

> **Anaerobic respiration** does not use oxygen to release energy from food.

▶▶▶ ACTIVITY 5.2

To demonstrate water vapour as a product of respiration

Method

1 Breathe out onto a clean mirror or a piece of glass (such as a microscope slide).

2 As a control, do not breathe onto a second mirror or piece of glass.

3 Test any liquid that forms by placing a piece of blue, dry cobalt chloride paper on it.

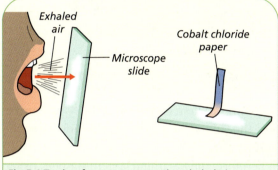

Fig 5.4 Testing for water vapour in exhaled air

Result

● A liquid forms on the mirror or glass that was breathed on. This liquid turns dry (blue) cobalt chloride a pink colour. This shows the liquid is water.
● No liquid forms on the other piece of glass.

Conclusion

This shows that we breathe out water vapour.

Respiration and breathing

Respiration is a chemical process which releases energy from glucose. It occurs in all living cells and requires oxygen while it produces carbon dioxide and water.

Breathing is the way in which animals take in oxygen and release carbon dioxide and water. Breathing takes place using the breathing system.

The breathing system

When we breathe in oxygen passes from our lungs into our bloodstream. The blood then carries oxygen to all the cells of our body. Oxygen is used in each cell to carry out respiration.

Carbon dioxide and water vapour are produced by respiration in each body cell. They enter the bloodstream and are carried by the blood to our lungs.

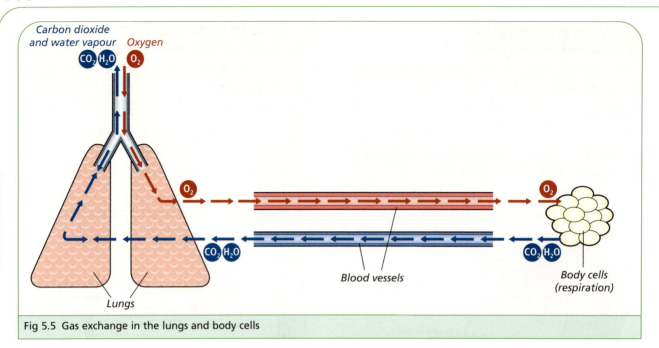

Fig 5.5 Gas exchange in the lungs and body cells

In the lungs carbon dioxide and water vapour pass from the blood into the lungs and are then breathed out.

Therefore, the breathing system supplies oxygen for respiration and gets rid of the waste products of respiration. For this reason, the breathing system is also called the respiratory system.

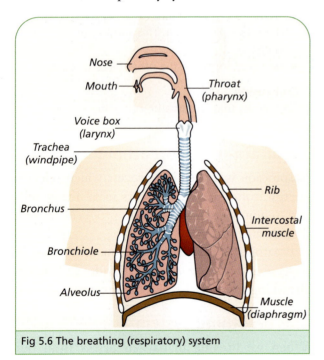

Fig 5.6 The breathing (respiratory) system

The functions of the parts of the breathing system

Nose

We are supposed to breathe in (inhale) through our nose. The reasons for doing this are:

● Firstly, hairs and mucous in the nose trap dirt particles and bacteria.

● In addition, air is warmed and made more moist as it passes through the nose. This later helps oxygen to pass from the lungs into the bloodstream.

Trachea or windpipe

The windpipe is made from rings of rigid cartilage (similar to the cartilage in our ears). The cartilage keeps the windpipe open, even when we are gasping for breath.

The windpipe carries air to and from the lungs.

Bronchus and bronchioles

Each bronchus (plural bronchi) carries air between the windpipe and a lung. The two bronchi subdivide many times to form tiny tubes called bronchioles. Bronchioles carry air to and from the air sacs or alveoli.

Some of the tiny bronchioles are made of muscle. If this muscle contracts it causes the bronchioles to become narrow. As a result, less air passes to and from the lungs. This condition is called asthma.

Alveolus

Each lung contains millions of tiny air sacs called alveoli (singular alveolus). Each alveolus has a thin lining and is surrounded by a large number of tiny blood vessels called capillaries. The function of the alveoli is gas exchange.

Oxygen passes from the air in the alveolus into the blood vessels. At the same time carbon dioxide and water pass from the blood vessels into each alveolus.

As a result of gas exchange in the alveoli, the composition of inhaled air is different to the composition of exhaled air. The approximate figures are given in Table 5.1.

Table 5.1 The composition of inhaled and exhaled air

Substance	% in inhaled air	% in exhaled air
Nitrogen	78	78
Oxygen	21	16
Carbon dioxide	0.04	4
Water vapour	Variable	Much higher than in inhaled air

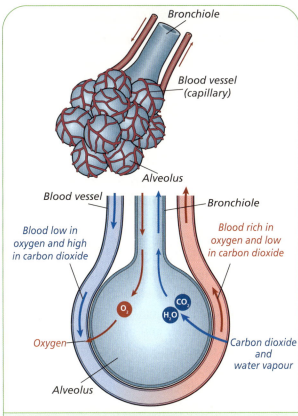

Fig 5.7 Gas exchange in an alveolus

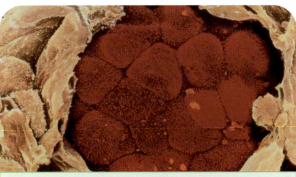

A cluster of cancer cells (in red) growing in an alveolus

The diaphragm and intercostal muscles

The diaphragm is a sheet of muscle that forms the base of the chest. The intercostal muscles are located between the ribs. All these muscles help to cause air to move into or out of the lungs.

Exercise and breathing rate

▶▶▶ ACTIVITY 5.3

To demonstrate the effect of exercise on the rate of breathing

Method

1 Count the number of times per minute you breathe in (inhale) while at rest.

2 Repeat step 1 three times.

3 Add the three values together and divide by three to calculate the average number of breaths per minute at rest.

4 Exercise vigorously for two minutes (e.g. step up and down onto a chair repeatedly or run on the spot).

5 Immediately after exercising count the number of times per minute you breathe in.

6 Continue to record the number of inhalations per minute until the rate returns to the average resting rate.

7 Record how long it takes for the breathing rate to return to normal after exercise.

8 Compare the rate of breathing at rest with the rate after exercise.

Sample result

Table 5.2		
Number of inhalations per minute at rest	Average number of inhalations per minute at rest	Number of inhalations per minute after exercise
20		62
22	21	35
21		21

Conclusion

As shown by the sample results in Table 5.2, the rate of breathing increases after exercise.

BIOLOGY

Explanation of the link between exercise and breathing rate

Breathing is necessary to take in and release the gases needed for respiration. Respiration occurs in all the cells of the body in order to provide them with energy.

When we exercise, our cells need greater supplies of energy. As a result the rate of respiration increases. This causes us to breathe faster in order to allow for the increases in gas exchange.

MANDATORY ACTIVITY 6

To compare the carbon dioxide levels of inhaled and exhaled air

Method

1 Place equal amounts of clear limewater in two test tubes, as shown in Figure 5.8.

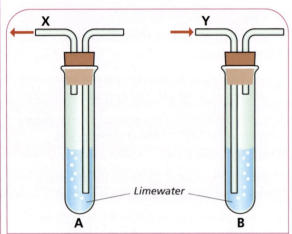

Fig 5.8 Testing for carbon dioxide in inhaled and exhaled air

2 Suck air in through tube X and hold your breath for as long as possible.

3 Breathe air out through tube Y.

4 Repeat steps 2 and 3 until the limewater in one of the test tubes turns milky.

Result
The limewater in tube B turns milky first.

Conclusion
We breathe out (exhale) more carbon dioxide than we breathe in (inhale).

● Note that the limewater in test tube A will eventually turn milky if you continue to breathe in through tube X. This is because there is a small amount of carbon dioxide in inhaled air.

The effects of smoking

Smoking is bad for our health. This fact is recognised by the health warning printed on all packets of cigarettes. In addition, smoking is banned in an increasing number of locations and advertisements for cigarettes are strictly controlled.

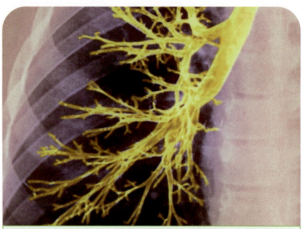

X-ray of a lung showing a bronchus and many bronchioles

Smoking has the following effects:

● Smoke clogs up the tiny hairs in the nose. This allows more dirt particles to enter the lungs.

● Smoke irritates the nose and bronchioles. This causes increased mucous to form which leads to 'smoker's cough'.

● Smoking results in increased lung infections such as pneumonia and bronchitis.

● Smoking increases the risk of getting lung cancer and other cancers.

● Gases in cigarette smoke enter our blood. Some of these gases reduce the ability of our blood to carry oxygen. As a result, our heart must pump harder and faster. This strains the heart and often leads to heart attacks.

● If a pregnant woman smokes then the chemicals in her blood will enter the baby's body. This can affect the development of the baby.

A normal lung (on the left) and a smoker's lung

ACTIVITY 5.4

To show the effect of smoking on the lungs

Method

1 Set up the apparatus as shown in Figure 5.9.

2 The filter pump draws air (containing smoke) through the cotton wool.

3 As a control, set up the apparatus without a cigarette.

Result

● The cotton wool becomes discoloured due to tar and other impurities present in cigarette smoke.

● The cotton wool in the control does not change colour.

Conclusion

There are many impurities present in cigarette smoke.

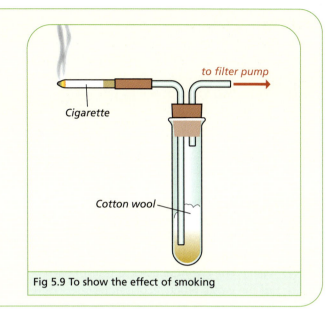

Fig 5.9 To show the effect of smoking

 Key Points

● Respiration is the release of energy from food.

● Aerobic respiration needs oxygen to release energy from food.

● The equation for aerobic respiration is:

food + oxygen $\longrightarrow$ energy + carbon dioxide
+ water vapour

● Carbon dioxide turns limewater milky.

● Cobalt chloride turns pink in the presence of water.

● Anaerobic respiration does not need oxygen to release energy from food.

● To show the effect of exercise on breathing rate compare the number of breaths taken at rest with the number taken after exercise.

● Exercise increases the rate of breathing in order to supply extra oxygen and remove the extra waste products formed during increased respiration.

● Respiration is the release of energy from glucose. Breathing involves taking gases in and out of the body.

● The parts of the breathing system and their functions are:

Part	Function
Nose	Cleans, warms and moistens air
Trachea (windpipe)	Allows air to pass to and from the lungs
Bronchi	Allow air to pass to and from the lungs
Bronchioles	Allow air to pass to and from the lungs
Alveoli	Enable gas exchange
Diaphragm and intercostal muscles	Cause air to move in or out of the lungs

● In the lungs:
  – Oxygen passes from the alveoli into the blood.
  – Carbon dioxide and water pass from the blood into the alveoli.

● Breathing out through limewater causes it to turn milky faster than breathing in through limewater. This shows there is more carbon dioxide in exhaled air than in inhaled air.

● Smoking damages health by:
  – clogging hairs in the nose
  – increasing mucous production in the breathing system
  – increasing the risk of lung infections and cancers
  – increasing the risk of heart attacks
  – damaging the growth of a baby in the womb.

BIOLOGY

? Questions

1 Answer the following with reference to a Bunsen burner which is burning:
  (a) Name the fuel used by the Bunsen burner.
  (b) Name the gas needed for the fuel to burn.
  (c) Name two forms of energy produced.
  (d) Name two waste products.
  (e) State the energy conversion that takes place.

2 (a) What is respiration?
  (b) Where does respiration take place in (i) humans, (ii) plants?
  (c) Why must living things respire?
  (d) Name the waste products formed in aerobic respiration.
  (e) Name one chemical used to test for each of the waste products of aerobic respiration.

3 (a) What is the difference between aerobic respiration and anaerobic respiration?
  (b) Name one living thing which carries out (i) aerobic, (ii) anaerobic respiration.

4 Rewrite the following and fill in the blanks:
  (a) Respiration is the release of _____ from _____. Aerobic respiration requires _____. Anaerobic respiration does not require _____.
  (b) Respiration is essential for all living things because it gives them _____. The gas that is normally used in respiration is _____ while the gas that is produced is _____.
  (c) The process of aerobic respiration is given by the equation:

  Glucose + _____ ⟶

  Energy + _____ _____ + _____

5 (a) Name the chemical used to test for the presence of carbon dioxide.
  (b) State the colour of this chemical when:
   (i) carbon dioxide is not present
   (ii) carbon dioxide is present.

6 (a) Name the chemical used to test for the presence of water.
  (b) State the colour of this chemical when:
   (i) water is not present
   (ii) water is present.

7 What is the difference between respiration and breathing?

8 (a) Name the parts of the breathing system labelled A, B, C, D and E.

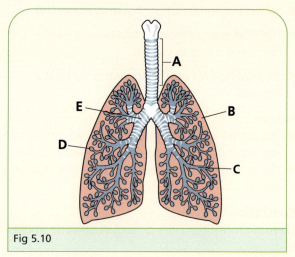

Fig 5.10

  (b) Name the material of which part A is made.
  (c) In which labelled part does gas exchange take place?
  (d) Name the gas taken in during breathing and used by the body.
  (e) Name one waste gas the body gets rid of by breathing.

9 (a) Name the gas that passes from the lungs into the bloodstream.
  (b) Name the gas that passes from the bloodstream into the lungs.
  (c) Name any muscle that helps to get air in and out of the lungs.

10 Arrange the following parts into the correct order:
  bronchiole, nose, trachea, alveolus, bronchus, throat.

11 Copy out the table and use the following values to complete it:
  16, 0.04, 78, 21, 4, 78

Substance	% in inhaled air	% in exhaled air
Nitrogen		
Oxygen		
Carbon dioxide		

? Questions

12 State a reason for each of the following:
   (a) There are tiny hairs in the nose.
   (b) Cobalt chloride paper turns pink when we exhale onto it.
   (c) The lungs are surrounded by the ribs.
   (d) We have two bronchi.
   (e) People with asthma find it hard to breathe.
   (f) Exercise causes us to breathe faster.
   (g) The alveoli have very thin walls.
   (h) The lungs have huge numbers of capillaries.
   (i) Breathing problems are often more common in cities than in country areas.

13 An experiment was carried out by Sinéad using the apparatus shown in Figure 5.11.

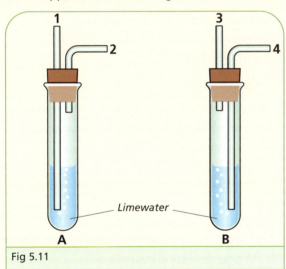

Fig 5.11

   (a) Through which numbered tube in apparatus A should Sinéad inhale?
   (b) Through which numbered tube in apparatus B should she exhale?
   (c) Which limewater would turn milky the fastest?
   (d) Give a reason for your answer to part (c).

14 Suggest one reason in each case why smoking causes damage to
   (a) the breathing system
   (b) the heart
   (c) an unborn baby.

15 It is suggested that different brands or types of cigarettes are more or less harmful to the lungs. Design an experiment to test the visible effects of different cigarettes on the lungs.
   (a) List the apparatus you will require.

   (b) Suggest any reason why different cigarettes would be more or less harmful.
   (c) State two factors that will be kept constant in each experiment.
   (d) Suggest two risks associated with your experiment.
   (e) What precautions could you take to reduce the risks that you mentioned?
   (f) How will you measure the effect of the different cigarettes?

Examination Question

16 (a) Name the parts of the breathing system labelled X and Y in the diagram.

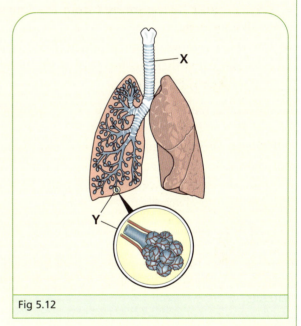

Fig 5.12

   (b) Complete the sentence below using words from the list below.

   > Oxygen
   > Carbon dioxide
   > Hydrogen

   There is more _____ in exhaled air than in inhaled air.
   (c) A balance of exercise and rest promotes good health. Name one activity which has a harmful effect on the breathing system.
   (d) How does gas exchange take place in the structures labelled Y?
   (Adapted from JC, HL, 2006; JC, OL, 2006)

Chapter 6 ●●●

THE CIRCULATORY SYSTEM

Introduction

The human body needs a transport system to move materials from one place to another. For example:

● We need to move food from the intestines to all the body cells.
● Oxygen must be transported from the lungs to all body cells.
● We must remove waste products from our cells and carry them to the part of the body that will get rid of them.

All these tasks (and many more) are carried out by our circulatory or transport system.

The circulatory system is made up of our blood and a system of tubes or blood vessels. In addition, our heart is needed to pump the blood through the blood vessels.

Blood

Blood is made up of four parts or components: plasma, red blood cells, white blood cells and platelets.

Plasma

Plasma is the liquid part of blood. It is a pale yellow colour and is mostly made of water.

Plasma transports many dissolved chemicals around the body. Examples of these chemicals are useful materials such as foods and hormones, along with wastes such as carbon dioxide, salts and urea.

Chemical reactions in all our body cells produce heat. Plasma transports heat from one part of the body to another. By doing this it plays an important role in maintaining our body temperature at 37°C.

> **The function of plasma is to transport chemicals and heat.**

If we are too hot extra blood is sent to our skin (especially to the face causing it to go red). This allows more heat to pass out of our body and we cool down.

When we are ill our body temperature may rise. This helps to destroy the bacteria or, especially, the viruses that are causing us to be ill.

However, high temperatures can have damaging effects on the body. This is why the high temperature should not last too long.

Plasma also carries red blood cells, white blood cells and platelets.

Red blood cells

Red blood cells are made in the jelly-like centre of bones, called bone marrow. They contain a red dye or pigment called haemoglobin. To make haemoglobin we need iron.

Haemoglobin – and therefore red blood cells – carries oxygen.

Oxygen enters our red blood cells in the alveoli of the lungs. It attaches to haemoglobin in the red cells. When the blood reaches cells in other parts of our body (such as our muscles or the brain) haemoglobin releases the oxygen into these cells.

> **The function of red blood cells is to transport oxygen.**

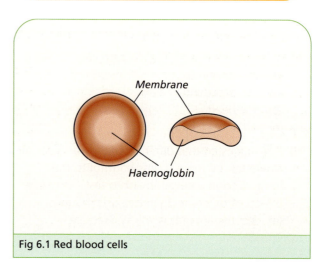

Membrane

Haemoglobin

Fig 6.1 Red blood cells

Red blood cells are round. They are tiny cells (about five million are contained in a drop of blood).

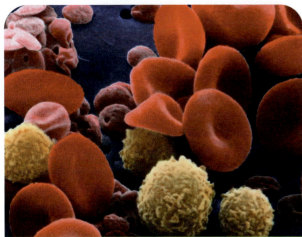

Red blood cells, white blood cells (in yellow) and platelets (in pink)

White blood cells

White blood cells are also made in bone marrow. They are larger than red cells and can change shape so that they have no definite shape.

We have much fewer white cells than red cells (which is why our blood is red in colour).

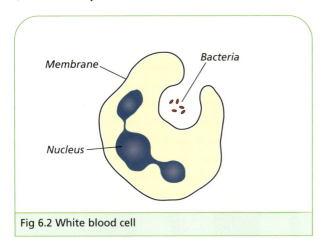

Membrane

Bacteria

Nucleus

Fig 6.2 White blood cell

White blood cells fight infection. Some do this by surrounding micro-organisms (such as bacteria and viruses) and destroying them.

Some white blood cells form proteins called antibodies. Antibodies help to destroy micro-organisms which have entered the body. The importance of antibodies can be seen by realising that AIDS is a condition in which antibodies are not produced.

> The function of white blood cells is to fight infection.

Platelets

Platelets are formed in the bone marrow when large cells break down into smaller pieces.

Platelets help to form blood clots. In this way, they help to prevent loss of blood. Blood clots also prevent micro-organisms entering the body.

> The function of platelets is to clot the blood.

Fig 6.3 Platelets

Functions of blood

The functions of blood are a combination of the functions of the parts of blood. These functions and the parts of blood that carry them out are given in Table 6.1.

Table 6.1 The functions of the parts of the blood	
Part of blood responsible	**Function**
Plasma	Transports chemicals such as foods, wastes, hormones
Plasma	Maintains body temperature
Red blood cells	Transport oxygen
White blood cells	Fight infection
Platelets	Form blood clots

Blood vessels

There are three main types of blood vessels: arteries, veins and capillaries.

Arteries and veins

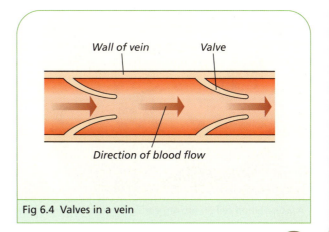

Wall of vein   Valve

Direction of blood flow

Fig 6.4 Valves in a vein

41

Table 6.2 The work of arteries and veins	
Arteries	**Veins**
1 Arteries carry blood away from the heart. (Remember A is for artery and A is for away.)	1 Veins carry blood to the heart
2 As a result of carrying blood away from the heart there is a strong flow of blood in arteries. The blood in arteries is under high pressure	2 The blood flow or pressure in a vein is low
3 In order to withstand the high pressure, the walls of arteries are very thick and strong	3 The low pressure means that veins can have thinner, weaker walls than arteries
4 There is no danger of the blood in an artery flowing backwards. This means there is no need for valves in an artery	4 To prevent blood flowing backwards in veins, they have valves at regular intervals (as shown in Figure 6.4)

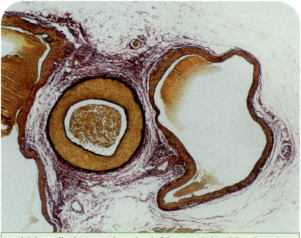

A thick-walled artery (on the left) containing blood and a thin-walled vein containing two clumps of blood cells

Capillaries

Capillaries are tiny blood vessels. They are found between arteries and veins. There are a huge number of capillaries in the human body, but the best way to see them is by looking at your eyes in a mirror.

The walls of a capillary are very thin. This allows materials to pass in and out of capillaries.

For example, in our intestines food passes into blood capillaries. This food is carried by the bloodstream to all the cells of the body. The food then leaves the capillaries to enter the body cells.

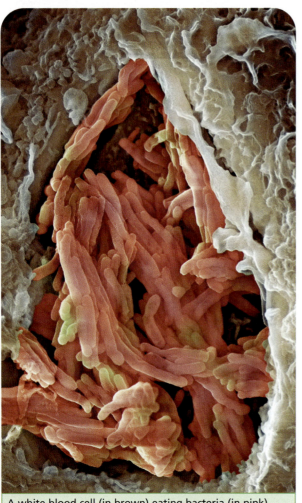

A white blood cell (in brown) eating bacteria (in pink)

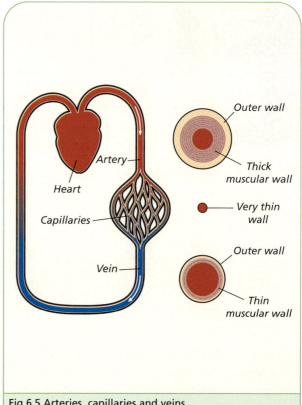

Fig 6.5 Arteries, capillaries and veins

The heart

The heart is about the size of a clenched fist. It is located in the left-hand side of the chest.

The heart is made of a special type of muscle called cardiac muscle. Cardiac muscle is very strong and does not tire easily.

The heart contracts in order to pump blood around our body in blood vessels.

When we are resting the average rate of an adult heartbeat is 70 beats per minute (bpm). When we exercise the heart beats faster. This causes blood and the materials it carries to move faster around our bodies.

> **The function of the heart** is to pump blood around the body.

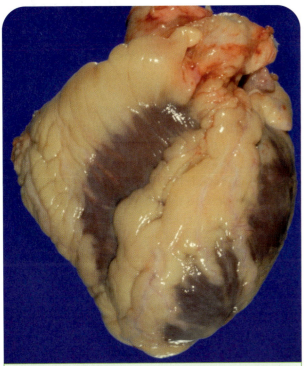

A human heart showing some blood vessels at the top (note that this heart is covered in yellow fat)

Structure of the heart

The heart contains four chambers. The top two are the right and left atrium (plural atria) and the bottom two are the right and left ventricles.

The two sides of the heart are separated by a muscular wall called the septum. When a child is born with a hole in the heart the hole is usually found in the septum.

Valves in the heart make sure that blood can only flow in one direction. In this way they are similar to valves in a car tyre or football (which only let air pass in, but not out).

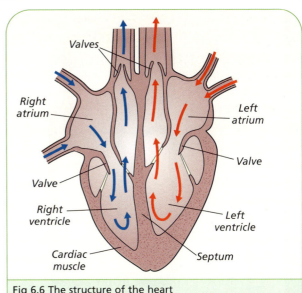

Fig 6.6 The structure of the heart

Blood flow through the heart

1. Blood from the arms, legs and other parts of the body enters the right atrium of the heart. This blood is low in oxygen.
2. The right atrium contracts to pump the blood down through a valve into the right ventricle.
3. When the right ventricle contracts the valve shuts to prevent the blood from going back into the right atrium. As a result blood is pumped out of the heart to the lungs.
4. In the lungs the blood gains oxygen (and also loses carbon dioxide and water vapour).
5. Blood from the lungs flows back into the left atrium of the heart. This blood is now rich in oxygen.
6. The left atrium contracts to pump the blood through a valve and into the left ventricle.
7. The left ventricle contracts, the valve snaps shut and blood is forced out of the heart and all around the body.
8. Eventually this blood will lose oxygen to the body cells and it will return to the heart in the right atrium. The cycle then starts all over again.

The two blood circuits

The circulatory system consists of two circuits. In one circuit, the lung circuit, blood flows from the heart to the lungs and back to the lungs.

In the second, longer, body circuit, blood flows from the heart to the rest of the body and back to the heart again. The two circuits are shown in Figure 6.7.

BIOLOGY

BIOLOGY

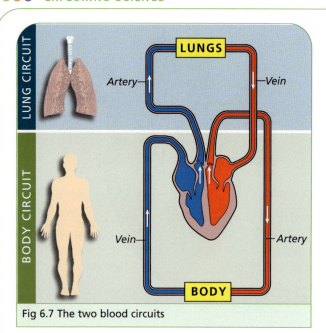

Fig 6.7 The two blood circuits

The ventricles of the heart

The right ventricle pumps blood from the heart to the lungs. This is a reasonably short distance and so the muscular walls of the right ventricle are fairly thin.

The left ventricle pumps blood from the heart all around the body. This is a very long distance and so the muscular walls of the left ventricle are very thick.

Table 6.3 **The differences between the ventricles of the heart**		
	Right ventricle	**Left ventricle**
Pumps blood to	Lungs	Rest of body
Size of wall	Thin	Thick
Oxygen content	Low	High
Carbon dioxide content	High	Low

Blood vessels attached to the heart

There are four major blood vessels attached to the heart, as seen in Figure 6.8.

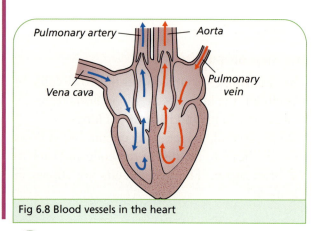

Fig 6.8 Blood vessels in the heart

Blood flow through the blood vessels in the heart

1 Deoxygenated (low oxygen) blood enters the right atrium through a vein called the vena cava.
2 The blood vessel that takes deoxygenated blood from the heart to the lungs is the pulmonary artery.
3 Oxygenated (oxygen-rich) blood flows from the lungs to the heart through the pulmonary vein.
4 Oxygenated blood flows from the heart through a large artery called the aorta.

Most of the arteries in the body carry oxygenated blood. The pulmonary artery is an exception to this rule as it carries deoxygenated blood.

Most of the veins in the body carry deoxygenated blood. The pulmonary vein is an exception as it carries oxygenated blood.

Pulse rates

On average the adult heart beats 70 times a minute. This figure can change depending on age, health and fitness.

As the blood surges down the arteries its pressure can be felt in areas of the body where the arteries are close to the surface, for example at the wrist and in the neck. The surge of blood in the arteries in these locations is called a pulse. The pulse rate is used to measure the rate of heartbeat.

Why we need exercise

● When we exercise, the cells in the body (especially in the muscles) need increased supplies of food and oxygen. In addition, the cells need to get rid of extra amounts of carbon dioxide and heat. As a result exercise causes the heart to beat faster and our pulse rate increases.
● As well as causing our heart to beat faster, exercise also causes the heart to contract more strongly. In the same way that exercise increases the strength of the muscles in our arms and legs, it also increases the strength of our heart.
● Exercise also helps to reduce weight. This means there is a lower demand for oxygen in the body.

For these reasons moderate exercise is good for the heart and the circulatory system in general.

ACTIVITY 6.1

To demonstrate the effects of exercise and rest on pulse rates

Method

1 Count the number of pulses in your wrist or neck per minute while at rest.

2 Repeat step 1 three times.

3 Add the three values together and divide by three to calculate the average pulse rate per minute at rest.

4 Exercise vigorously for two minutes (e.g. step up and down onto a chair repeatedly or run on the spot).

5 Immediately after exercising count the number of pulses per minute.

6 Continue to record the number of pulses per minute until the rate returns to the average resting rate.

7 Record how long it takes for the pulse rate to return to normal after exercise.

8 Compare the pulse rate at rest with the rate after exercise.

Results

Table 6.4		
Pulse rate per minute at rest	Average pulse rate per minute at rest	Pulse rate per minute after exercise
74		139
72	72	92
70		72

Conclusion

As shown by the sample results in Table 6.4 the pulse rate increases after exercise.

Heart disease

Heart disease is a major cause of illness and death in Ireland. It is mainly caused by small arteries that supply blood to the heart becoming clogged (with a fatty substance called cholesterol).

The risk of heart disease can be reduced by:

- Exercising regularly.
- Eating a healthy diet (especially by eating less fats and salt).
- Not smoking.

 Key Points

- Blood is composed of plasma, red blood cells, white blood cells and platelets.

- Plasma is the liquid part of blood and it transports chemicals and heat around the body.

- Red blood cells contain haemoglobin which transports oxygen.

- White blood cells fight infection.

- Platelets help to form blood clots.

- Arteries carry blood away from the heart.

- Veins carry blood to the heart.

- Capillaries allow materials to pass in and out of the bloodstream.

- The heart pumps blood around the body.

- The right ventricle pumps blood to the lungs, the left ventricle pumps blood to the rest of the body.

- The vena cava takes deoxygenated blood from the body to the right atrium.

- The pulmonary artery takes deoxygenated blood from the right ventricle to the lungs.

- The pulmonary vein takes oxygenated blood from the lungs to the left atrium.

- The aorta takes oxygenated blood from the left ventricle to the body.

- Each heartbeat causes a pulse of blood in the arteries.

- Exercise increases the pulse (and heartbeat) rate.

- To show the effect of exercise on pulse rate:
  - Count the number of pulses per minute at rest.
  - Count the number of pulses per minute after exercise.
  - Compare the two values.

- Heart disease can be avoided by:
  - exercise
  - a healthy diet
  - not smoking.

BIOLOGY

? Questions

1 (a) Why do humans need a circulatory system?
   (b) Name (i) the substance that circulates, (ii) the three types of tubes through which it circulates, (iii) the organ that causes circulation.

2 Copy out and complete the following table.

Substance	Carried from	Carried to
Oxygen		All body cells
	All body cells	Lungs
Glucose		All body cells
		Kidneys

3 (a) Name the four components of blood.
   (b) Give one function for each named component.
   (c) Name any two components of blood and say where they are formed in the body.

4 Write out and complete the following:
   (a) The normal temperature of the human body is _____. Our temperature may _____ due to exercise or _____. High temperatures are an attempt by the body to destroy _____.
   (b) The red pigment in human blood is called _____. The element _____ is needed to form this pigment. The function of this pigment is to transport _____.
   (c) Antibodies are made by _____ blood cells and are necessary to fight _____.
   (d) Blood clots help to reduce the loss of _____ from the body. They also help to prevent the entry of _____ into the body.

5 Figure 6.9 shows three blood vessels A, B and C.

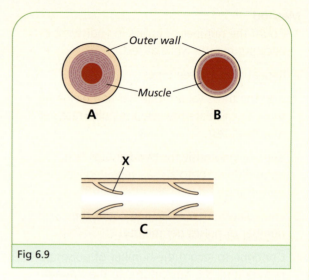

Fig 6.9

   (a) Name the types of blood vessels shown as A, B and C.
   (b) Name the structure labelled X.
   (c) Explain why structure X is needed in some blood vessels.

6 (a) In terms of blood flow, what is the difference between an artery and a vein?
   (b) Why can we feel a pulse in some arteries but not in veins?
   (c) Name the type of blood vessels that connect arteries and veins.
   (d) What is the function of the blood vessels named in part (c)?

7 Suggest a reason for each of the following:
   (a) Our blood is red in colour even though it contains white blood cells.
   (b) A lack of red blood cells often means a person has much less energy.
   (c) People who cannot produce white blood cells may get many infections.
   (d) We do not normally bleed to death when we get a cut.
   (e) Cardiac muscle should not get tired.
   (f) Our heart rate increases when we run.
   (g) There are valves in our heart.
   (h) Capillaries have very thin walls.
   (i) The walls of the left ventricle are stronger than the walls of the right ventricle.
   (j) Blood is pumped to the lungs.

Questions

8 Match the following functions with the structure that carries them out.

Function	Structure
Carry oxygen	Platelets
Pumps blood	Capillaries
Take blood to the heart	White blood cells
Kill bacteria	The heart
Connect arteries and veins	Veins
Form blood clots	Red blood cells

9 Figure 6.10 shows a human heart.

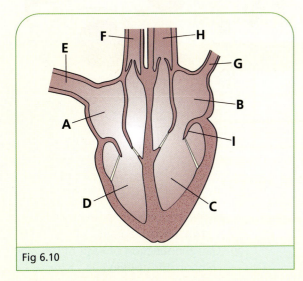

Fig 6.10

(a) Name the chambers labelled A, B and C.
(b) Name the structure labelled I.
(c) What is the function of the structure labelled I?
(d) Copy out this diagram (make it about the size of your clenched fist).
(e) On your diagram show by arrows the direction of blood flow in the blood vessels labelled E, F, G and H.
(f) Is the vessel labelled H an artery or a vein?
(g) For each of the chambers A, B, C and D say if the blood in them is oxygen-rich or oxygen-poor.

10 The pulse of a man at rest was taken three times. The values obtained were 66, 70 and 68 pulses per minute. The man then jogged on the spot for three minutes. His pulse rate was measured each minute after he stopped jogging. The values obtained were 116, 84 and 76 pulses per minute.
(a) Name one place in the body where the pulse rate may be felt.
(b) What causes a pulse?
(c) What was the average pulse rate of the man at rest?
(d) Why did the pulse rate increase after exercise?
(e) Why did the man's pulse rate fall after he stopped the exercise?
(f) Suggest one benefit of exercise.
(g) Suggest one problem that might result from having a high pulse rate.

11 Copy out and complete the following table:

Blood vessel	Blood flows		State of blood
	From	To	
Aorta		Body	Oxygenated
Pulmonary artery	Heart		
	Lungs	Heart	
Vena cava			Deoxygenated

12 Copy out and complete the following to describe the flow of blood:

Body → Vena _____ → Right _____ → _____ ventricle → Pulmonary _____ → _____ → _____ vein → _____ atrium → _____ _____ → _____ → Body.

? Questions

13 Design an investigation to find out if pulse rates depend on age.

Write a report of your method under the following headings:
(a) Method used to determine pulse rates.
(b) In selecting groups to test:
    (i) How many would be in each group?
    (ii) What feature would the members of each group share?
    (iii) What feature would be different for the groups selected?
    (iv) Apart from age, what other factors might effect pulse rates?

Examination Questions

14 (a) Blood consist of white blood cells, red blood cells and platelets in a liquid called plasma. Blood is carried around the body in arteries, veins and capillaries.
    (i) Give one function of blood.
    (ii) Give one difference between veins and arteries.
  (b) The heart pumps blood around the body.
    (i) Name the organ at which the blood arrives after it leaves the chamber marked X.
    (ii) Name the chamber labelled Y in the diagram.
    (JC, OL, Sample Paper)

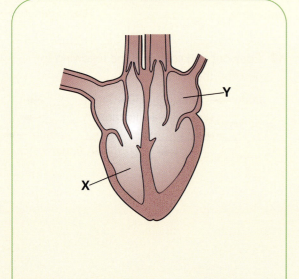

Fig 6.11

15 (a) Blood helps transport food and other materials around the body. It also helps fight infection.
    (i) Name the liquid part of the blood that helps to transport materials.
    (ii) Name the blood cells that help fight infection.
  (b) Blood is a liquid tissue. The diagram shows blood viewed through a microscope.

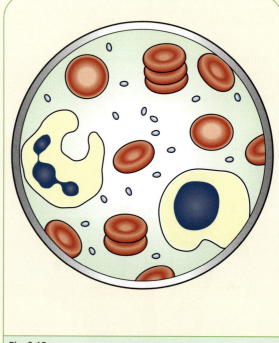

Fig 6.12

    (i) Name any two components of blood shown in the diagram.
    (ii) Give the function of each of the components of blood you have named.
    (iii) Why has the left ventricle got a thicker wall than the right ventricle of the heart?
    (Adapted from JC, HL, 2006; JC, OL, 2006)

BIOLOGY

? Questions

16 (a) The diagram shows the human heart. Why has the left ventricle got a thicker wall than the right ventricle?

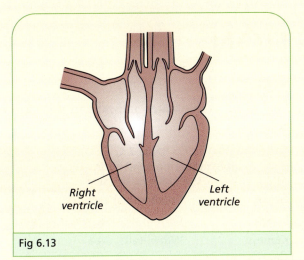

Right ventricle Left ventricle

Fig 6.13

(b) The diagram shows a person's pulse rate being taken.

Fig 6.14

(i) What causes a person's pulse?

(ii) How is a person's pulse rate measured using this method?

(iii) An athlete's resting pulse rate is 58. After 10 minutes of strenuous exercise the pulse rate was 120. After resting for 5 minutes the pulse rate reduced to 63. Clearly account for the rise and fall in the pulse rate experienced by the athlete.
(JC, HL, 2006)

BIOLOGY

Chapter 7

EXCRETION

Introduction

Many chemical reactions take place in the human body in order for it to work properly. Most of these reactions produce waste products that could be damaging to the body in large amounts. The body needs to remove waste products before they cause problems. The removal of this waste is called excretion.

> **Excretion** is the removal from the body of the waste products of chemical reactions in the body.

The main organs of excretion are the lungs, the kidneys and the skin. The main substances excreted by these organs are given in Table 7.1.

Lungs

The lungs excrete carbon dioxide and water vapour. These wastes are produced by respiration in all the living cells of the body. The way in which the lungs work is explained in Chapter 5.

Kidneys

The kidneys excrete water, salts and a nitrogen-containing compound called urea. The combination of these products is called urine.

The water and salts are produced from the liquids and salt-containing foods that we eat. Urea is produced when we break down excess proteins in the liver.

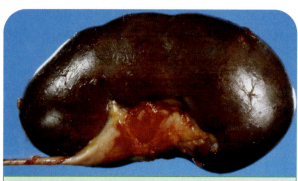

A human kidney and ureter

Skin

The skin excretes water and salts in the form of sweat.

Table 7.1 Substances excreted by the lungs, the kidneys, the skin	
Organ	**Substance(s) excreted**
Lungs	Carbon dioxide, water vapour
Kidneys	Water, salts, urea
Skin	Salts, water

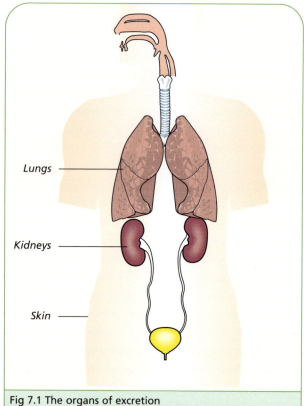

Lungs

Kidneys

Skin

Fig 7.1 The organs of excretion

Urea

Before studying how urine is formed it is necessary to understand urea.

Protein is essential for a healthy body. However, if we eat too much protein we cannot store it in the body. Excess protein can become poisonous.

To prevent it from damaging the body excess protein is taken in the bloodstream to our liver. In the liver protein is broken down to form a compound called urea. Urea is carried in the bloodstream from the liver to the kidneys.

The urinary system

The urinary system consists of the kidneys (and the blood vessels entering and leaving them), the ureters, the bladder and the urethra.

The parts of the urinary system

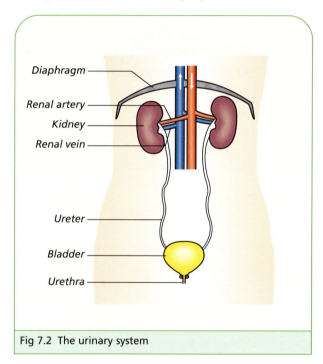

Fig 7.2 The urinary system

Renal arteries

The renal arteries carry blood into the kidneys. The blood in the renal arteries contains excess water and salts, along with urea that was made in the liver.

Kidneys

The kidneys remove excess water, salts and urea from the blood. These compounds are combined to form urine.

Renal veins

The renal veins take blood from the kidneys and carry it back towards the heart. The blood in the renal veins contains no waste materials.

Ureters

The ureters are tubes which carry urine from the kidneys to the bladder.

Bladder

The bladder stores urine until it can be released from the body.

Urethra

The urethra is a tube which takes urine from the bladder and passes it out of the body. The urethra passes urine out through the penis in a male and from the vagina in a female.

Filtration in the kidneys

Blood enters each kidney through the renal artery. In the kidney waste substances are removed from the blood by **filtration**.

Waste substances such as excess water, salts and urea are not needed by the body. The salts and urea dissolve in the water to form urine. Urine passes from the kidneys to the bladder through the ureters.

By reabsorbing different amounts of water the kidneys help to control the amount of water in the body. For example, when we are thirsty the kidneys reabsorb lots of water and help to retain water in the body. This means we produce less urine.

However, if we drink large amounts of water the kidneys reabsorb very little of it and so the extra water is excreted as urine.

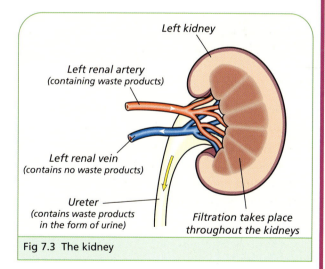

Fig 7.3 The kidney

The blood that leaves the kidneys in the renal veins does not contain waste products.

If the kidneys do not work properly then waste products build up in the blood. To prevent damage to the body the person's blood is filtered by a machine in order to remove waste products. This process is called dialysis.

Skin

Excess water and salts are carried to the skin by numerous arteries. As the blood passes through capillaries in the skin, some of the water and salts pass out of the blood. The combination of water and salts forms sweat in sweat glands in the skin.

Sweat passes to the surface of the skin through sweat ducts. Sweat is excreted from the skin through sweat pores.

As well as removing waste products from the body, sweat also helps to cool the body when it evaporates.

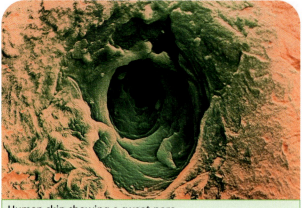

Human skin showing a sweat pore

🔑 Key Points

- Excretion is the removal from the body of the waste products of chemical reactions in the body.

- The main organs of excretion and their products are as follows:
  - Lungs excrete carbon dioxide and water.
  - Kidneys excrete water, salts and urea.
  - Skin excretes water and salts.

- Urea is:
  - made in the liver
  - formed from excess proteins
  - removed from the blood in the kidneys.

- The kidneys remove water, salts and urea from the blood to form urine.

- The ureters carry urine to the bladder.

- The bladder stores urine.

- The urethra carries urine out of the body.

- The kidneys:
  - filter water, salts and urea from the blood
  - form urine
  - control the amount of water in the body.

- The skin excretes water and salt in the form of sweat.

❓ Questions

1 (a) What is meant by excretion?
   (b) Why is excretion necessary?

2 (a) Name the organs of excretion.
   (b) Draw a diagram of the body to show the location of the excretory organs.
   (c) Name the main materials excreted by each of the excretory organs.

3 Apart from being an organ of excretion give another function for the lungs.

4 Figure 7.4 shows the human urinary system.

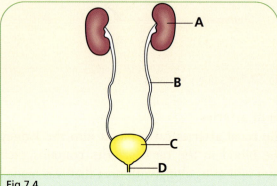

Fig 7.4

   (a) Name each of the parts labelled A, B, C and D.
   (b) Name the waste product produced in structure A.
   (c) What is the function of the part labelled C?

5 (a) Where in the body is urea made?
   (b) Name the type of food from which urea is made.
   (c) In what part of the body is urea removed from the blood?
   (d) How does urea get from where it is made to where it is removed from the blood?
   (e) Suggest one problem that might result from a failure to form urea.

BIOLOGY

? Questions

6 Write out and complete the following:

(a) Excretion is the removal of _____ products of _____ reactions from the body. The lungs excrete _____ _____ and _____. The kidneys excrete _____, _____ and _____.

(b) Urea is made from the breakdown of _____ in the _____. Urea passes to the kidneys in the _____ arteries. In the kidneys it becomes part of a waste product called _____.

(c) Urine is made in the _____, stored in the _____ and excreted through a tube called the _____.

7 Give a reason for each of the following:

(a) The human body has two ureters.

(b) Kidney failure can damage the body.

(c) We produce more carbon dioxide during the day than we do at night.

(d) There are more waste products in the renal arteries than in the renal veins.

(e) Our skin may taste salty after exercise.

8 Figure 7.5 represents the human urinary system.

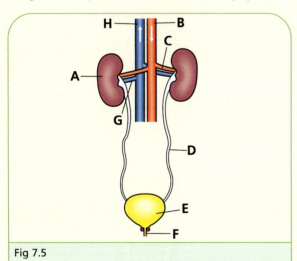

Fig 7.5

(a) Name the parts labelled A, B, C, D, E, F, G and H.

(b) Name the process that takes place in structure A that is responsible for the formation of urine.

(c) Name one substance in each case that is found in higher concentrations in (i) C compared to G, (ii) B compared to H, (iii) D compared to G.

(d) To what organ (not shown on the diagram) are B and H both connected?

9 (a) Name the substance excreted by the skin.

(b) Name two compounds contained in this substance.

(c) Name a structure in the skin responsible for excretion.

(d) Give two reasons why humans sweat.

Examination Question

10 Excretion is important for the removal of cellular wastes from the body. The urinary system has an important role in excretion from the body.

(a) Name two substances excreted from the human body.

(b) Identify the parts of the urinary system labelled A, B and C in the diagram below.

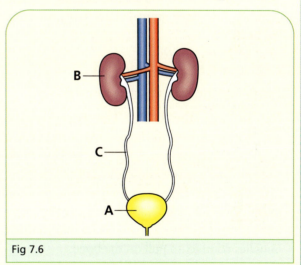

Fig 7.6

(c) Give one function of the part of the urinary system labelled A.

(d) Name a human organ of excretion other than an organ of the urinary system. (Adapted from JC, OL, Sample Paper)

BIOLOGY

Chapter 8 ●●●

THE SKELETAL AND MUSCULAR SYSTEMS

Introduction

Plant cells have cell walls which give them support. Animal cells do not have cell walls.

Some animals, such as jellyfish and earthworms, have no skeleton. They are soft-bodied animals.

Some animals have a skeleton on the outside (e.g. crabs). Advanced animals have bony, internal skeletons.

Functions of the skeleton

The main functions of the skeleton are:

1 **To support the body**. The bones of the skeleton act to keep the body upright and to give it shape. Without a skeleton our bodies would collapse like a tent without the tent poles.
2 **To allow movement**. A joint is where bones meet. Some joints are not able to move but many others are moveable. The bones are moved when muscles contract to pull on the bone.
3 **To protect the internal parts of the body**. The skull protects the soft material of the brain. The backbone is made of 33 bones (called vertebrae) which surround and protect the nerves in the spinal cord. Each vertebra can move slightly to allow us to bend our back.

The ribs protect the lungs and heart.

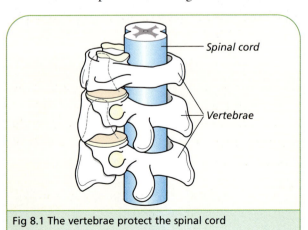

Fig 8.1 The vertebrae protect the spinal cord

The parts of the human skeleton

The major bones in the human body are shown in Figure 8.2.

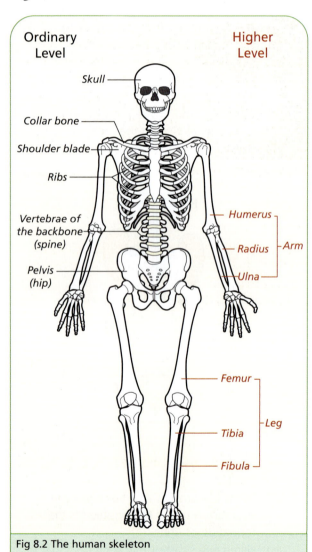

Fig 8.2 The human skeleton

Joints

Bones meet at joints. Some joints, such as those in the skull, do not move. These joints help to protect the body. However, most joints allow some amount of movement.

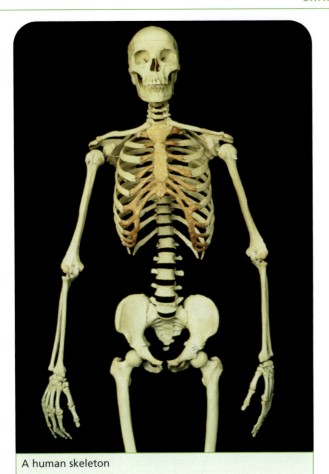

A human skeleton

Muscles

Muscles are made of protein. Muscles have the ability to move by contracting (i.e. getting smaller). Each muscle is attached to two different bones across a joint.

When a muscle contracts it pulls on one of the bones to which it is attached. This causes the bone to move, while the other bone does not move.

> **The function of muscles and joints** is to allow movement.

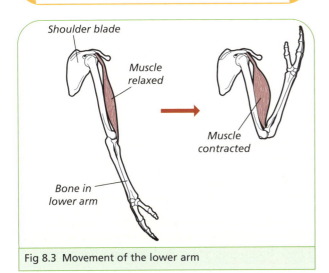

Shoulder blade

Muscle relaxed

Muscle contracted

Bone in lower arm

Fig 8.3 Movement of the lower arm

For example, in the upper arm a muscle (the biceps) is attached to the shoulder blade and to one of the bones in the lower arm. When this muscle contracts it pulls the lower bone in the arm upwards.

Tendons

Tendons attach muscle to bone. When a muscle contracts the pull of the muscle is transferred to the bone through a tendon.

To allow them to work properly tendons do not stretch. This is why they are sometimes damaged if we exercise too strenuously when we have not warmed up.

Tendons can be seen in the arm just below the wrist (especially if you clench your fist). They can also be felt in the underside of the thigh close to the knee.

The Achilles tendon connects the muscle in the calf of the leg to the heel bones. It is often injured by sportspeople.

> **Tendons** connect muscle to bone.

Ligaments

Ligaments are slightly elastic fibres. They attach bone to bone. Ligaments prevent joints from being damaged by reducing the amount of movement between the bones at a joint.

> **Ligaments** connect bone to bone.

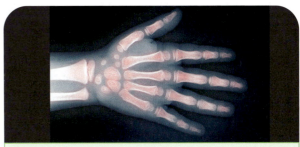

A coloured X-ray of a child's hand

Antagonistic pairs

Muscles can contract but they cannot make themselves elongate. Each muscle needs a second muscle to stretch it again. This means that muscles work in pairs, with each one being elongated when its partner contracts.

> **Antagonistic pairs** are pairs of muscles which carry out opposite functions.

BIOLOGY

For example, the biceps contracts to pull up the forearm. A second muscle called the triceps then contracts to pull down the forearm (while at the same time elongating the biceps).

By carrying out opposite effects the biceps and triceps form an antagonistic pair of muscles.

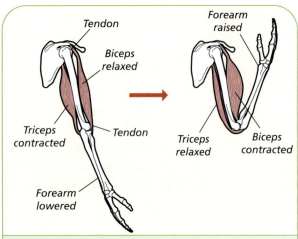

Fig 8.4 The action of the biceps and triceps

Table 8.1 Effects of contracting the biceps and triceps			
Muscle	State of muscle	Effect on forearm	Effect on second muscle
Biceps	Contracts	Raised	Elongates triceps
Triceps	Contracts	Lowered	Elongates biceps

Types of joints

There are two main types of joints: fixed joints and freely moving joints.

Fixed joints

In fixed (also called fused or immovable) joints there is no movement between the bones. The bones in the skull of an adult form a number of fixed joints.

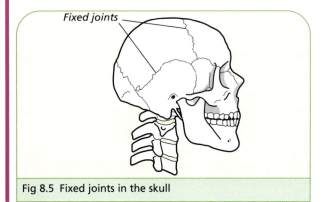

Fig 8.5 Fixed joints in the skull

Freely moving joints

Freely moving joints normally contain a liquid called synovial fluid. This liquid lubricates the joint and allows the joint to move more freely. These freely moving joints are called synovial joints.

Very often in synovial joints the ends of the bones are covered with a tough pad of cartilage. This protects the end of the bone and reduces friction in the joint.

Synovial joints are divided into two different types: ball and socket joints and hinge joints.

Ball and socket joints

Ball and socket joints are located in the shoulders and hips. They allow movement in all directions.

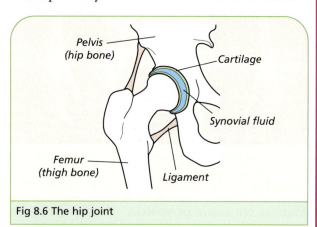

Fig 8.6 The hip joint

Hinge joints

Hinge joints are located in the elbow and knee. They allow movement in one direction only (similar to the hinge on a door).

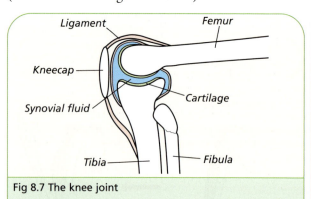

Fig 8.7 The knee joint

In many sports, especially football, it is common to suffer ligament damage, especially in the knee. Ligaments on the inner and outer sides of the knee are called medial ligaments. A more severe type of injury occurs when internal ligaments, which cross over inside the knee, are damaged. These ligaments are called cruciate ligaments.

Key Points

- The functions of the skeleton are:
  - support
  - allow movement
  - protection.

- A joint is where bones meet.

- Most joints allow movement to take place.

- Muscles are protein structures that can contract.

- Muscles cause bones to move.

- Tendons connect muscle to bone.

- Ligaments connect bone to bone.

- An antagonistic pair is two muscles which pull in opposite directions.

- Joints can be:
  - fixed (where there is no movement between the bones), e.g. the skull
  - freely movable or synovial which are of two types:
    (i) ball and socket joints which allow movement in all directions, e.g. shoulders and hips
    (ii) hinge joints which allow movement in one direction only, e.g. elbows and knees.

Questions

1 (a) Give three functions of the skeleton.
  (b) Explain briefly how each function is carried out.

2 Name the bones in each of the following cases:
  (a) The bones that form the backbone.
  (b) The bones that protect the brain.
  (c) The bone at the base of the spine to which the legs are connected.
  (d) The bone located between the top of the chest and the shoulder at the front of the body.
  (e) The bones that protect the lungs.

3 Write out and complete the following:
  (a) The skeleton _____ soft inner parts of the body from damage. It also gives our body its _____. The skeleton _____ the body and prevents it from collapsing. The _____ protect the spinal cord.

  (b) _____ contract to move _____ at joints. Muscles are made of _____ and connect to _____ on either side of a _____. An example of a muscle is the _____ in the upper arm. When the biceps _____ it causes the lower arm to be raised.

4 Name the bones labelled A, B, C, D, E and F on the diagram of the human skeleton.

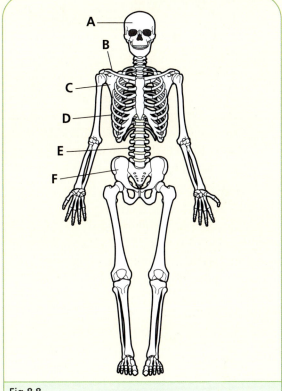

Fig 8.8

5 Give a reason for each of the following.
  (a) Plants do not have a skeleton.
  (b) A skeleton is not much use without muscles.
  (c) A muscle must be attached to a bone.
  (d) The backbone is made of 33 bones rather than a single bone.
  (e) Muscles can make themselves shorter.

BIOLOGY

BIOLOGY

? Questions

6 The diagram shows the arm of a human.

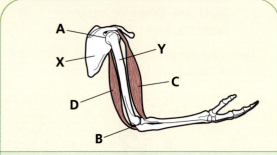

Fig 8.9

(a) Name the bones labelled X and Y.
(b) Name the type of joints found at A and B.
(c) Name the muscles labelled C and D.
(d) Name the muscle which contracts to
 (i) straighten the arm, (ii) bend the arm at the elbow.

7 Give a reason for the following:
(a) Muscles must work in pairs.
(b) People who have their cartilage removed may suffer from stiffness in the joint.
(c) Tendons do not stretch.
(d) The shoulder is a more flexible joint than the elbow.
(e) Some joints contain synovial fluid.

8 (a) What is a synovial joint?
(b) State two ways by which friction is reduced in a synovial joint.
(c) Why is there no synovial fluid in the skull?

9 (a) What is an antagonistic pair?
(b) Name any antagonistic pair.
(c) Give one function for each member of the antagonistic pair you named.

Examination Questions

10 (a) Name the bone of the human skeleton found between the hip and the knee.
(b) Name an organ that is protected by the skull.
 (Adapted from JC, OL, 2006)

11 The diagram shows the structure of an elbow. Name bone A and identify the type of moveable joint.
 (JC, HL, 2006)

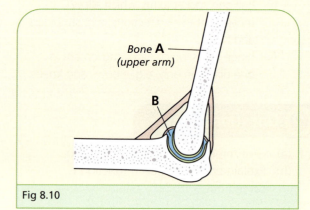

Fig 8.10

Chapter 9

THE SENSORY SYSTEM

Introduction

It is important for humans to be able to be made aware of changes in the world around us. Also we have to be able to respond to any changes of which we become aware.

For example, when we pour a cup of tea we need to be aware that the liquid is close to the top of the cup. Then we need to stop pouring and put down the teapot.

Our sensory system makes us aware of what is happening in and around us. It also allows us to respond to these changes.

Stimulus and response

Anything in our surroundings that causes us to take an action is called a **stimulus** (plural stimuli). The action we take is called a **response**.

For example, the sound of a car horn is a stimulus. Our response might be to jump out of the way of the car.

Sense organs

Our sensory system is made up of five senses. Each sense is associated with a different sense organ.

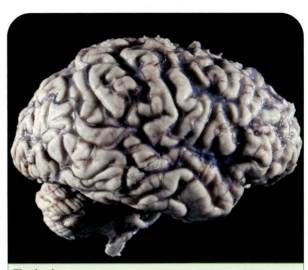
The brain

Table 9.1 Senses and sense organs	
Sense	**Sense organ**
Sight	Eyes
Hearing	Ears
Smell	Nose
Touch	Skin
Taste	Tongue

The central nervous system

The central nervous system consists of the brain and the spinal cord. The brain is a complex structure containing many millions of nerve cells. The spinal cord contains many nerves which take messages to and from the brain.

The rest of the nervous system contains nerves which run to and from the central nervous system.

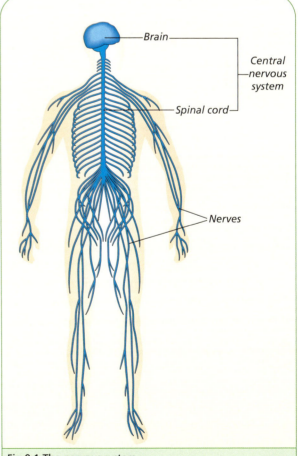
Fig 9.1 The nervous system

Sense organs and the central nervous system

Some nerves carry messages from the sense organs to the brain. In this way, the sense organs allow us to gather information from our surroundings. Messages are carried along a nerve as electrical impulses.

When the brain receives information from the sense organs it interprets the message, memorises it and decides on the correct response.

The brain then sends a message along different nerves to a muscle (or a group of muscles). In this way, the brain is responsible for causing us to respond to a stimulus.

For example, when pouring a cup of tea the eye detects that the cup is almost full of tea. This is the stimulus. A message is sent along a nerve from the eye to the brain.

The brain decides to stop pouring tea. It sends a message along another nerve, down the spinal cord and out to muscles in the arm or hand. We respond by raising the teapot.

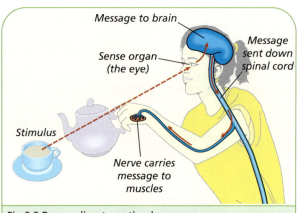

Fig 9.2 Responding to a stimulus

Sensory and motor nerves

A nerve cell is also called a neurone.

A sensory nerve carries messages to the brain or central nervous system. This means that sensory neurones carry impulses from the sense organs towards the brain. Sensory neurones make our central nervous system aware of a stimulus.

> **Sensory nerves** carry messages to the brain.

A motor nerve carries a message away from the brain or central nervous system. This means that motor neurones carry impulses from the brain (or central nervous system) to a muscle. Motor neurones cause us to carry out a response.

> **Motor nerves** carry messages away from the brain.

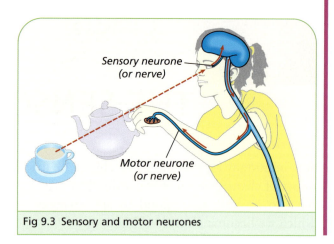

Fig 9.3 Sensory and motor neurones

The eye

The functions of the parts of the eye

Iris
The iris is the coloured part at the front of the eye. Its function is to expand or contract to control the amount of light entering the eye.

Pupil
The pupil is the black circle in the middle of the iris. Its function is to allow light to enter the eye. The pupil changes size due to changes in the iris.

In bright light the pupil is small (to prevent too much light entering the eye). In dim light the pupil enlarges (to allow more light into the eye).

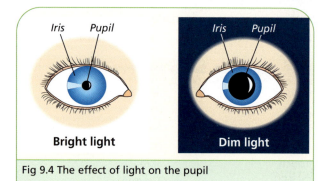

Fig 9.4 The effect of light on the pupil

Lens
The lens is a flexible structure. It changes shape depending on whether we are looking at a near or a far object. The function of the lens is to focus light on the retina.

If the lens does not take up the correct shape the object we are looking at will appear blurred. In such cases glasses or contact lenses are needed to allow us to see clearly.

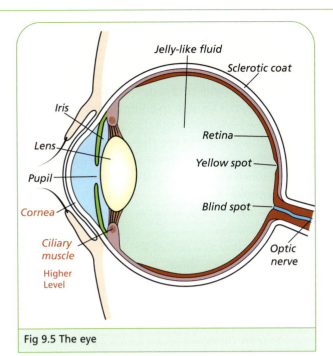

Fig 9.5 The eye

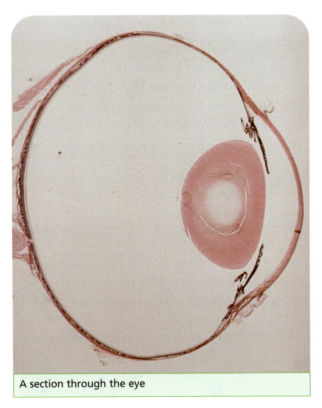

A section through the eye

Retina

The retina is a light-sensitive layer at the back of the eye. It contains millions of receptor cells which are sensitive to light. The function of the retina is to absorb light and to allow us to see.

Optic nerve

The optic nerve carries messages from the eye to the brain. It provides the link between the eye and the central nervous system.

Cornea

The cornea is a transparent section of the sclerotic coat. The function of the cornea is to allow light to pass into the eye.

Ciliary muscle

The ciliary muscle surrounds the lens. When the ciliary muscle contracts or relaxes it causes the lens to change shape.

Key Points

- The sensory system makes us aware of our surroundings and allows us to respond.

- A stimulus causes us to take an action called a response.

- The sense organs and their senses are:
  - eyes for sight
  - ears for hearing
  - nose for smell
  - skin for touch
  - tongue for taste.

- The central nervous system is made up of the brain and spinal cord.

- Some nerves carry messages from the sense organs to the brain.

- The brain decides how to respond to the stimulus.

- Different nerves carry messages from the brain to muscles to cause a response.

- Sensory nerves carry messages to the brain.

- Motor nerves carry messages from the brain.

- The functions of the main parts of the eye are:
  - The iris controls the amount of light entering the eye.
  - The pupil lets light into the eye.
  - The lens focuses light on the retina.
  - The retina absorbs light to allow us to see.
  - The optic nerve carries messages from the eye to the brain.
  - The cornea allows light to enter the front of the eye.
  - The ciliary muscle changes the shape of the lens.

BIOLOGY

BIOLOGY

❓ Questions

1 Rewrite the table to match the following sense organs and their senses:

Sense organ	Sense
Ear	Touch
Tongue	Sight
Skin	Smell
Eye	Taste
Nose	Hearing

2 Write out and complete the following:
   (a) Our _____ system makes us aware of changes in our surroundings. Anything that causes us to take an action can be called a _____. The action we carry out due to being stimulated is a called a _____.

   (b) Traffic lights turning red are an example of a _____. The response to the red lights should be to _____ the car.

   (c) The central nervous system consists of the _____ and _____ _____. Messages in the nervous system are carried in the form of _____.

3 Arrange the following in the order in which they work in the body:
   1 = muscle, 2 = nerve to the brain, 3 = sense organ, 4 = brain, 5 = nerve from the brain.

4 What is the main difference between sensory and motor nerves?

5 Figure 9.6 shows the eye.

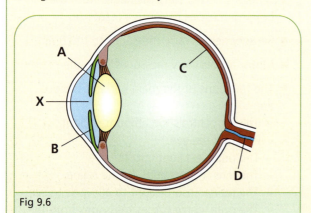

Fig 9.6

   (a) Name the parts of the eye labelled A, B, C and D.
   (b) Name and state the function of the part labelled X.
   (c) Name two other sense organs in the human body.

6 Give a reason for each of the following:
   (a) If the spinal cord is damaged the person may not be able to control those parts of the body below the damaged area.
   (b) The lens in the eye changes shape.
   (c) There is less electrical activity in the brain when we are asleep than when we are awake.
   (d) A damaged optic nerve may result in blindness.
   (e) The coloured part at the front of the eye can change shape.

7 (a) Draw a diagram of the eye and label on it the following parts:
       the iris, the pupil, the lens, the retina, the optic nerve.
   (b) Match the following parts of the eye with their functions.

Part of eye	Function
Retina	Lets light into the eye
Iris	Carries message to the brain
Lens	Controls light entering the eye
Pupil	Absorbs light
Optic nerve	Helps form a clear image on the retina

8 (a) Show the following parts on a diagram of the eye: (i) the cornea, (ii) the ciliary muscle.
   (b) Give the function for (i) the cornea, (ii) the ciliary muscle.

Examination Question

9 (a) Name the parts of the eye labelled X and Y in the diagram.
   (b) State the function of the part labelled X.

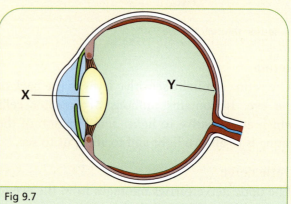

Fig 9.7

(Adapted from JC, OL, Sample Paper; JC, HL, 2006)

Chapter 10 •••

THE REPRODUCTIVE SYSTEM

Introduction

Reproduction is the production of new individuals. Reproduction is necessary to replace those who die. It may also result in an increase in the number of individuals.

Humans reproduce sexually. This means that a nucleus from a sperm joins with the nucleus from an egg. This process is called fertilisation.

Fertilisation results in a single cell called a zygote. During pregnancy the zygote grows in the uterus (womb) of the mother to form an embryo. As it gets older the embryo is called a foetus. The embryo or foetus can be called a baby. At birth the foetus passes out of the mother's uterus.

The male reproductive system

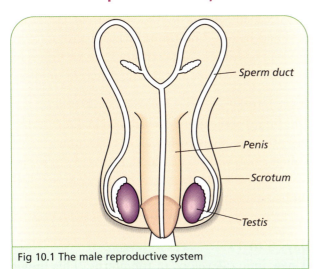

Fig 10.1 The male reproductive system

The functions of the main parts of the male reproductive system

Testis

The testis (plural testes) makes sperm.

The testes start to make sperm between the ages of 12 and 14. This is the age of sexual maturity (or puberty) in boys.

Other changes also take place in a boy's body at puberty. These include:

- a rapid growth spurt
- the enlargement of the penis and testes
- the deepening of the voice
- the growth of hair on the body.

Sperm are the male sex cells (also called the male gametes). Each sperm cell is tiny and they are produced in huge numbers by the testes.

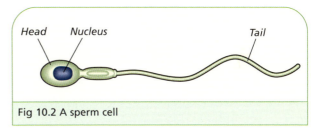

Fig 10.2 A sperm cell

Scrotum

The scrotum is a sac in which the testes are held. The scrotum allows the testes to be kept at a temperature just below body temperature. This allows sperms to be made successfully.

Sperm ducts

Two sperm ducts carry sperm from the testes to the penis.

A number of glands are located beside the sperm ducts. These glands produce a liquid called seminal fluid in which the sperm are released. The combination of sperm and seminal fluid is called semen.

A single sperm penetrating an egg

63

Penis

The sperm ducts join to form a tube called the urethra. Sperm pass through this tube, which is located in the centre of the penis.

The penis allows semen (or sperm) to pass out of the male body and into the body of the female.

The female reproductive system

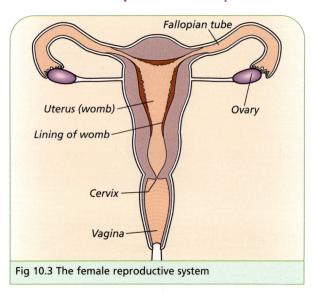

Fig 10.3 The female reproductive system

The functions of the main parts of the female reproductive system

Ovary

The ovaries produce eggs.

The ovaries start to make eggs at puberty. This occurs between the ages of 10 and 13 in girls. Other changes taking place in the girl's body at this time include the enlargement of the pelvis, breasts, vagina and uterus along with the growth of hair on parts of the body.

Eggs are the female sex cells or gametes. Each egg is larger than a sperm cell. Normally one egg is formed each month in the female body from puberty until about 50 years of age. At this stage, called the menopause (or the change of life), the female stops forming eggs.

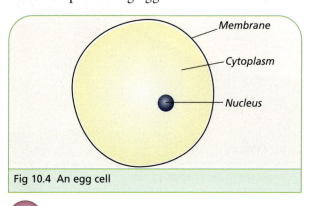

Fig 10.4 An egg cell

Fallopian tube

The fallopian tube collects the egg from the ovary and carries it to the uterus.

If sperm are present one of them may join with the egg in the fallopian tube. If there are no sperm present the egg dies.

Uterus

The uterus or womb is the area in which a baby (or embryo) will develop.

Cervix

The cervix is the opening or neck of the uterus. Sperm pass through the cervix in order to reach an egg.

Vagina

The vagina is a muscular tube into which the penis releases sperm. It forms the birth canal when the baby passes down the vagina at childbirth.

The menstrual cycle

In females one menstrual cycle takes place about every 28 days between puberty and the menopause. The menstrual cycle does not take place during pregnancy (i.e. when a baby is developing in the uterus).

The main events in the menstrual cycle are outlined below. The times given in this account are average times. These timings can be different in different females and during different months.

Days 1 to 5

The lining of the uterus (which had built up during the previous menstrual cycle) breaks down. This lining, along with some blood, is passed out of the body through the vagina. This process is called **menstruation** or having a period.

During these days an egg matures in the ovary.

Days 6 to 13

A new lining develops in the uterus. This lining will be needed to nourish a developing baby if the female becomes pregnant.

The egg continues to develop in the ovary.

Day 14

The egg is released from the ovary. This is called **ovulation**. The egg can survive for two days in the fallopian tube.

> **Ovulation** is the release of an egg from the ovary.

Days 15 to 28

The lining of the uterus remains in place until it breaks down on the first day of the next menstrual cycle.

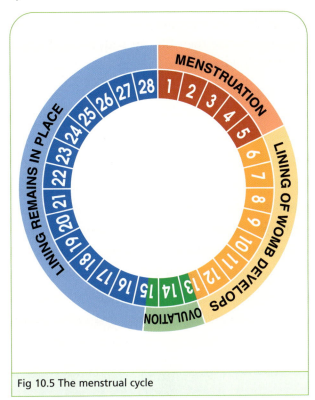

Fig 10.5 The menstrual cycle

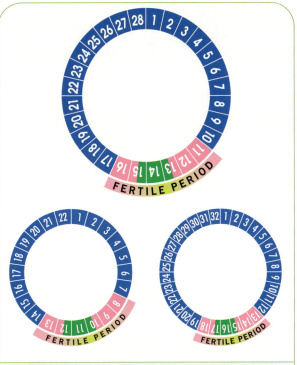

Fig 10.6 The fertile period depends on the length of the menstrual cycle

The fertile period

The fertile period is the days in the menstrual cycle when a female is most likely to become pregnant if she has intercourse.

Sperm can stay alive in the female reproductive system for up to three days. This means that pregnancy can occur if the female has intercourse three days *before* ovulation.

The egg can also stay alive for two days. This means that pregnancy can occur if the female has intercourse two days *after* ovulation.

Therefore, the fertile period lasts from day 11 to day 16 in a typical menstrual cycle. However, because menstrual cycles can be longer or shorter in different females, the fertile period is different for every female.

In some cycles the fertile period could start as early as day 8 or 9, while in other cycles it could last until day 18 or beyond.

Sexual intercourse

Sexual intercourse (which is also called copulation) takes place when the erect penis of the male is placed in the vagina of the female.

The movement of the penis in the vagina causes semen to be released from the penis. This is called ejaculation.

Path of the sperm

Normally over 100 million sperm are released into the vagina. The sperm swim through the cervix and into the uterus. They then swim from the uterus towards the fallopian tube.

If there is no egg present in the fallopian tube the sperm die within three days.

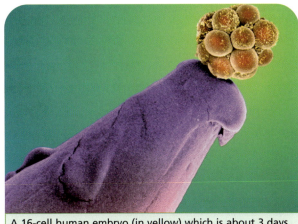

A 16-cell human embryo (in yellow) which is about 3 days old (shown on the point of a pin)

Fertilisation

After ovulation the egg (which is the female gamete or sex cell) is pushed along by tiny hairs in the fallopian tube. Many sperm (which are the male gametes or sex cells) swarm around the egg in the fallopian tube. Soon the head of one of the sperm will enter the egg.

Fertilisation takes place when the nucleus of the sperm joins or fuses with the nucleus of the egg. This takes place in the fallopian tube. A single cell called a zygote is formed as a result of fertilisation.

> **Fertilisation** is the joining of the male and female gametes to form a zygote.

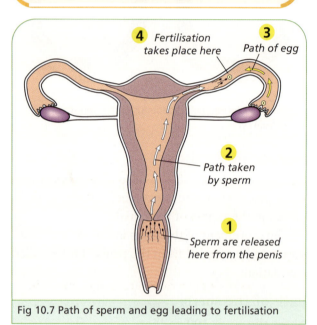

Fig 10.7 Path of sperm and egg leading to fertilisation

Pregnancy

Once the zygote has formed it soon goes through many cell divisions to form an embryo.

Within a few days of fertilisation the embryo (or young baby) becomes attached to the lining of the uterus. This attachment is called implantation.

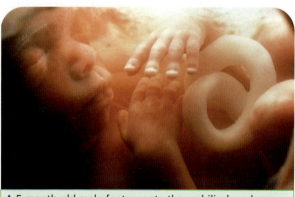

A 5-month-old male foetus: note the umbilical cord near his hands

> **Implantation** is the attachment of the embryo to the lining of the uterus.

Soon after implantation the baby becomes surrounded by a membrane called the amnion. This membrane then fills up with a liquid called amniotic fluid.

The amniotic fluid acts as a shock absorber to protect the baby during pregnancy.

> **Pregnancy** is the length of time the baby spends developing in the uterus.

Pregnancy normally lasts from implantation until birth. A normal pregnancy lasts 9 months or about 38 weeks.

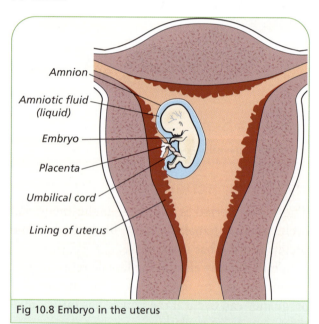

Fig 10.8 Embryo in the uterus

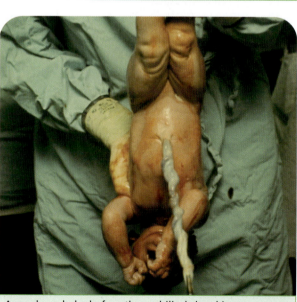

A newborn baby before the umbilical chord is cut

The placenta

The placenta forms early in the pregnancy. It attaches to the lining of the uterus. The baby's blood passes through the umbilical cord to and from the placenta.

The umbilical cord attaches to the baby at the navel (belly button). The blood of the baby and the mother do not mix in the placenta.

The function of the placenta is to allow food and oxygen to pass from the mother's blood into the baby's blood. It also allows waste products to pass in the reverse direction (i.e. from the baby to the mother).

Along with these useful functions the placenta also allows harmful substances to pass into the baby. These substances include alcohol, smoke and drugs (both legal and illegal).

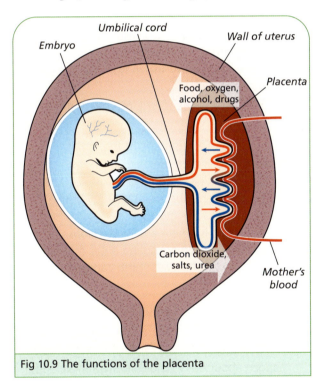

Fig 10.9 The functions of the placenta

The placenta's function is to allow materials to pass between the mother and the baby in the uterus.

Birth

Towards the end of pregnancy muscles in the uterus begin to contract. These contractions are called labour. Over the course of about 12 hours the contractions happen more often.

At some point these contractions cause the amnion to burst. The release of liquid through the

vagina is called 'the breaking of the waters'. The cervix gradually widens during these contractions.

The contractions become stronger and the baby is pushed head first out through the cervix and the vagina. The umbilical cord is clamped (to prevent loss of blood from the baby) and cut. The baby soon starts to breathe through its lungs for the first time.

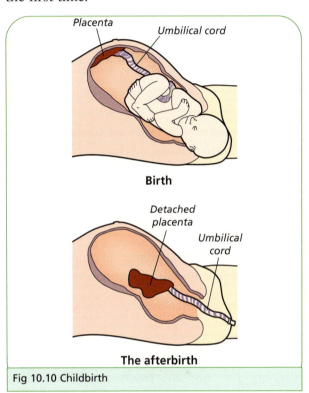

Fig 10.10 Childbirth

The uterus continues to contract after the baby is born. After about 20 minutes these contractions push the placenta and the remains of the umbilical cord out of the vagina. These materials are called the afterbirth.

Growth of the baby

The baby may feed on breast milk produced by the mother. This milk is full of the ideal nutrients that a young baby needs. In addition, it contains many substances (antibodies) which help to protect the baby from infection.

The remains of the umbilical cord fall away from the baby's navel after about seven days.

The different systems in the baby's body develop and become co-ordinated over the first few months of life. For example, after the first month the baby begins to hold its head up. Then it begins to push itself up on its hands. Soon after this it starts to turn to locate sounds and then begins to make gurgling noises.

BIOLOGY

The teeth start to emerge from the gums at around 6 months. Within the first year the baby begins to eat solid foods. It is then said to be weaned. Most babies begin to walk and say one or two words at about one year old.

The final organs to develop are the sex organs. These develop fully at puberty.

Contraception

Some couples want to control the number of children they have or to control how soon after each other their children are born. Others wish to have sexual intercourse without the female becoming pregnant.

These couples may use contraception as a method of birth control, or family planning, in order to prevent unwanted pregnancies.

> **Contraception** is the prevention of fertilisation or pregnancy.

There are two main types of contraception: preventing fertilisation and preventing implantation.

Preventing fertilisation

Natural methods

Natural methods of preventing the sperm from reaching the egg are based around avoiding intercourse during the female's fertile period. These methods aim to predict or detect the time of ovulation.

Artificial methods

Artificial methods of contraception include stopping the female from producing eggs. This can be achieved by the female taking the contraceptive pill.

Other artificial methods involve preventing the sperm from reaching the egg. These methods include the use of a condom which covers the top of the penis, a cap which covers the cervix, chemical creams or foams which kill sperm, and medical operations in which the sperm ducts or fallopian tubes are cut and sealed.

Preventing implantation

Some pills prevent pregnancy by stopping the embryo from attaching (or implanting) in the uterus. A plastic coil inserted in the uterus also acts in this way.

Some methods of contraception are more successful than others. In addition couples may decide that a number of these methods are not acceptable to them. This may be because they do not agree with them (i.e. morally) or because of their religious beliefs. For example, some people believe that preventing implantation is the same as destroying a life.

Key Points

- Reproduction is the production of new individuals.

- Puberty is when a person becomes sexually mature.

- The functions of the parts of the male reproductive system are as follows:
  - The testes make sperm.
  - The scrotum keeps the testes at the correct temperature.
  - Sperm ducts carry sperm from the testes to the penis.
  - The penis passes sperm into the body of the female.

- The functions of the parts of the female reproductive system are as follows:
  - The ovaries produce eggs.
  - The fallopian tubes transfer eggs from the ovary to the uterus.
  - The uterus is where a baby develops.
  - The vagina is where sperm enter the female body. It also forms the birth canal when a baby is born.

- The menopause is the age at which a female stops producing eggs.

- The menstrual cycle is a series of changes that happen about every 28 days in the female body.

- The menstrual cycle takes place:
  - between puberty and the menopause
  - only when a female is not pregnant.

- In each menstrual cycle:
  - the lining of the uterus breaks down and passes out of the body (a period occurs)
  - a new lining forms in the uterus and a new egg develops in the ovary
  - an egg is released from the ovary (ovulation)
  - the lining of the uterus remains in place until the next menstrual cycle.

- The fertile period is the days on which pregnancy is most likely to take place.

- Sexual intercourse takes place when the erect penis of the male is placed in the vagina of the female.

- Ejaculation is the release of semen from the penis.

- Fertilisation is the fusion of the male and female gametes to form a zygote.

- Fertilisation takes place in the fallopian tube.

- Implantation is the attachment of the embryo to the lining of the uterus.

- Pregnancy is the length of time the baby spends developing in the uterus.

- During pregnancy:
  - the baby is surrounded by a membrane called the amnion
  - amniotic fluid surrounds and protects the baby
  - the umbilical cord connects the baby and the placenta
  - the placenta allows substances to be exchanged between the mother and the baby.

- Birth is caused by contractions of muscles in the uterus of the mother.

- The afterbirth consists of the placenta and part of the umbilical cord.

- The early growth of a baby is helped by breast milk from the mother.

- Contraception is the prevention of fertilisation or pregnancy.

- Contraceptive methods include:
  - avoiding intercourse at certain times
  - using a condom, a cap or chemicals to kill sperm
  - surgery to close off the sperm ducts or fallopian tubes.

- Contraceptive methods which prevent implantation include certain pills or inserting a coil in the uterus.

? Questions

1 (a) What is meant by reproduction?
  (b) Give two reasons why reproduction is necessary.
  (c) Name the type of reproduction carried out by humans.

2 Figure 10.11 shows the male reproductive system.

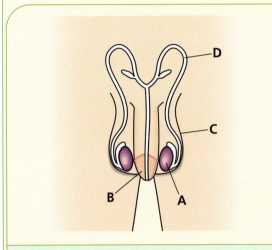

Fig 10.11

  (a) Name the parts labelled A, B, C and D.
  (b) In which labelled part (i) are sperm produced? (ii) is the temperature below body temperature?

3 (a) What is meant by puberty?
  (b) Give the age range at which puberty normally occurs in (i) males, (ii) females.
  (c) List three changes that takes place during puberty in (i) a male, (ii) a female.

4 (a) What is meant by gametes?
  (b) Name the gametes formed in (i) males, (ii) females.
  (c) Draw labelled diagrams of the human male and female gametes.
  (d) How do male gametes move?
  (e) How do female gametes move?

5 Figure 10.12 shows the female reproductive system.
  (a) Name the parts labelled A, B, C and D.
  (b) In which labelled part (i) are eggs made? (ii) does fertilisation take place? (iii) are sperm released by the penis?

? Questions

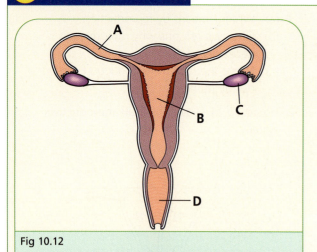

Fig 10.12

6 Write out and complete the following:
(a) Sperm are produced in the _____. Eggs are produced in the _____. Sperms and eggs are sex cells or _____. During _____ sperm are released from the _____ into the _____ of the female.
(b) The release of an egg from the ovary is called _____. This normally takes place on the _____ day of the _____ cycle. The time in the menstrual cycle during which fertilisation is most likely to occur is called the _____ _____.
(c) At fertilisation a _____ and an _____ unite to form a single cell called a _____. The single cell goes through many cell divisions to form an _____. The attachment of the embryo to the lining of the uterus is called _____.

7 Figure 10.13 shows a typical menstrual cycle.

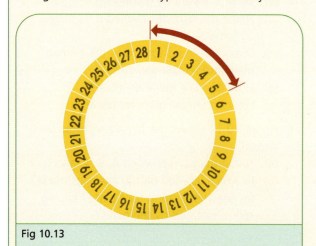

Fig 10.13

(a) What happens between days 1 and 5 of the cycle?

(b) Copy out this cycle and mark on your copy the days on which the following are most likely: (i) ovulation, (ii) the female may become pregnant, (iii) the egg dies, (iv) a new egg starts to develop.

8 Suggest a reason for the following:
(a) A woman is most likely to become pregnant during the middle of her menstrual cycle.
(b) Sperm are not formed if the testes are located inside the male body.
(c) During pregnancy the woman does not have a period.
(d) A baby in the womb does not breathe using its lungs.
(e) If a pregnant woman smokes or drinks her baby may be harmed.
(f) Breast-fed babies tend to get less infections than bottle-fed babies.
(g) A placenta is essential for the development of a baby.
(h) Sperm have tails.

9 Draw a diagram of the female reproductive system and mark on it where the following events take place.
(a) The entry of sperm
(b) Ovulation
(c) Fertilisation
(d) Implantation
(e) The formation of the placenta
(f) Birth.

10 Name a structure in each case found between
(a) the testis and the penis
(b) the placenta and the baby
(c) the ovary and the uterus
(d) the uterus and the vagina.

11 (a) What is meant by each of the following terms?
(i) ovulation
(ii) ejaculation
(iii) fertilisation
(iv) implantation
(v) pregnancy
(vi) contraception
(vii) menopause.
(b) Which of the above events are associated with
(i) males only?
(ii) females only?
(iii) males and females?

BIOLOGY

? Questions

12 Figure 10.14 shows the reproductive system of a pregnant female.
  (a) Name the parts labelled A, B, C, D, E and F.
  (b) Name two useful substances that may pass to the baby through structure C.
  (c) Name two harmful substances that may pass to the baby through structure C.
  (d) Name two substances that pass from the baby through structure C.

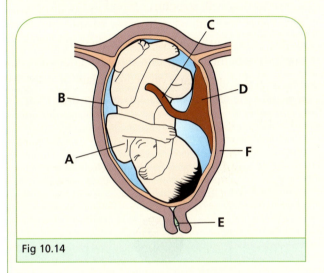

Fig 10.14

13 (a) What is meant by contraception?
  (b) State one reason why a couple might want to use contraception.
  (c) Name one method of contraception in each case which
      (i)   prevents the sperm and egg from joining
      (ii) prevents the zygote from developing.

Examination Question

14 The diagram shows the female reproductive system during the fertile period of the menstrual cycle.

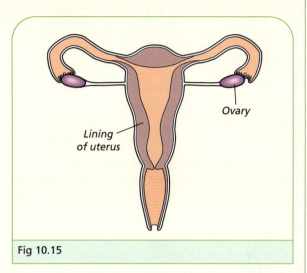

Lining of uterus

Ovary

Fig 10.15

  (a) What happens in the ovary during this time?
  (b) What happens to the lining of the uterus during this time?
      (JC, HL, 2006)

BIOLOGY

Chapter 11

GENETICS

Introduction

Humans have many traits or characteristics in common. For example, humans have one nose, two eyes, fingernails and many other shared features.

However, humans also show differences or variations. For example, humans may differ in their height, eye colour, length of eyelashes, accents, ability to speak a language or interest in a particular sport.

Some of these variations are passed on from parents to their children, while others develop during life.

> **Genetics** is the study of how traits or characteristics are inherited.

Inherited characteristics

Inherited characteristics are those that are passed on from parents to their children. These characteristics are controlled by structures in the nucleus of each cell called **genes**.

Examples of inherited characteristics include height, eye colour, length of eyelashes, shape of face and the ability to produce substances such as saliva, tears or enzymes. All these characteristics are controlled by genes.

Genetics is the study of how genes pass from one generation to the next.

Non-inherited characteristics

Non-inherited characteristics are not inherited at birth. They develop or are learned or acquired during life.

Examples of non-inherited characteristics are an accent, the ability to speak one or more languages, an interest in a sport, adding and subtracting numbers, writing, reading, riding a bicycle or using a computer. These characteristics are not controlled by genes.

Chromosomes

Chromosomes are structures found in the nucleus of each plant and animal cell. Most of the time they cannot be seen in a nucleus because they are stretched out into extremely long, thin threads. These threads are wound around each other like a ball of wool. In this state the chromosomes appear as a dark mass of material.

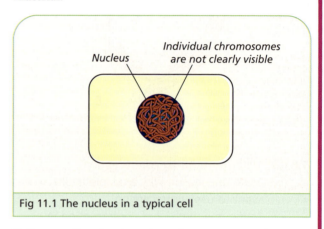

Fig 11.1 The nucleus in a typical cell

When cells divide, the chromosomes become shorter and thicker and can be seen using microscopes. At this time each chromosome looks like a tiny, thin thread.

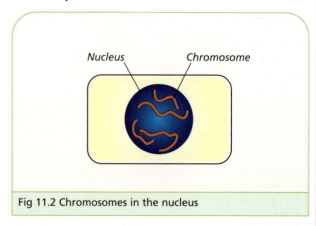

Fig 11.2 Chromosomes in the nucleus

Each chromosome is made of a chemical called DNA (deoxyribonucleic acid) and protein. The DNA is present as a pair of spirals. These are wrapped around some of the protein and held in place by surrounding protein.

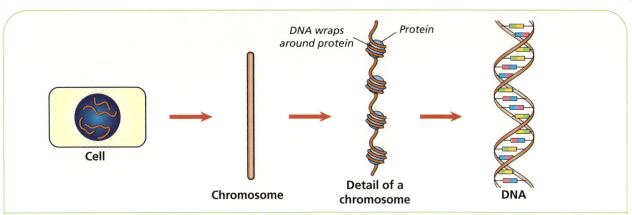

DNA wraps around protein   Protein

Cell

Chromosome

Detail of a chromosome

DNA

Fig 11.3 The structure of a chromosome

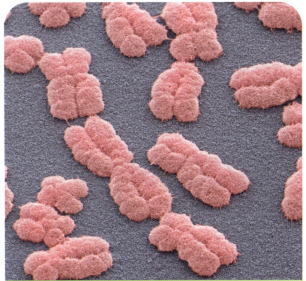

Pairs of human chromosomes at huge magnification. Chromosomes are present in the nucleus of every cell of the body

Genes

Genes are short sections of DNA. Genes are therefore located on chromosomes. Each human cell contains about 30 000 genes.

The proteins produced by genes can cause the production of inherited characteristics.

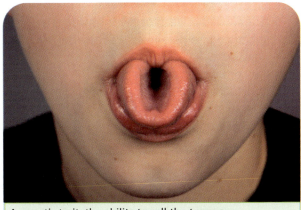

A genetic trait: the ability to roll the tongue

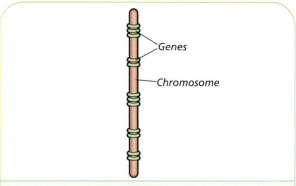

Genes

Chromosome

Fig 11.4 Genes on a chromosome

Chromosome numbers

Each cell (except sperm and egg cells) in the body of every human contains 46 chromosomes. The 46 chromosomes consist of two sets of 23 chromosomes.

The special type of cell division that forms sperm and eggs results in the number of chromosomes being halved. This means that sperm and eggs (the sex cells or gametes) only carry 23 chromosomes.

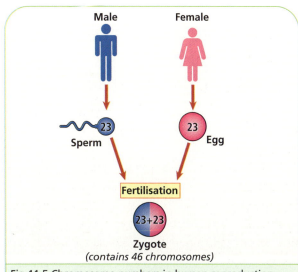

Male   Female

23   23
Sperm   Egg

Fertilisation

23+23

Zygote
(contains 46 chromosomes)

Fig 11.5 Chromosome numbers in human reproduction

BIOLOGY

At fertilisation the chromosomes from the sperm and egg unite to form a zygote. This means the zygote has the normal number of 46 chromosomes. Half of these chromosomes came from the mother and the other half came from the father.

A different type of cell division takes place in the zygote (and in the baby and adult formed from each zygote). This cell division means that each new cell has the same number of chromosomes as the parent cell.

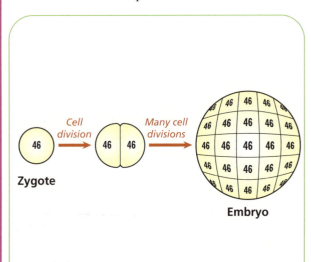

Fig 11.6 Cell division in human cells

As a result, all the cells in a person's body will have 46 chromosomes. In other words, there are two sets of 23 chromosomes in each human cell.

This means that the person who develops from each zygote inherits half of his or her chromosomes (and half of his or her genes) from each parent.

This is why most children have some characteristics that resemble their mother and some that resemble their father.

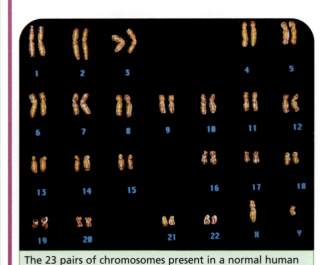

The 23 pairs of chromosomes present in a normal human male cell

Key Points

- Genetics is the study of how traits or characteristics are inherited.
- Inherited characteristics are controlled by genes and are passed on from parents to their children.
- Non-inherited characteristics are not controlled by genes but are learned or acquired during life.
- Chromosomes are located in the nucleus of a cell.
- Chromosomes are made of DNA and protein.
- Genes are located on chromosomes.
- Genes are short sections of DNA which control the production of a protein.
- Human cells have 46 chromosomes made up of two sets of 23 inherited from each parent.

Questions

1 (a) List three characteristics found in all humans.
   (b) List three characteristics which may differ between one human and another.

2 (a) Explain why some characteristics are said to be inherited characteristics.
   (b) Name five inherited characteristics in a human.

3 (a) Explain why some characteristics are said to be non-inherited characteristics.
   (b) Name five non-inherited characteristics in a human.

4 (a) What controls inherited characteristics?
   (b) How do non-inherited characteristics develop?

5 State whether each of the following is an inherited or a non-inherited characteristic or trait:
   (a) production of saliva
   (b) toenails
   (c) ability to play the guitar
   (d) speaking English
   (e) colour of hair
   (f) forming teeth
   (g) swimming
   (h) mathematics
   (i) production of sweat.

? Questions

6 Suggest a reason for each of the following:
  (a) A child may look like its parent.
  (b) The children of tall people are usually tall.
  (c) The children of musically talented parents are not born with these talents.
  (d) Genes do not control all our abilities.

7 (a) Name the chemicals of which chromosomes are made.
  (b) Where in a cell are chromosomes located?
  (c) Why are chromosomes not clearly visible in most cells?

8 (a) Where in a cell are genes located?
  (b) Name the chemical of which genes are made.
  (c) What is the function of a gene?

9 Figure 11.7 shows a cell during cell division as seen under the microscope.

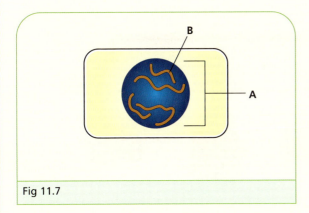

Fig 11.7

  (a) Name the structures labelled A and B.
  (b) Name the sub-units of structure B which control the production of proteins.
  (c) Draw a diagram to show this cell as it might appear when it is not taking part in cell division.

10 Say how many chromosomes are present in each of the following human cells:
  (a) a cheek cell
  (b) a sperm
  (c) an egg
  (d) a zygote
  (e) a gamete
  (f) a cell in the arm of a baby
  (g) a sex cell.

11 Suggest a reason for each of the following:
  (a) Chromosomes are not always clearly seen in a cell.
  (b) DNA is important for passing on traits.
  (c) The zygote has twice as many chromosomes as a sex cell.
  (d) Some of our characteristics are similar to our mother and some are similar to our father.
  (e) All normal body cells (not the sex cells) have 46 chromosomes.

Examination Question

12 Eye colour, hair texture and many other human characteristics are controlled by genes.
  (a) Name the structures in the nuclei of our cells where genes are located.
  (b) Name the substance that genes are made of.
  (JC, HL, 2006)

BIOLOGY

Chapter 12 ⬤⬤⬤

STRUCTURE AND TRANSPORT IN FLOWERING PLANTS

Introduction

There are many different types of plants. These include algae (such as seaweed), mosses and ferns (both of which live in damp places), conifers (which produce seeds in cones) and flowering plants.

Flowering plants are the most advanced type of plants. They include oak, ash and chestnut trees, grasses, cereals (such as wheat, oats and barley) and flowers (such as daffodils, tulips, daisies and dandelions).

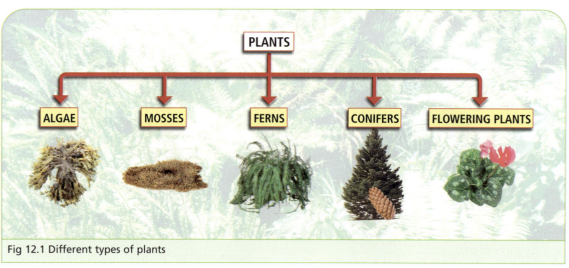

Fig 12.1 Different types of plants

Moss (green) growing on the bark of a tree

The main features of flowering plants are as follows:

- They are normally divided into roots, stems and leaves.
- They have special cells (or tissues) that allow them to transport water, minerals and food throughout the plant.
- They produce flowers (which are not always brightly coloured).
- They reproduce using seeds.
- The seeds are enclosed in fruits.

Structure of typical flowering plant

Each of the main parts of a flowering plant has its own special function. The roots are located below the ground. The shoot is composed of the parts of the plant that are above the ground (i.e. stems, leaves and flowers).

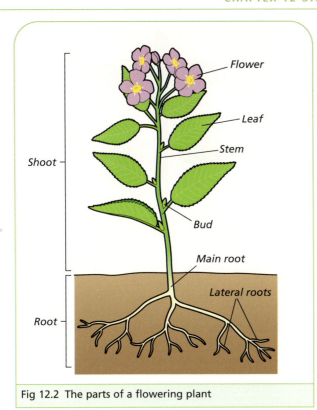

Fig 12.2 The parts of a flowering plant

Stems

- Stems support and hold up the leaves and flowers.
- They transport water and minerals from the roots to the leaves and also transport food from the leaves to the roots.
- Some stems store food (e.g. potatoes).

Leaves

- Leaves make food in a process called photosynthesis.
- They allow gases to pass in and out (in daylight carbon dioxide passes in and oxygen passes out).
- They allow water vapour to pass out of the plant.
- Some leaves store food (e.g. lettuce and cabbage).

Fig 12.3 The parts of a leaf

Root

- The roots anchor and support the plant in the soil.
- In addition, roots take in (or absorb) water and minerals (such as nitrogen, phosphorus and potassium) from the soil. The water and minerals pass up through the stem to the leaves where they are used by the plant.
- Some plants (e.g. carrots and turnips) store food in their roots.

Buds

- A bud is a potential growth point. New leaves or flowers may develop from buds.

A fern

A bud on a tree in winter

BIOLOGY

Flowers

The function of a flower is to produce seeds so that the plant can reproduce. (Flowers are dealt with in more detail in Chapter 14, Plant Reproduction.)

Hawthorn tree in blossom

Transport in flowering plants

Water transport

Water is carried in tiny tubes throughout a plant. These tubes allow a continuous flow of water from the roots to the leaves of a plant. The tiny tubes appear as the veins in a leaf.

Tiny openings (called stomata) on the lower surfaces of leaves allow water to evaporate from leaves into the air. This loss of water is called transpiration.

> **Transpiration** is the loss of water vapour from a plant.

As water is drawn out of the leaves into the air it pulls a column of water up through the tubes in the leaves, stem and roots.

The plant must absorb water from the soil in order to replace the water that passes up from the roots. Water passes from the soil into the tiny (invisible) roots of a plant.

This water will eventually pass from the roots to the stem. It will be drawn up the stem and into the leaves. From the leaves it will pass out into the air. The flow of water through a plant is called the transpiration stream.

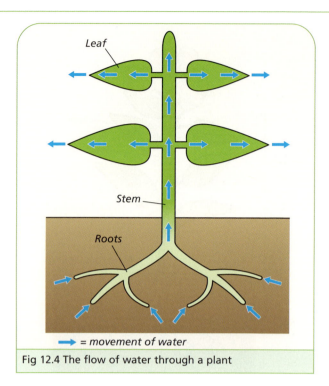

Leaf

Stem

Roots

→ = movement of water

Fig 12.4 The flow of water through a plant

> **The transpiration stream** is a flow of water from the roots to the leaves of a plant.

Functions of transpiration

- Transpiration supplies water to the leaves of plants for photosynthesis.
- It supplies minerals to the leaves of plants. This happens because minerals dissolved in water enter the roots. These minerals are carried to all parts of the plant, dissolved in water.
- Transpiration helps to cool plants (in the same way that sweating cools the human body).

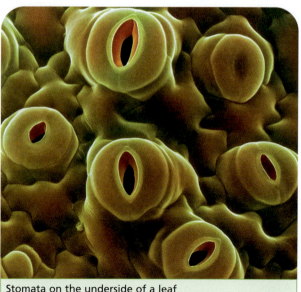

Stomata on the underside of a leaf

ACTIVITY 12.1

To show the path of water through a plant

Method

1 Add some food dye to water in a beaker.

2 Place a stalk of celery with some leaves attached into the beaker.

3 Leave it in a bright, warm room for a few days.

4 Remove the celery and rinse off the coloured water.

5 Observe the pattern of the coloured water in the stalk and leaf.

6 Cut sections through the stalk and note the pattern of the coloured water in the stalk.

Result

The coloured water is seen as lines of colour in the stalk and leaves.

Conclusion

This shows that water passes up through bundles of plant tissue in the stalk and leaves of celery.

Fig 12.5 Examining the flow of water in a plant

 ACTIVITY 12.2

To show that water is taken in by the roots and passes up and out of a plant

Method

1 Set up the apparatus as shown in the diagram.

2 The oil prevents water from evaporating.

3 Use masking tape to mark the top of the water level in the test tube.

4 Leave the apparatus in as warm a place as possible for about a week.

5 Check the level of the water in the test tube.

Result

The water level in the test tube goes down.

Conclusion

Water is absorbed by the roots, rises up through the stem and is lost from the leaves due to transpiration.

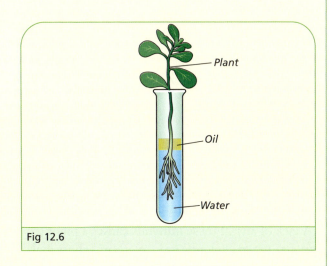

Fig 12.6

▶▶▶ **ACTIVITY 12.3**

To show that water evaporates from the surface of a leaf by transpiration

Method

1 Water a potted plant.

2 Set up the apparatus as shown in the diagram.

3 Leave the plant in a bright, warm room for a few hours (or overnight).

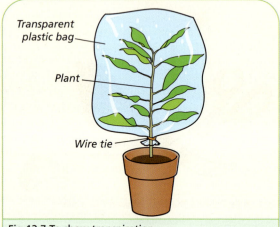

Fig 12.7 To show transpiration

Result

● Note that drops of liquid form inside the plastic bag.

● When dry cobalt chloride paper (blue) is touched against the liquid drops, it changes to a pink colour. This indicates that the liquid is water.

Conclusion

Water is lost from the leaves of a plant.

Mineral transport

Numerous minerals such as nitrogen, phosphorus and potassium are dissolved in water in the soil. Plants need these (and many other minerals) to allow them to grow properly.

A lack of any of these minerals causes a plant to lose its green colour and to show abnormal growth. Fertilisers are applied to the soil to add minerals when necessary.

These minerals pass into the roots of plants dissolved in water. They are transported around the plant as the water passes through the plant.

Food transport

Food is made in the leaves of a plant by photosynthesis. Some of this food is used by the leaves (for example to produce energy or new cell walls or for storage).

However, most of the food passes from the leaves to the other growing parts of the plant such as the stem, roots or flowers.

Food is carried or transported around the plant by a different type of tube to those that transport water.

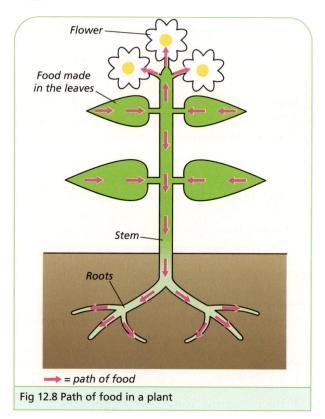

➡ = path of food

Fig 12.8 Path of food in a plant

Xylem and phloem

Plants have two types of transport (or vascular) tissue: xylem and phloem. Both of these tissues form continuous tubes throughout the plant, i.e. they are found in the roots, stems, leaves and flowers. They both carry liquid around the plant.

● **Xylem** is the tissue that carries water and dissolved minerals in plants. Generally water is carried up from the roots to the other parts of the plant, especially the leaves.

Xylem often forms a hard tissue in plants, e.g. it forms the wood in trees.

● **Phloem** is the tissue that carries food in plants. Generally the food is in the form of the sugar called sucrose (or table sugar) dissolved in water.

Phloem is a softer tissue than xylem.

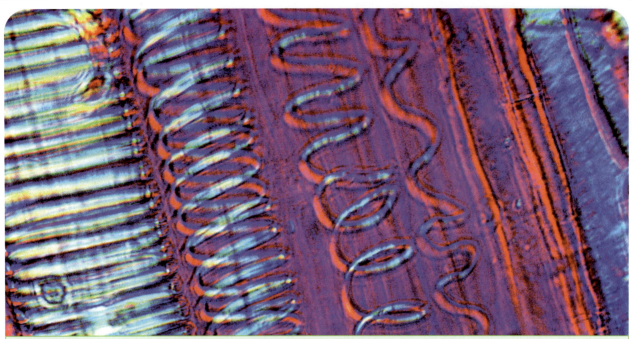

Xylem of a tree: note the spirals are supporting material around the xylem tubes

 ## Key Points

- Flowering plants are the most advanced type of plants.

- The main functions of the parts of a flowering plant are as follows:
  - Roots anchor the plant and take in water and minerals.
  - Stems support and transport materials to the upper parts of the plant.
  - Leaves make food, exchange gases and allow water to pass out.
  - Buds are future growth points.
  - Flowers produce seeds for reproduction.

- Transpiration is the loss of water from a plant.

- Most transpiration takes place through openings called stomata on the underside of leaves.

- The flow of water from the roots to the stems and into the leaves is called the transpiration stream.

- Transpiration:
  - supplies water to the leaves for photosynthesis
  - supplies minerals to the leaves
  - helps cool plants.

- To show the path of water through a plant:
  - Place a piece of celery in coloured water.
  - Observe the path taken by the coloured water as it rises through the plant.

- To show that water passes from the roots to the stem and out of the leaves:
  - Place the roots of a young plant in water.
  - Cover the water with a layer of oil.
  - Note that the level of water goes down over time.

- To show that water evaporates from the surface of a leaf by transpiration:
  - Cover a well-watered pot plant with a plastic bag.
  - Test the droplets of liquid with cobalt chloride paper (blue if dry, pink if water is present).

- Minerals enter a plant dissolved in water.

- Minerals pass around the plant dissolved in water.

- Food is transported from the leaves to the other growth areas in a system of tubes.

- Xylem transports water and minerals.

- Phloem transports food.

? Questions

1 Name the following:
   (a) Five types of plants
   (b) The most advanced type of plant
   (c) Five flowering plants.

2 Write out and complete the following:
   (a) Flowering plants are divided into an underground _____ system and a shoot system which is _____ the ground. The shoot system is made up of an upright _____, green _____ and _____ for reproduction.
   (b) Flowering plants have special tubes to transport _____, minerals (such as _____ and _____) and _____ around the plant. Flowering plants reproduce by means of _____ which are contained inside _____.
   (c) The roots of a plant help to _____ it in the soil. The _____ provides support for the aerial parts of the plant. The leaves make _____ and allow the gas _____ to pass out, while the gas _____ _____ passes in.

3 Rewrite the following so that the parts match the functions:

Parts	Functions
Leaf	Absorbs water and minerals
Flower	Carries out photosynthesis
Bud	Future growth point
Root	Supports the leaves and flowers
Stem	Carries out reproduction

4 (a) Name the parts labelled A, B, C, D and E on the diagram.
   (b) Which of these parts is associated with (i) making seeds? (ii) taking in water? (iii) exchanging gases?
   (c) Name one part of a plant that is not labelled on the diagram.

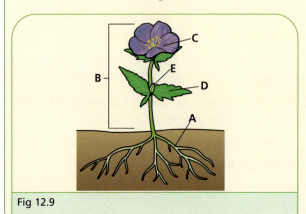

Fig 12.9

5 (a) Name the process by which water is lost from a plant.
   (b) From what part of a plant is water mostly lost?
   (c) Name the openings through which water is lost.
   (d) What is the transpiration stream?
   (e) What causes water to pass out of a plant?

6 (a) State one benefit of transpiration to a plant.
   (b) State one disadvantage of transpiration to a plant.

7 Name three parts of a plant in the correct sequence through which water passes during the transpiration stream.

8 A plant was set up as shown in the diagram.

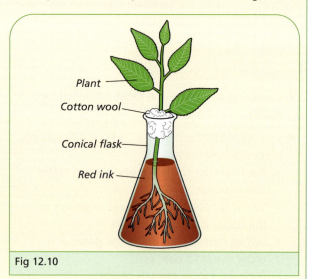

Plant
Cotton wool
Conical flask
Red ink

Fig 12.10

   (a) What will happen to the level of water in the conical flask after a few days?
   (b) What will you notice in the plant after a few days?
   (c) Name the process that causes the results observed in parts (a) and (b).

9 Give a reason for each of the following:
   (a) Plants produce flowers.
   (b) Fertiliser is spread on some soils.
   (c) Water passes out of leaves into the air.
   (d) Water rises up against gravity in plants.
   (e) Cobalt chloride may turn pink.
   (f) When showing the path of water through a plant, a dye is added to the water.
   (g) Plants have xylem.
   (h) Plants have phloem.

Questions

10 The apparatus shown in the diagram was set up and left for a few days.

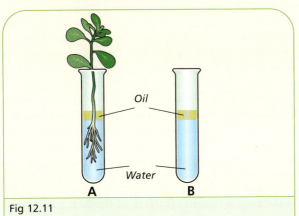

Fig 12.11

(a) What will happen to the water level in tube A after a few days?
(b) Why was the oil placed on the water?
(c) Draw diagrams to show the level of water in tube A and tube B after a few days.
(d) Explain the reasons for the changes (if any) in tubes A and B.
(e) Name the chemical used to test for water.

11 Shane set up the apparatus shown below in order to investigate transpiration.

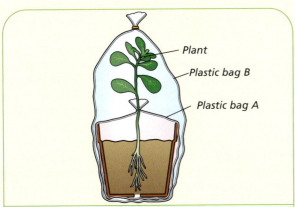

Fig 12.12

(a) Why was plastic bag A used?
(b) What changes would you expect to see after a few days in both plastic bags?
(c) Why should bag B be transparent?
(d) What chemical is used to allow Shane to conclude that transpiration has taken place?

12 (a) What function have xylem and phloem got in common?
(b) What is the function of xylem?
(c) What is the function of phloem?

Examination Question

13 (a) The plant in the test tube was allowed to stand in the laboratory for a few days to investigate the transport of water in the plant.

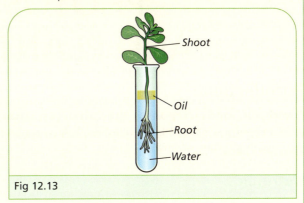

Fig 12.13

(i) Which part of the plant takes in water?
(ii) What would you notice about the level of water in the test tube after a few days?
(iii) Why is it necessary to put oil on the surface of the water in the test tube?
(b) Water evaporates from cells in the leaves of plants and exits the leaves by way of tiny pores in their leaves.

Fig 12.14

(i) What is this process called?
(ii) How would you test the drops of liquid inside the plastic bag covering the shoot of the plant shown in the diagram to show that the drops are water?
(c) What is the main function served by phloem tissue in a plant?
(Adapted from JC, OL, 2006; JC, HL, Sample Paper; JC, HL, 2006)

BIOLOGY

Chapter 13

PHOTOSYNTHESIS AND PLANT RESPONSES

Introduction

Nutrition and response are two of the characteristics of life. Plants display each of these characteristics.

Nutrition

The sun is the source of most of the energy on Earth. It provides heat to keep our planet warm. It also provides energy in the form of light.

Plants use solar (sun) energy to make food in a process called photosynthesis. In this way they convert energy from one form (light) to another form (the chemical energy in food).

Response

Plants also respond to features in their surroundings. For example, the shoots of a plant respond to light by slowly growing towards it. In addition, the roots of a plant respond to gravity by slowly growing down towards gravity.

Photosynthesis

One of the main differences between plants and animals is that plants can make their own food, while animals cannot do so. Plants make food by a process called photosynthesis.

> **Photosynthesis** is the way in which green plants make food.

Factors needed for photosynthesis

1 Carbon dioxide

Plants take carbon dioxide from the air into the leaves. The carbon dioxide passes into the leaves through tiny openings called stomata on the lower surface of the leaf.

2 Water

Water from the soil enters the roots and passes up through the stem to the leaves. The veins in the leaf carry water to all parts of the leaf.

3 Light

Light from the sun is absorbed by the leaves. Light provides the energy needed to form food.

Leaves have large flat leaf blades that enable them to absorb as much light as possible.

4 Chlorophyll

Chlorophyll is a green dye or pigment found in the leaves (and some other parts) of a plant. It absorbs light and allows photosynthesis to take place.

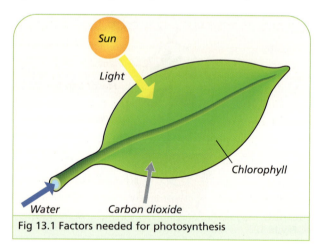

Fig 13.1 Factors needed for photosynthesis

Products of photosynthesis

1 Glucose

The food made by a plant is a carbohydrate or sugar called glucose. Glucose made in a leaf may be used in the following ways:

- to provide energy in respiration
- stored as starch in the leaves or transported out of the leaf through the veins, midrib and petiole to be stored as starch in other parts of the plant

- to form cellulose for new cell walls
- converted to fat or protein by the plant.

2 Oxygen

Oxygen is formed in photosynthesis. The oxygen made in this way may be used as follows:

- to provide energy in respiration in the leaf
- released out of the leaf and into the air.

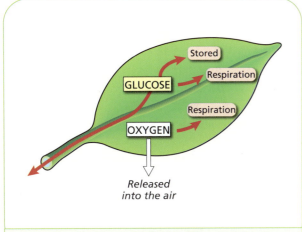

Fig 13.2 The products of photosynthesis and their uses

Equation for photosynthesis

Photosynthesis can be summarised by the word equation shown below:

carbon dioxide	sunlight	glucose
+	→	+
water	chlorophyll	oxygen

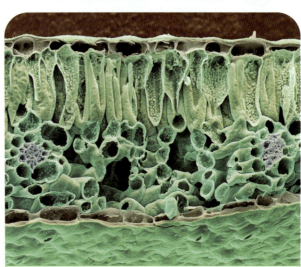

The inside of a leaf showing long vertical cells (where photosynthesis takes place), transport cells (in blue) and stomata on the lower surface

---

★ MANDATORY ACTIVITY 7

To show that starch is made by a plant in photosynthesis

Method

1. Place a plant in the dark for two days. This allows all the starch to move out of the leaves to other parts of the plant. The leaves are said to be destarched.

2. Cover some of the leaves of the plant with aluminium foil. This blocks out light and prevents photosynthesis. These leaves will act as a control.

3. Place the plant in strong light for a few hours (or overnight if necessary). This allows photosynthesis to take place.

4. Test some of the covered leaves and some of the uncovered leaves for starch, as outlined on the following page.

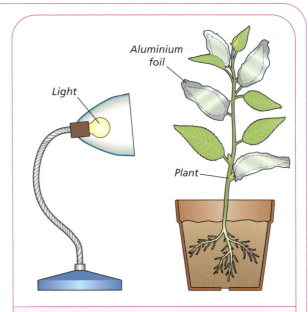

Fig 13.3 To show that starch is made in photosynthesis

BIOLOGY

To test leaves for starch

Method

1 Place the leaves in boiling water for a few minutes. This kills and softens the leaves.

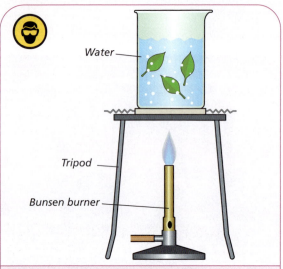

Fig 13.4 Boiling leaves in water

2 Soak the leaves in hot alcohol for about 10 minutes. This removes the chlorophyll from the leaves turning them a cream colour.

Chlorophyll is insoluble in water but soluble in alcohol. The alcohol turns green as chlorophyll dissolves in it.

The alcohol is heated by placing it in hot water as shown. The Bunsen burner should be turned off at this stage as alcohol catches fire very easily.

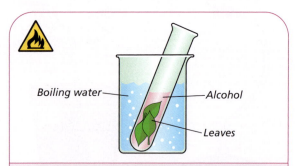

Fig 13.5 Removing chlorophyll from leaves

3 Place the leaves back into boiling water for a few seconds. This washes off the alcohol which makes the leaves brittle. The boiling water resoftens the leaves.

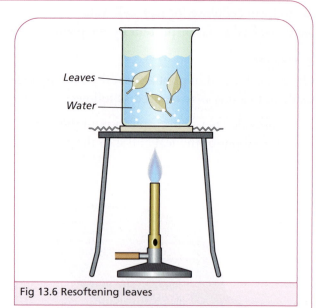

Fig 13.6 Resoftening leaves

4 Place the leaves on a white surface (a tile or white paper) and cover them with iodine solution (which is a red-yellow colour).

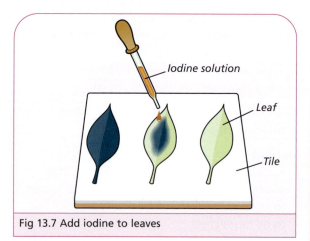

Fig 13.7 Add iodine to leaves

Results

1 The leaves that were uncovered (and which could carry out photosynthesis) turned a blue-black colour when iodine was added. This shows that starch was present.

2 The leaves which were covered with aluminium foil and which were in the dark (and could not carry out photosynthesis) remained a red-yellow colour when iodine was added. This shows that starch was not present.

Conclusion

Starch is produced by plants in photosynthesis (i.e. in the light).

Leaf starch test: the blue leaf was in the light (starch present), the brown leaf was in the dark (no starch)

Plant responses

Plants respond to factors or stimuli such as light and gravity by changing their growth. The change in growth of a plant in response to these factors is called a tropism.

A **tropism** is the change in growth of a plant in response to an outside stimulus.

Phototropism

Phototropism is the way in which a plant changes its growth in response to light.

The shoots of a plant respond by growing towards light. This allows the leaves to absorb as much light as possible. This, in turn, allows them to make more food.

Phototropism: the plants on the left were grown in normal light, the plants in the middle were grown in the dark, and the plants on the right were exposed to light from the right-hand side

▶▶▶ **ACTIVITY 13.1**

To investigate phototropism

Method

1 Divide a cardboard box (an old shoebox) into three sections as shown.

   Remove some cardboard from the side of the first section and from the top of the second section. Leave all the cardboard in the third section.

2 Place wet cotton wool into three petri dishes. Sprinkle small seeds (such as cress or mustard seeds) onto the cotton wool in each dish.

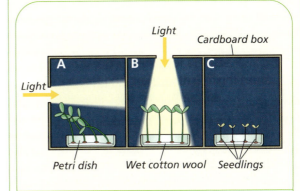

Fig 13.8 Investigating phototropism

3 Place one petri dish in each section of the box as shown.

4 Leave the box in a bright, warm room for about one week (check that the cotton wool remains wet throughout this time).

Results

1 The seedlings in dish A grew towards the light.

2 The seedlings in dish B grew straight up towards the light.

3 The seedlings in dish C were smaller and less green.

Conclusion

Plants (seedlings) grow towards the light.

Geotropism

Geotropism is the way in which a plant changes its growth in response to gravity.

The shoots of a plant respond by growing away from gravity. This allows the stem to grow upright so that the leaves get more light.

The roots of a plant grow towards gravity. This allows them to grow down into the soil where they can anchor the plant and absorb more water and minerals.

Geotropism: note how the root grows down in response to gravity

▶▶▶ ACTIVITY 13.2

To investigate geotropism

Method

1 Soak some broad bean or pea seeds in water for a day or two. This will allow them to absorb water and grow (germinate) better.

2 Line a beaker with blotting paper or filter paper.

3 Support the paper with damp compost, as shown in Figure 13.9.

4 Place the soaked bean seeds between the paper and the glass. Make sure each of the seeds is placed in a different position.

5 Leave the jar in a warm, dark place for about one week. Make sure the compost is kept moist all the time.

Result

The seeds grow or germinate. In each case it is seen that the shoots grow upwards while the roots grow downwards.

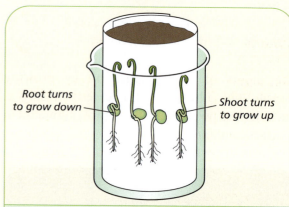

Root turns to grow down

Shoot turns to grow up

Fig 13.10 The positions of the roots and shoots after germination

Conclusion

Shoots grow upwards away from gravity while roots grow down in the direction of gravity.

Damp compost

Blotting paper

Soaked seed

Beaker

Fig 13.9 Setting up seeds to show geotropism

BIOLOGY

BIOLOGY

Key Points

- Photosynthesis is the way in which green plants make their food.

- Photosynthesis needs:
  - carbon dioxide
  - water
  - light
  - chlorophyll.

- Photosynthesis produces:
  - glucose
  - oxygen.

- The word equation for photosynthesis is:

carbon dioxide	sunlight	glucose
+	→	+
water	chlorophyll	oxygen

- To test a leaf for starch involves the following four steps:

Step	Reason
Boil the leaf in water	To kill and soften the leaf
Soak the leaf in hot alcohol	To remove the chlorophyll (green colour)
Place the leaf in boiling water	To resoften the leaf
Add iodine	To test for starch

- To show that starch is made by a plant in photosynthesis:
  - Put a plant into the dark for two days.
  - Cover some of the leaves with aluminium foil.
  - Leave the plant in bright light.
  - Test the covered and uncovered leaves for starch.
  - The uncovered leaves turn blue-black (starch made in photosynthesis).
  - The covered leaves remain red-yellow (no starch made as photosynthesis does not happen in the dark).

- A tropism is the change in growth of a plant in response to an outside stimulus.

- Phototropism is the way in which a plant changes its growth in response to light.
  - The shoots of a plant grow towards light.

- Phototropism can be investigated in the following way:
  - Grow some seedlings in lightproof boxes with holes cut in different positions.
  - The shoots of the seedlings are seen to always grow towards the light.

- Geotropism is the way in which a plant changes its growth in response to gravity:
  - The shoots of a plant grow away from gravity.
  - The roots of a plant grow towards gravity.

- Geotropism can be investigated in the following way:
  - Plant some seeds in different positions.
  - The shoots always grow upwards (away from gravity).
  - The roots always grow downwards (towards gravity).

Questions

1. (a) Name the main source of energy for the planet Earth.
   (b) Name two forms of energy that pass from this source to Earth.
   (c) Name the process that converts light energy into chemical energy.
   (d) Why does this process take place in plants but not in animals?

2. (a) What is photosynthesis?
   (b) Write out a word equation to represent photosynthesis.

3. (a) Name four factors needed by a plant in order to carry out photosynthesis.
   (b) Explain how plants get each one of these factors.
   (c) Which of the four factors is not available to animals?
   (d) In what part of a plant does photosynthesis normally take place?

4. (a) Name two substances formed in photosynthesis.
   (b) State two ways in which each of these two substances might be used in plants.

5. Give two reasons why photosynthesis is essential for life on Earth.

? Questions

6 Name a carbohydrate in each case that:
   (a)   is made in photosynthesis
   (b)   is stored by plants
   (c)   forms cell walls.

7 Write out and complete the following:
   (a)   Photosynthesis is the way in which _____ coloured plants convert a gas called _____ _____ and a liquid called _____ into a sugar called _____ and a gas called _____. This process requires energy in the form of _____ and a green dye called _____.

   (b)   A tropism is the response of a _____ to an outside stimulus. The shoots of a plant are those parts found above the _____. Shoots respond to light by _____ in the direction of the light. This response to light is called _____.

8 The following steps were carried out in an experiment:
   (a)   A plant was placed in a dark cupboard for 48 hours.
   (b)   A leaf of the plant was then tested for starch.
   (c)   The plant was placed in bright light for three hours.
   (d)   A leaf was removed from the plant and boiled in water for five minutes.
   (e)   The leaf was placed in hot alcohol for 10 minutes.
   (f)   The leaf was placed back into boiling water for two minutes.
   (g)   Iodine was added to the leaf.
   (h)   The leaf turned a dark blue colour.

   Give one reason for each of the eight steps (a) to (h) carried out in this experiment.

9 Give a reason for each of the following:
   (a)   Photosynthesis does not take place in the roots of a plant.
   (b)   The sun is essential for life on Earth.
   (c)   Most leaves have large flat leaf blades.
   (d)   Chlorophyll must be removed from a leaf before it is tested for starch.
   (e)   Shoots growing towards light benefit a plant.
   (f)   Most leaves turn blue-black when tested with iodine solution.

10 The apparatus shown in the diagram was set up to investigate the production of a gas in photosynthesis.

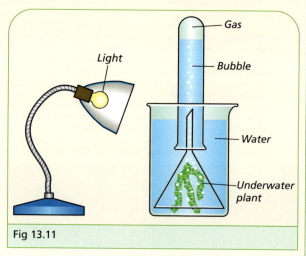

Fig 13.11

   (a)   Name the gas produced.
   (b)   Outline a test for this gas (if needed refer to Chapter 26, The Atmosphere).
   (c)   The rate of photosynthesis can be measured by counting the number of bubbles produced per minute.
      (i)   Suggest how you could vary the amount of light reaching the plant.
      (ii) How would the number of bubbles be affected by increasing the amount of light reaching the plant?
      (iii) What would happen to the bubbles if the apparatus were covered with a black sack?

11 The diagram shows two petri dishes of cress seeds grown under two different conditions.

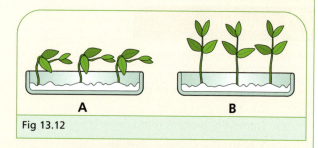

A   **B**

Fig 13.12

   (a)   Draw diagrams to show the conditions under which the seeds were grown.
   (b)   Name the process that caused the seeds in dish A to grow to the side.
   (c)   Suggest one way by which plants benefit from this response.

BIOLOGY

? Questions

12 (a) What is a tropism?

(b) What is meant by geotropism?

(c) Describe the response of
(i) shoots, and (ii) roots, to gravity.

(d) Why does the direction in which most seeds are sown not matter?

13 The diagram shows a germinating seed.

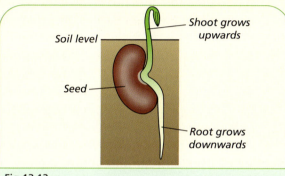

Fig 13.13

(a) Suggest one stimulus that causes the root to grow down.

(b) Suggest two stimuli that cause the shoot shown in the diagram to grow upwards.

(c) Draw the apparatus you would set up in order to observe the responses of a germinating seed to gravity.

Examination Questions

14 (a) (i) Complete the word equation for photosynthesis below.

(ii) Identify the substance X which must be present for photosynthesis to occur.

(b) What is the main function served by phloem tissue in a plant?

(c) The plant shown in Figure 13.14 was left in total darkness overnight and then exposed to strong sunlight for four hours. The leaf with the foil was removed from the plant and tested for starch.

Fig 13.14

(i) Clearly state the result you would expect from this test?

(ii) What conclusion can be drawn?
(Adapted from JC, OL, Sample Paper; JC, HL, Sample Paper; JC, HL, 2006)

15 The seedlings in the flower pot drawn in Figure 13.15 were grown in a closed box which had a window to let light in at one of the points X, Y or Z.

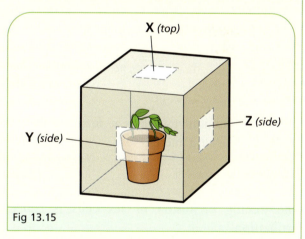

Fig 13.15

(a) Was the window at X, Y or Z?

(b) Give a reason for your previous answer.

(c) Why is the growth response helpful to the plants?

(d) What is the correct name of the growth response of the seedlings observed in this investigation?
(Adapted from JC, OL, 2006; JC, HL, Sample Paper)

BIOLOGY

Chapter **14** ○○○

PLANT REPRODUCTION

Introduction

Reproduction is the formation of new individuals. It is one of the characteristics of life.

There are two types of reproduction: sexual and asexual reproduction. This chapter will deal mainly with sexual reproduction in flowering plants.

The flower of a plant contains the structures of sexual reproduction. The process of plant sexual reproduction involves the following stages:

1 pollination
2 fertilisation
3 seed and fruit formation
4 seed and fruit dispersal
5 germination.

Sexual and asexual reproduction

Sexual reproduction

> **Sexual reproduction** involves two sex cells joining together.

The cells that join together at fertilisation are called sex cells or gametes.

> **A gamete** is a sex cell.

The joining or fusion of sex cells is called fertilisation.

As a result of sexual reproduction, the offspring have features or characteristics of both the male and female parents. This means that the offspring are not identical to the parents. This is the main advantage of sexual reproduction. Humans reproduce only by sexual reproduction.

Asexual reproduction

> **Asexual reproduction** means that new individuals are formed from only one parent.

Asexual reproduction does not involve sex cells or gametes. Fertilisation is not part of the process of asexual reproduction.

In asexual reproduction the offspring are identical to the parent. This can be a major disadvantage for asexual reproduction.

Plants reproduce by both sexual and asexual reproduction.

Example of asexual reproduction in a plant

Strawberry (and buttercup) plants produce special stems called runners which grow from the base of the parent plant.

At a certain distance from the parent plant the runner forms new roots and a new shoot. In this way new plants are produced from the original plant.

Runners are a method of asexual reproduction. Apart from having runners, plants use many different forms of asexual reproduction.

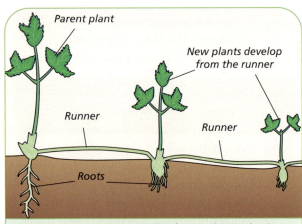

Fig 14.1 Runners as an example of asexual reproduction

For example, daffodil bulbs increase in number each year, spider plants produce new plants at the end of projecting stalks. People often take a cutting from plants such as geraniums, busy Lizzies or hedges.

Sexual reproduction in plants

The structure of a flower

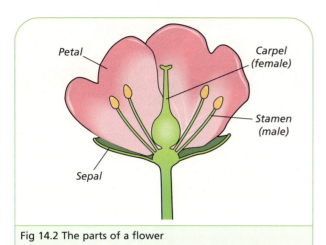

Fig 14.2 The parts of a flower

The functions of the parts of a flower

Sepals

The sepals protect the flower when it is a bud, i.e. before the petals open out and the flower blooms.

Fig 14.3 Sepals protecting a bud

Petals

Petals protect the internal parts of the flower. In many plants the petals are brightly coloured in order to attract insects to the flower.

The green sepals enclose the red petals of a poppy flower

Carpel

The carpel is the female part of the flower. It produces an egg cell which in turn will produce the female gamete. Some flowers have more than one carpel.

Stamen

The stamen is the male part of the flower. The stamen produces pollen grains which in turn will produce the male gametes. Most flowers have a large number of stamens.

A lily: note pollen grains on the tops of the brown stamens, the pink tipped carpel and white petals

Structure of the carpel

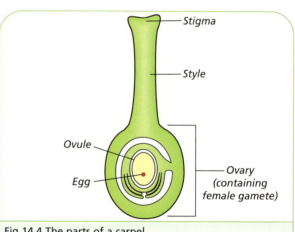

Fig 14.4 The parts of a carpel

Stigma

The stigma is the place where pollen grains will land.

Style

The style connects the stigma to the ovary. In some plants the style is very short.

Ovary

The ovary contains one or more ovules. Each ovule produces an egg. The egg produces a number of nuclei, one of which is the female gamete.

BIOLOGY

BIOLOGY

Structure of the stamen

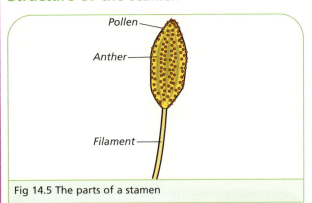

Fig 14.5 The parts of a stamen

Filament

The filament is a stalk that supports the anther. It ensures that the anthers are located high up in the flower so that the pollen can leave the flower more easily.

Anther

The anther makes pollen grains. These produce a number of nuclei, some of which become the male gametes.

Pollination

> **Pollination** is the transfer of pollen from a stamen to a carpel.

In order for sexual reproduction to take place the male gamete (located in the pollen) must reach the female gamete (located in the ovary).

Plants cannot move from place to place and therefore they depend on external agents to transfer the pollen to a carpel. The agents used for pollination are wind and insects.

The flowers of different plants are adapted to allow either wind or insect pollination. The main differences between wind and insect pollinated flowers are given in Table 14.1. When using this table refer also to the structure of the flowers shown in Figures 14.6 and 14.7.

A variety of pollen grains from different plants

Wind pollination

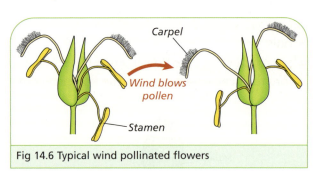

Fig 14.6 Typical wind pollinated flowers

Insect pollination

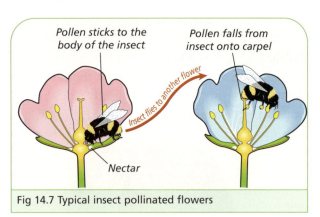

Fig 14.7 Typical insect pollinated flowers

Table 14.1 Differences between wind- and insect-pollinated flowers		
	Wind-pollinated flowers	**Insect-pollinated flowers**
Petals	Small and often green coloured No scent or nectar	Large and brightly coloured Have scent and nectar (a sweet liquid)
Pollen	Large amounts of small pollen grains	Small amount of larger pollen grains
Stamens	Located mostly outside the petals	Located inside the petals
Carpels	Located outside the petals	Located inside the petals
Examples	Grasses, oak and hazel trees	Buttercups, roses, dandelions

Insect pollination: note the pollen baskets on the bees legs

Fertilisation

> **Fertilisation** is the joining of the male and female gametes to form a zygote.

Pollen are transferred by wind or by an insect to the top of the carpel. The pollen then forms a tube which grows down through the carpel.

The male gamete passes down the pollen tube to the female gamete located in the lower part of the carpel. Fertilisation takes place in the base of the carpel.

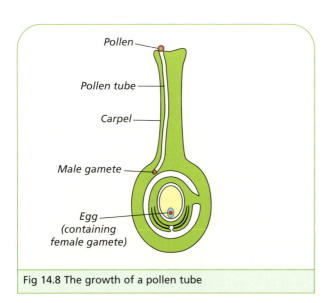

Fig 14.8 The growth of a pollen tube

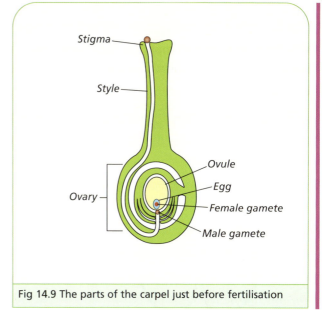

Fig 14.9 The parts of the carpel just before fertilisation

Fertilisation results in the formation of a single cell called a zygote. The zygote will grow to form a seed.

A seed consists of a young plant surrounded by a food supply. The seed or seeds (which are often known as pips) become surrounded by another food supply which is called the fruit.

Pollen tubes growing from pollen grains

Seed (and fruit) formation

Once fertilisation is complete the ovule forms a seed. Each seed starts off as a zygote surrounded by a food supply.

The zygote grows to form a baby plant called an embryo. The embryo consists of a radicle (which will form the future roots of the plant) and a plumule (which will form the future shoot of the plant).

After fertilisation, the ovule swells with food (such as starch and oils) and forms a hard, tough outer wall called the testa. The testa forms the coat of the seed.

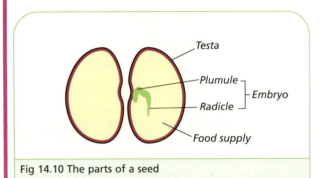

Fig 14.10 The parts of a seed

After fertilisation the ovary forms the fruit which surrounds and protects the seeds. If the ovary had many ovules then the fruit will contain many seeds. This is the case in an orange or a tomato.

If the ovary contained only a single ovule then the fruit will contain only a single seed. This is the case in peaches and plums (where the seed is called a stone).

Pollen being blown by the wind from the flowers of an alder tree

Seed (and fruit) dispersal

Dispersal is the carrying of the seed (and the surrounding fruit) as far as possible from the parent plant.

The main benefit of dispersal is that the young plants do not have to compete for scarce resources with the parent plant.

This means that dispersal reduces competition between the seedlings and the parent plant for light, space, water and minerals.

Plants use four main methods of seed dispersal: wind, animal, self and water dispersal.

Wind dispersal

Seeds that are wind dispersed are often small and light, e.g. orchid seeds.

Some wind dispersed seeds are larger but have special devices so that the wind can carry them longer distances.

For example, dandelions have hairy tufts which act like parachutes. Ash and sycamore seeds have wings which allow them to spiral down like helicopters.

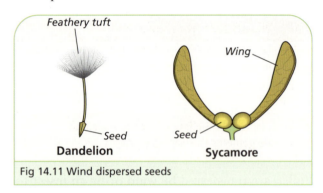

Fig 14.11 Wind dispersed seeds

Animal dispersal

Animals such as birds, bats and squirrels disperse seeds in the following two ways:

1 They swallow the fruits (and seeds) and pass out the seeds some time later.
2 The fruits or seeds may stick to the animal and be carried away to fall off some time later.

Fruits that are eaten are usually juicy and include strawberries, blackberries and tomatoes. The fruit is digested by the animal but the seeds (pips) are passed out in the faeces or droppings.

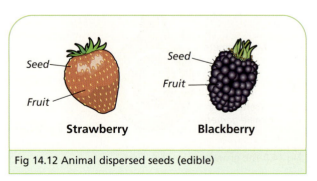

Fig 14.12 Animal dispersed seeds (edible)

Sticky seeds include burdock and cleavers or goosegrass.

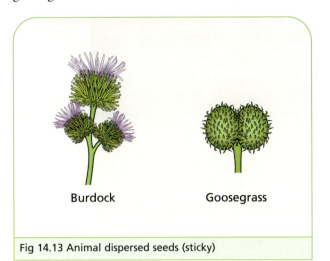

Fig 14.13 Animal dispersed seeds (sticky)

Self dispersal

Self dispersal usually involves the fruit (pod) bursting open when it is ripe. In this way the seeds are flung away as far as possible. Examples of this type of dispersal are peas, beans, lupins and gorse plants.

Fig 14.14 Self dispersed seeds

Water dispersal

Some fruits or seeds are able to float. This allows them to be carried away by streams, rivers and ocean currents. Examples include coconuts, water lilies and alder.

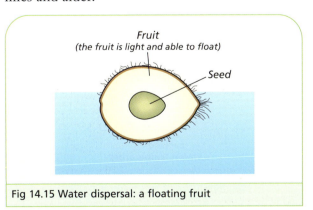

Fig 14.15 Water dispersal: a floating fruit

Germination

Germination is the growth of a seed to form a new plant.

Once the seeds have been dispersed they will germinate (or sprout) provided the conditions are suitable.

If the conditions are not suitable the seeds will not germinate. In this case they remain dormant or resting in the soil.

The conditions necessary for germination are:

● water
● oxygen
● a suitable temperature, i.e. warmth.

The condition which often prevents germination in winter is the low temperature. Once the temperature rises in spring seeds can germinate.

The main events in germination

When the conditions are suitable the seed absorbs water. The young root or radicle absorbs food from the seed.

The root bursts through the seed coat and grows out of the seed. The food supply in the seed is also used to allow the young shoot to grow out of the seed.

During germination the seed shrivels up as its food supply is used to allow the shoot and root to develop.

Once the shoot emerges above the ground it forms green leaves and begins to make food for itself.

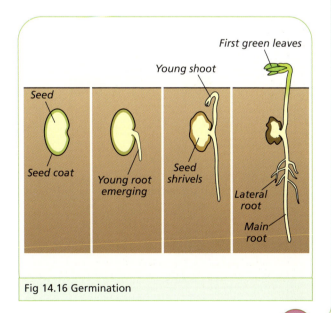

Fig 14.16 Germination

Stages of germination of a soya bean plant

 MANDATORY ACTIVITY 8

To investigate the conditions necessary for germination

Method

1 Set up four test tubes as shown in the diagram.

- Tube A has five seeds on wet cotton wool and is in a warm place.
- Tube B has five seeds on dry cotton wool and is in a warm place.
- Tube C has five seeds covered by boiled, cooled water.

Boiling removes the oxygen from the water. A layer of oil is placed on top of the water to prevent oxygen passing back into the water.

- Tube D has five seeds on wet cotton wool and is placed in a cold place (such as a fridge).

2 Leave tubes A, B and C in a warm room. Put tube D in a cold fridge.

3 Observe the seeds each day for about one week.

Result

1 The seeds in test tube A germinate.

2 The seeds in test tubes B, C and D do not germinate. These seeds were missing water, oxygen and a suitable temperature, respectively.

Conclusion

This shows that water, oxygen and a suitable temperature are all needed for germination.

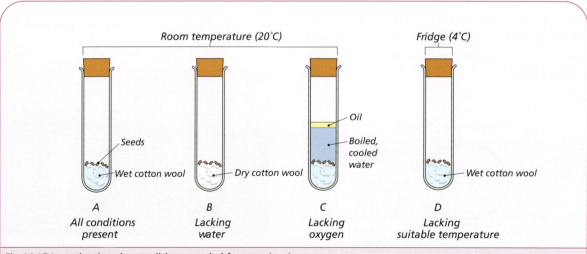

Fig 14.17 Investigating the conditions needed for germination

 **Key Points**

- Reproduction is the formation of new individuals.

- The flower is an organ of sexual reproduction.

- Sexual reproduction in plants involves:
  - pollination
  - fertilisation
  - seed and fruit formation
  - seed and fruit dispersal
  - germination.

- Sexual reproduction involves two sex cells joining together.

- A gamete is a sex cell.

- Asexual reproduction means that new individuals are formed from only one parent.

- Strawberries produce runners which allow them to reproduce asexually.

- The functions of the parts of a flower are:
  - Sepals protect the flower when it is a bud.
  - Petals protect the reproductive parts and may attract insects.
  - Carpels produce egg cells which contain the female gametes.
  - Stamens produce pollen grains which contain the male gametes.

- A carpel consists of:
  - a stigma for pollen grains to land on
  - a style which connects the stigma to the ovary
  - an ovary which contains one or more ovules (each ovule contains an egg cell).

- A stamen consists of:
  - a filament which supports the anther
  - an anther which produces pollen.

- Pollination is the transfer of pollen from a stamen to a carpel.

- The two methods of pollination are wind and insect pollination.

- Flowers are adapted for a particular type of pollination in the following ways:
  - Wind pollination: Small, green petals with no nectar or scent; large amounts of pollen; stamens and carpels are mostly outside the petals.
  - Insect pollination: Large, brightly coloured petals with nectar and a scent; smaller amounts of pollen; stamens and carpels are mostly inside the petals.

- Fertilisation is the joining of the male and female gametes to form a zygote.

- A pollen tube allows the male gamete to pass down to the female gamete in the carpel.

- After fertilisation the ovule forms a seed.

- A seed consists of:
  - a plumule which will form the shoot of the adult plant
  - a radicle which will form the roots of the adult plant (the plumule and radicle form the young plant or embryo)
  - a food supply around the embryo
  - a seed coat called the testa.

- The ovary forms the fruit, which may have one or more seeds.

- Dispersal is the carrying of the seed or fruit as far as possible from the parent plant.

- Seed dispersal reduces competition between seedlings and the parent plant.

- The main methods of dispersal are:
  - wind
  - animal
  - self
  - water.

- Germination is the growth of a seed to form a new plant.

- The conditions necessary for germination are:
  - water
  - oxygen
  - a suitable temperature, i.e. warmth.

- To investigate the conditions necessary for germination, place seeds in four tubes so that:
  - tube A has water, oxygen and a suitable temperature (germination takes place)
  - tube B has no water (no germination)
  - tube C has no oxygen (no germination)
  - tube D is in a low temperature (no germination).

BIOLOGY

? Questions

1 (a) What is meant by reproduction?
   (b) What would happen if living things did not reproduce?

2 (a) Name the two main types of reproduction.
   (b) Name one plant in each case involved in each type of reproduction.
   (c) Which type of reproduction is involved in each of the following?
      (i) the formation of offspring which are identical to the parent
      (ii) two cells joining together
      (iii) the production of a zygote
      (iv) human reproduction
      (v) the formation of runners in strawberries
      (vi) gametes
      (vii) one parent only
      (viii) fertilisation.

3 Rearrange the following into the order in which they occur in plant reproduction:
   (a) germination
   (b) pollination
   (c) seed dispersal
   (d) fertilisation
   (e) seed formation.

4 What is meant by each of the following terms?
   (a) sexual reproduction
   (b) asexual reproduction
   (c) gamete
   (d) pollination
   (e) fertilisation
   (f) dispersal
   (g) germination.

5 Figure 14.18 shows the structure of a typical flower.

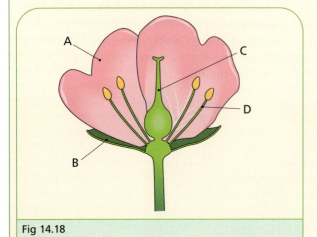

Fig 14.18

(a) Name the parts labelled A, B, C and D.
(b) Why are the structures labelled A sometimes brightly coloured?
(c) Which letter represents (i) a male structure? (ii) a female structure?
(d) What is produced at the top of structure D?
(e) What is produced in structure C?
(f) What is the function of a flower?

6 Rewrite the following table so that the parts match the functions.

Part	Function
Sepal	Forms female gametes
Petal	Organ of sexual reproduction
Carpel	Forms pollen
Stamen	Protects the flower as a bud
Flower	Attracts insects

7 Figure 14.19 shows a flower.

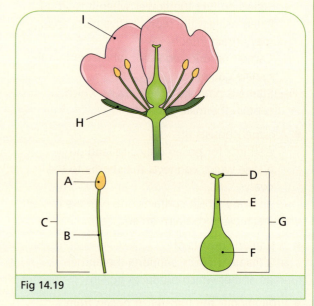

Fig 14.19

(a) Name all the parts labelled in the diagram.
(b) Name the labelled part associated with the following:
   (i) The production of pollen
   (ii) Protecting the flower in bud form
   (iii) Enclosing the ovule
   (iv) Where pollen land
   (v) Attracting insects
   (vi) Containing the female gamete
   (vii) The region through which the pollen tube grows
   (viii) The location for fertilisation
   (ix) Losing pollen in pollination
   (x) Gaining pollen in pollination.

BIOLOGY

? Questions

8 Give one function for each of the following:
- (a) stamen
- (b) filament
- (c) stigma
- (d) style
- (e) ovary
- (f) pollen
- (g) ovule
- (h) egg cell.

9 (a) What is meant by pollination?
(b) Name the two main types of pollination.
(c) Which type of pollination is associated with each of the following?
- (i) large petals
- (ii) huge amounts of pollen
- (iii) small flowers
- (iv) stamens and carpels enclosed within petals
- (v) scented flowers
- (vi) flowers that do not produce nectar.

10 In Figure 14.20, A shows a primrose flower and B shows the flower of a grass.

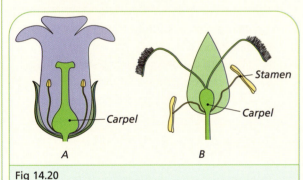

Fig 14.20

(a) State the type of pollination carried out by each of the two plants.
(b) Give two reasons based on the diagram to support each of your answers.

11 (a) What is meant by fertilisation?
(b) Name the part of the flower in which fertilisation takes place.
(c) After pollination is completed, name the structure that allows the male gamete to get to the female gamete.
(d) Name the cell formed as a result of fertilisation.
(e) Arrange the following words into the correct order in which they occur in reproduction: seed, gametes, zygote, fertilisation.

12 (a) Name the structures that are formed from the following parts after fertilisation:
(i) gametes, (ii) ovule, (iii) ovary.
(b) Name the structures that are formed from the following parts after germination:
(i) radicle, (ii) plumule, (iii) seed.

13 Figure 14.21 represents a typical seed.

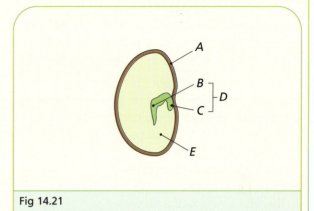

Fig 14.21

(a) Name the parts labelled A, B, C and D.
(b) Name two types of materials that are likely to be found in part E.
(c) Seeds are enclosed in fruits. Give one function for the fruit of a named plant.
(d) Explain why some fruits have many seeds while others have only a single seed.

14 (a) What is meant by dispersal?
(b) Give one major advantage to a plant for seed dispersal.
(c) Choose a plant from the box below which disperses its seeds using: (i) animals, (ii) wind, (iii) self dispersal, (iv) water.

> blackberry, water lily, bean, dandelion

15 (a) Name the part of a plant in which seeds are formed.
(b) Explain how the seeds shown in Figure 14.22 are dispersed.

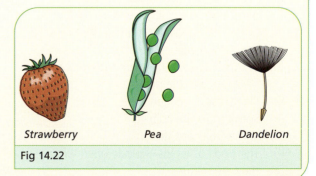

Strawberry   Pea   Dandelion

Fig 14.22

? Questions

16 (a) What is meant by germination?
   (b) State the three conditions needed for seeds to germinate.
   (c) Draw diagrams of a seed to show the process of germination.

17 When investigating the conditions needed for germination:
   (a) What conditions must be present for the seeds to germinate?
   (b) How do you show that water is necessary?
   (c) How are the seeds deprived of oxygen?
   (d) Why are some seeds placed in a fridge?

18 Write out and complete the following:
   (a) _____ is the formation of new individuals. The two types of reproduction are _____ and _____. Strawberries reproduce asexually by forming _____. Sexual reproduction involves the _____ of sex cells to form a _____.
   (b) The function of a flower is _____. When a flower is in bud it is enclosed by _____. The male structures in a flower are called _____ while the female structures are _____.
   (c) In plant reproduction pollen is transferred from the _____ to a _____ in a process called _____. The joining of the male and female _____ is called _____. To prevent _____ plants are _____ by one of four methods: wind, _____, _____ or _____ _____. The conditions needed for _____ to germinate are _____, _____ and the correct _____.

19 Give a reason for each of the following:
   (a) Buttercups have bright yellow petals.
   (b) Wind pollinated plants produce much more pollen than insect pollinated plants.
   (c) Grasses do not produce scented flowers.
   (d) Seeds do not germinate in winter but may germinate in the spring.
   (e) Seeds do not germinate in dry summers.
   (f) Alder trees have light seeds.
   (g) Dandelion seeds are more widely dispersed than ash seeds.
   (h) In some flowers the reproductive parts hang outside the petals.
   (i) Plants form a pollen tube.
   (j) Some fruits are edible.
   (k) Some fruits are sticky.
   (l) The pods of lupins explode open when they are mature.

Examination Questions

20 (a) Name the part of the flower labelled A in the diagram.
   (b) Give one reason why insects are attracted to flowers.
   (JC, OL, Sample Paper)

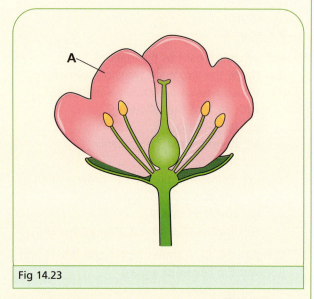

Fig 14.23

21 The diagram shows the fruit and the seed of the dandelion.
   (a) How are dandelion seeds dispersed?
   (b) Why is seed dispersal important?
   (JC, OL, Sample Paper)

Fig 14.24

Questions

22 (a) Plants produce a wide variety of seed types which need to be dispersed (scattered) in order to avoid competition.

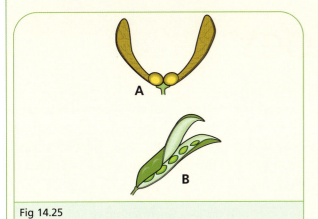

Fig 14.25

(i) Identify how the seeds A and B in the diagram are dispersed.

(ii) Name one resource that seeds must compete for with the parent plant.

(b) A number of cress seeds were set up as shown in the diagram and left for a few days to investigate the conditions necessary for germination. Test tubes A, B and D were kept in the laboratory at room temperature. Test tube C was placed in the fridge at 4°C.

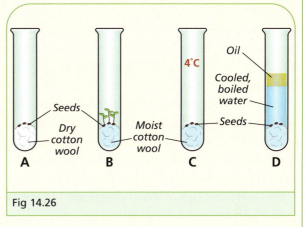

Fig 14.26

(i) Why do only the seeds in test tube B germinate?

(ii) Why is the water in test tube D boiled before use?

(iii) Explain why the seeds in test tube C failed to germinate.

(iv) Why is this investigation considered to be a 'fair test'?

(JC, OL, 2006)

BIOLOGY

Chapter **15** ○○○

ECOLOGY

Introduction

There are billions of living things on the Earth. Every one of these living things is affected in some way by other plants and animals and by their surroundings.

For example, plants may need animals for pollination or to disperse their seeds. Animals need plants as a source of food and oxygen. Both plants and animals depend on the weather, fresh water supplies and many other factors.

The surroundings of living things are known as their environment. Living things are affected by their environments. The number of living things depends on their environment along with their food supply and the presence or absence of other living things.

> **Ecology** is the study of the relationships between plants, animals and their environment.

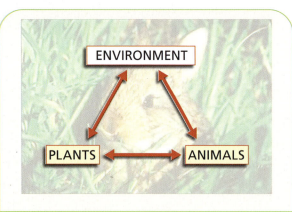

Fig 15.1 The interactions in ecology

In order to simplify the study of ecology, the world is divided into similar groups of plants, animals and environments. These groupings are called **ecosystems**. Examples of major ecosystems are deserts, tropical rainforests, grasslands and seashores.

Habitat

Ecosystems are large areas containing similar types of environments and living things. Normally an ecosystem is too large to study. Instead a small, local part of the ecosystem called a habitat is studied.

> **A habitat** is the area where a plant or animal lives.

There are a wide variety of habitats that can be studied. These include a grassland, a hedgerow, a rocky seashore, a woodland, a local park, a school field or a pond.

This chapter will deal with one sample habitat – a grassland. For examination purposes it is necessary to study only **one** habitat.

The rabbit's large ears help to catch sound from any direction

Feeding relationships in a habitat

One of the main lessons of ecology is that all living things depend on each other and on their environment for survival. This means that if any one part of a habitat is altered it will cause effects on many other parts of the habitat.

One of the main ways in which living things depend on each other is for food.

BIOLOGY

Producers

Plants that contain chlorophyll carry out photosynthesis and make their own food. They are called producers because they produce food.

> **Producers** are plants that make their own food.

Examples of producers are grasses, dandelions, buttercups, daisies and nettles.

Producers: daisies, buttercups and clover

Consumers

Animals do not make their own food. Instead they get their food by eating (or consuming) plants or other animals. For this reason animals are called consumers.

> **Consumers** are animals that get their food by eating plants or other animals.

Depending on what they eat, consumers can be placed into three different groups: herbivores, carnivores and omnivores.

Herbivores

Animals that eat plants only are called herbivores. Examples of herbivores are rabbits, sheep, slugs and snails.

> **A herbivore** is an animal that eats plants only.

Herbivore: a garden snail

Carnivores

Animals that eat other animals only are called carnivores. Examples of carnivores or flesh eaters are foxes, hawks and ladybirds.

> **A carnivore** is an animal that eats other animals only.

Carnivore: a fox hunting its prey

Omnivores

Animals that eat both plants and animals are called omnivores. Examples of omnivores are badgers, thrushes, blackbirds and humans.

> **An omnivore** is an animal that eats both plants and animals.

BIOLOGY

Omnivores: a pair of badgers

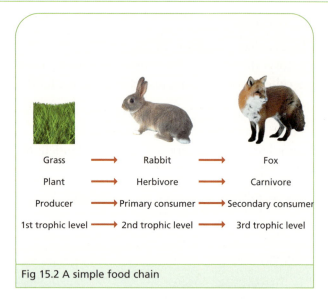
Fig 15.2 A simple food chain

Decomposers

A range of small animals (such as woodlice and earthworms), bacteria and fungi feed on dead plant and animal remains. These organisms break down the dead remains and are called decomposers.

> **Decomposers** are organisms that feed on dead plants and animals.

Decomposers are of great value as they release chemicals back into the environment. This allows other organisms to use these chemicals.

Food chains

A food chain is a simple way of explaining how energy and nutrients pass from one living thing to another. The arrows in a food chain show the direction in which the energy and nutrients pass.

An example of a simple food chain is where grass is eaten by a rabbit and the rabbit is then eaten by a fox. This food chain is shown as follows:

> grass ⟶ rabbit ⟶ fox

This food chain has three feeding levels. The grass is the first feeding level. Normally the first feeding level is a plant or a producer.

Rabbits are the second feeding level. They are herbivores and are called the primary (first) consumers.

The fox is the third feeding level. It is a carnivore and is called the secondary (second) consumer.

All the details of this food chain are provided in Figure 15.2.

> **A food chain** is a list of organisms in which each organism is eaten by the next one in the chain.

A more realistic way to show a food chain is to refer to the decomposers. These act on dead plants and animals, or on the waste products (such as urine and faeces) of animals. If the decomposers are included, the food chain given in Figure 15.2 is shown as:

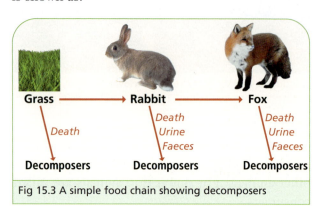
Fig 15.3 A simple food chain showing decomposers

Examples of other food chains are given in Figure 15.4.

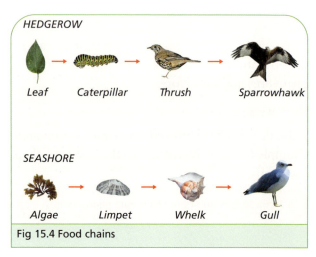
Fig 15.4 Food chains

In the food chains shown the sparrowhawk and the kestrel are both carnivores and occupy the fourth feeding level.

Adaptation

Living things must have structures or habits that allow them to survive in their habitat. They are said to have adaptations that suit them to their environment.

Adaptations shown by grassland organisms

- Rabbits have large ears so that they can hear foxes.
- Foxes' coat colour allows them to blend in with brown foliage so that they are not easily seen.
- Plants such as primroses grow and produce flowers early in the spring. This means they can get more light and grow better because the other plants in their habitat have not produced their leaves.

Primroses bloom early in spring to get as much light as possible

- Caterpillars are adapted by having a similar colour to the plants they feed on. This allows them to be well camouflaged. In addition, caterpillars have strong mouth parts so they can bite and chew leaves.
- Butterflies have long tubular mouthparts so they can drink nectar (sugary water) from flowers.

A butterfly drinking nectar: note the long tubular tongue entering the flower

> **An adaptation** is a structure or habit that helps an organism to survive in its habitat.

Competition

In any habitat a number of animals seek to find space, food, water and partners. In the same way many plants are trying to get light, space, water and minerals from the soil. These organisms compete to get these scarce resources.

Competition can occur between plants and animals of the same type. For example, grasses compete with each other for light and space. Foxes compete with each other for food.

Competition also occurs between different species. For example, grass and dandelions compete with each other for light. Robins and blackbirds compete with each other for food.

> **Competition** takes place when two or more organisms require something that is in short supply.

Table 15.1 Examples of competition		
Habitat	**Example**	**Resource required**
Grassland	Grass and dandelion	Light
	Robin and blackbird	Food

BIOLOGY

Food web

Food chains are simple representations of feeding relationships in a habitat.

For example, the simple food chain (grass ⟶ rabbit ⟶ fox) described earlier suggests that grass is only eaten by rabbits. It also suggests that rabbits are only eaten by foxes. Both of these situations are untrue.

A food web provides a more complete and realistic picture of the way in which organisms in a habitat feed. A food web consists of a number of interlinked food chains.

Remember that in all food webs the dead plants and animals are broken down by decomposers. An example of a food web is given in Figure 15.5.

> **A food web** consists of two or more interconnected food chains.

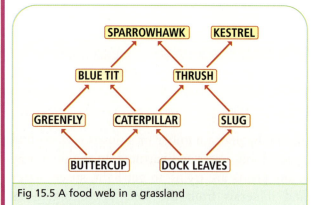

Fig 15.5 A food web in a grassland

Interdependence

All the organisms in a habitat depend on other organisms for their survival. In some cases animals depend on plants while in other cases plants depend on animals. Plants also depend on each other, while animals also depend on other animals. Both plants and animals depend on decomposers.

- One way in which animals depend on plants is for food. This means that rabbits depend on grass for food.
- Animals also depend on other animals for food. For example, foxes depend on rabbits for food.
- Plants may depend on animals for pollination (e.g. bees and butterflies) or for seed dispersal (e.g. birds and squirrels).

- Plants depend on each other for shelter. Some plants depend on others for support (e.g. ivy grows on other plants).
- Decomposers depend on dead plants and animals for food.
- Plants and animals depend on decomposers to break down dead material and release chemicals into the environment so that new forms of life can grow.

> **Interdependence** means that living things depend on each other for survival.

Table 15.2 Examples of interdependence	
Type of interdependence	Reason for interdependence
Animals depend on plants	Food, shelter (nests), oxygen
Animals depend on animals	Food, protection (adults protect young)
Plants depend on animals	Pollination, seed dispersal
Plants depend on plants	Shelter, support
Decomposers depend on plants and animals	Food
Plants and animals depend on decomposers	Release chemicals from dead material

Habitat study

The study of a local habitat is called fieldwork. If possible, the habitat should be visited during different seasons. This will show how the habitat changes over the course of a year.

When investigating any habitat it is important to follow the country code. This involves measures such as the following:

1 Get permission to enter private property.
2 Close gates behind you.
3 Do not disturb livestock.
4 Do not damage gates, fences or crops.
5 Do not light fires.
6 Do not leave litter.

Where possible you should 'leave only footprints and take only memories'.

The study of a habitat involves these five steps:

1 Draw a simple map of the habitat (or photograph it if possible).
2 Measure and record the environmental (non-living) features in the habitat.
3 Identify and list all the organisms in the habitat.
4 Collect samples of those organisms that cannot be identified in the habitat.
5 Estimate the number of each organism in the habitat.

Mapping the habitat

Draw a simple map to show the main features of the habitat. These features might include any walls, fences, ponds, large trees, hedges or paths that are present.

Photographs can be taken of the habitat if possible.

Measuring non-living features

Non-living features are also called abiotic features. Depending on the habitat being investigated the abiotic features that might be measured and the devices used to measure them are given in Table 15.3.

Identify the organisms

The organisms in the habitat form the biotic (or living) component of the habitat. If possible,

Table 15.3 Measuring environmental (abiotic) features	
Feature	**Devices**
Air temperature	Thermometer
Water temperature	Thermometer
Soil temperature	Soil thermometer
Light intensity	Light meter
Wind direction	Piece of ribbon and a compass
Soil pH	pH meter

organisms should be named and listed in the habitat. This may involve a teacher explaining the names of the organisms.

Organisms can be identified by using a suitable key (as described in Chapter 1, Living Things). Sometimes organisms can be named by comparing them to drawings or pictures in a book.

Plants and animals should only be removed from the habitat if they cannot be identified there and then.

Collecting organisms

Plants

Plants are easy to collect as they do not move away when approached. Samples of plants can be taken using a knife or a small trowel and placed in labelled bags.

APPARATUS	DIAGRAM	USED	COLLECTS
POOTER	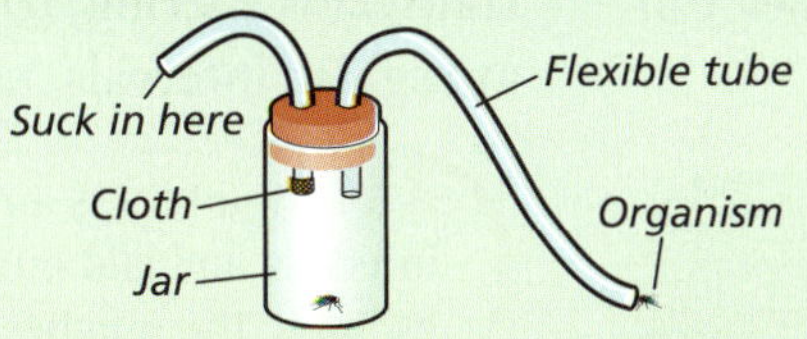 Suck in here — Flexible tube — Cloth — Organism — Jar	Suck the organism into a jar	Insects, spiders, small animals
SWEEP NET	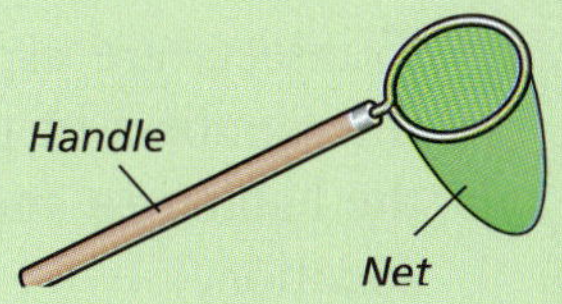 Handle — Net	Swept over long grass	Insects, caterpillars, beetles
BEATING TRAY (or upturned umbrella or sheet)	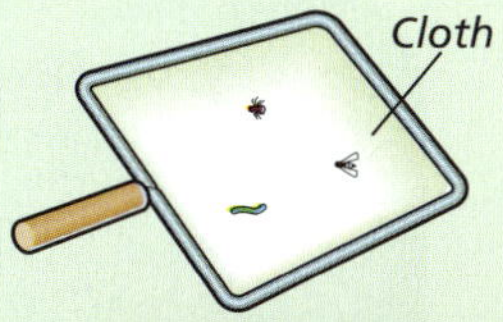 Cloth	Placed under bush or branch which is shaken	Insects, caterpillars, beetles, worms
PITFALL TRAP	 Rock — Glass or can	Placed in soil, overnight if possible	Crawling animals, e.g. snails, slugs, worms, woodlice

Fig 15.6 Apparatus used to collect animals

BIOLOGY

Only small sections of a plant should be removed. Remember to collect the leaves, twigs, flowers, seeds and fruits of each plant if they are present.

Animals

Some animals can be collected reasonably easily. For example, slow-moving animals such as slugs, snails and earthworms can be collected using a tin or glass jar.

Special apparatus is needed to collect fast moving animals. Examples of this type of apparatus are given in Figure 15.6.

> A pooter, a sweep net, a beating tray and a pitfall trap are devices used to collect animals.

Estimating the number of organisms in a habitat

It is not normally possible to count all the plants and animals in a habitat. Instead the number of organisms is calculated from samples taken at random in the habitat.

Two different methods are used to sample the number of organisms in a habitat: a quadrat or a line transect.

Quadrat

A quadrat is a square frame that is thrown at random in the habitat. Quadrats come in different sizes (e.g. their sides may measure 1 m, 0.5 m or 0.25 m). They may be made of wood, metal or plastic or they can be marked out using string or rope.

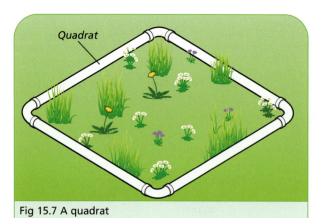

Fig 15.7 A quadrat

> A quadrat is a frame used to estimate the number of plants in a habitat.

 MANDATORY ACTIVITY 9A

To study a local habitat, using appropriate instruments and simple keys to show the variety and distribution of named organisms

Method

1 Visit the habitat and record the non-living features (as described in Table 15.3).

2 Identify as many of the plants and animals as possible in the habitat.

3 Take samples of any plants that you cannot name.

4 Collect animals using the appropriate equipment (as described in Figure 15.6).

5 Identify the plants and animals collected by using different keys (as described on pages 5 and 6 in Chapter 1).

6 Use a quadrat and/or a line transect to estimate the number and distribution of plants in the habitat (as described in Mandatory Activities 9B and 9C).

Line transect

In some habitats a quadrat is not suitable. For example, a hedgerow is vertical (i.e. it goes up into the air) while the rocky seashore may not be flat enough to use a quadrat properly.

In these cases a line transect can be used. A line transect is a string, rope or tape that is marked off at regular intervals, usually every 1 metre.

> A line transect is a rope marked at regular intervals and laid out across a habitat to estimate the number of plants present.

Transects are especially useful in showing how the numbers of plants change from one part of the habitat to another. These changes are called zonation.

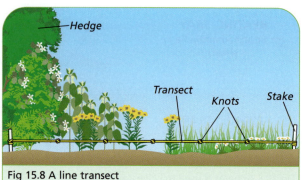

Fig 15.8 A line transect

MANDATORY ACTIVITY 9B

Using a quadrat to estimate the number and distribution of plants in a habitat

Method

1 Throw a pen or stone over your shoulder 10 times in different parts of a grass field.

2 Place a quadrat on the ground wherever the pen or stone lands.

3 Repeat this 10 times in different parts of the habitat.

4 Record the names of the plants in each quadrat.

5 Record the results as shown in Figure 15.9.

6 Calculate the percentage frequency of each plant in the habitat.

This is done by adding the number of quadrats in which the plant is found and expressing this as a percentage of the 10 quadrats (e.g. buttercups were found in 4 out of 10 quadrats, so their frequency is 40%).

7 The frequency of each type of plant can also be shown on a histogram (as in Figure 15.10).

TYPE OF PLANT	QUADRAT NUMBER										TOTAL	PERCENTAGE FREQUENCY
	1	2	3	4	5	6	7	8	9	10		
Grass	✔	✔	✔	✔	✔	✔	✔	✔	✔	✔	10	¹⁰/₁₀ = 100%
Buttercup	✘	✔	✔	✘	✔	✘	✔	✘	✘	✘	4	⁴/₁₀ = 40%
Daisy	✔	✔	✔	✔	✘	✘	✘	✔	✔	✔	7	⁷/₁₀ = 70%
Clover	✘	✘	✘	✘	✘	✔	✔	✔	✘	✘	3	³/₁₀ = 30%
Dandelion	✘	✔	✘	✔	✔	✘	✘	✔	✔	✘	5	⁵/₁₀ = 50%
Thistle	✘	✘	✘	✘	✘	✘	✔	✘	✘	✘	1	¹/₁₀ = 10%

Fig 15.9 Recording the names of plants in a quadrat study

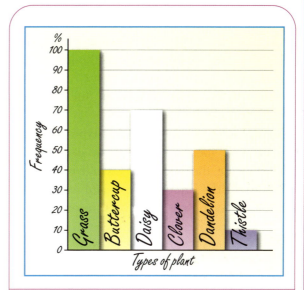

Fig 15.10 A histogram showing the percentage frequency of plants in a habitat

 MANDATORY ACTIVITY 9C

Using a line transect to estimate the number and distribution of plants in a habitat

Method

1 Stake one end of a line transect into the ground near the base of a hedge.

2 Lay the transect across the grassland and stake the second end of the transect in the ground some distance out from the hedge as shown in Figure 15.8.

3 Record the name of each plant that touches the transect in each metre section of the transect.

Results

The results can be presented as shown in Figure 15.11.

PLANT	METRE SECTION FROM HEDGE					
	1	2	3	4	5	6
Blackberry	✔	✔	✘	✘	✘	✘
Nettles	✘	✔	✔	✘	✘	✘
Ferns	✔	✘	✘	✘	✘	✘
Primroses	✘	✔	✔	✔	✘	✘
Daisy	✘	✘	✘	✘	✔	✔
Grasses	✘	✘	✘	✔	✔	✔

Fig 15.11 The results of a study using a line transect

BIOLOGY

BIOLOGY

Conservation

The number of humans on Earth has risen dramatically over the last two hundred years. All over the globe humans are increasingly damaging their environment in order to get supplies of fuel, water, energy and materials.

Habitats are being destroyed on a huge scale. For example, we knock down forests and woodlands for timber or for housing. Boglands are being drained for farming. Grassland is being used up for housing and factories. Lakes are being destroyed by pollution.

As habitats are destroyed the organisms living in these habitats are killed. In some cases the organisms may become extinct.

For example, birds called corncrakes are in danger of being wiped out due to the loss of hay meadows in which they breed. Some orchids are also threatened with extinction due to their habitats being destroyed.

The decrease in the number of types of plants and animals means there is a decrease in biodiversity (i.e. the different types of living things).

To prevent the loss of habitats and wildlife it is necessary to look after our natural resources.

> **Conservation** is the protection and wise management of natural resources.

The corncrake has become an endangered species due to intensive farming which has destroyed its natural habitat

The need for conservation

Conservation is necessary for the following reasons:

1 To prevent other organisms from being wiped out (made extinct). At present about one type of organism becomes extinct each day.
2 To maintain the balance of nature. The loss of any one type of organism can cause dramatic results for other types of organisms.
3 Future generations have the right to the same natural resources as are found at present. For example, the world would be a poorer place if elephants and tigers become extinct.
4 Plants are the source of many medicines. As plants become extinct we lose the ability to test them for new medications.
5 If our natural resources are not protected there is a danger that human lifestyles (and even our survival) would be at risk.

The only species on earth that threatens conservation is humans. It is up to us, both as a species and as individuals, to support conservation measures.

Humans can support conservation by making themselves aware of the issues, by joining conservation groups and by supporting groups in society that encourage proper conservation.

Pollution

> **Pollution** is the addition of harmful materials to the environment.

Pollution is mainly caused when humans add too much waste material to the environment. The materials which cause pollution are called pollutants. There are three main types of pollution: air, water and soil pollution.

Air pollution

The main source of air pollution is the burning of fossil fuels. These include coal, gas, turf, oils and petrol.

The main air pollutants and their effects on the environment are given in Table 15.4.

Water pollution

The main causes of water pollution are: badly treated sewage (toilet) waste, oil spills and the dumping of household, farming and industrial wastes.

When nutrients enter a water supply (such as a stream, river, pond or lake) they allow more algae to grow in the water. This may result in an overgrowth of algae, called an algal bloom. When the algae die they are decomposed by bacteria.

The bacteria use up all the oxygen in the water as they grow. This results in the death of all the animals and plants in the water.

The main types of water pollution are outlined in Table 15.5.

Pumping sewage into the sea causes water pollution

Table 15.4 Examples of air pollution		
Problem	**Pollutant**	**Effect**
Acid rain	Sulfur and nitrogen oxides from burning are dissolved in rainwater	Burns leaves on trees; kills plants and fish; dissolves marble, limestone and metal
Greenhouse gases	Increasing levels of carbon dioxide (and other gases)	Causing the Earth to warm up, resulting in floods and weather changes
Less ozone	Chemicals called CFCs (chloro-fluorocarbons) found in aerosols and refrigerators	CFCs destroy ozone (a form of oxygen) in the upper atmosphere, which lets in more ultraviolet radiation from the sun causing more skin cancers
Lead	Burning 'leaded' petrol	Harmful to the brains of young children
Dirt	Smoke and soot	Harmful to the lungs; blackens buildings
Smog	Particles in air that arise from burning	Harmful to the lungs; kills lichens

Table 15.5 Examples of water pollution		
Problem	**Pollutant**	**Effect**
Untreated sewage	Micro-organisms Nutrients	Cause diseases Remove oxygen
Excess slurry, fertiliser or silage leaks	Nutrients	Remove oxygen
Industrial waste	Poisonous chemicals Nuclear waste	Harmful to health May cause mutations or cancer
Oil spills	Oil	Kills birds and fish; destroys beaches

BIOLOGY

Soil pollution

Soil pollution is mainly caused by acid rain, overuse of fertilisers and slurry on the land and improper dumping of wastes. All of these factors have been discussed under the headings of air and water pollution.

Waste management

The large numbers of humans and the modern style of life produces huge amounts of waste materials. These wastes include dirty water, urine, faeces, plastics, packaging, along with agricultural and industrial wastes.

It is essential that we manage these wastes properly in order to prevent pollution and ensure that we conserve our environment.

Waste management includes composting, recycling, landfill and incineration, as outlined in Table 15.6.

Table 15.6 Methods of waste managment			
Method	**How it works**	**Advantages**	**Disadvantages**
Composting	Waste food, garden waste (e.g. grass clippings) and other organic waste decay and are then returned as compost to the soil	Reduces the amount of waste sent to landfill Returns nutrients to the soil	Can be dirty and smelly May attract insects and maggots
Recycling	Glass, plastics, paper and metals are returned to recycling centres and converted into new products	Reduces the amount of waste sent to landfill Reduces the use of scarce resources (such as trees to make paper)	Takes time and effort to sort out the wastes Some materials (such as paper and plastic) can only be recycled a limited number of times
Landfill	Waste is placed into holes in the ground (often called dumps)	Easy to use Wastes do not have to be sorted or processed	Unsightly, smelly, attracts gulls, rats and mice Wastes may leak into soil and cause pollution in nearby streams, rivers and lakes
Incineration	Waste is burned at high temperatures forming ash	Reduces the volume of waste that is sent to landfill The heat produced can be used to generate electricity	Poisonous gases may be released into the air The ash must be tested for safety before it is sent to landfill

Landfill sites attract gulls and vermin such as mice and rats

One of the main principles in waste management is known as the three Rs:

- Reduce the use of unnecessary goods and packaging (e.g. plastic bags).
- Reuse as many materials (e.g. glass) as possible.
- Recycle as much waste as possible (e.g. paper, glass, metals).

Observing these three rules will result in less waste being produced. This will help to reduce our waste management problems.

Problems of poor waste management

1 Wastes may contain micro-organisms that cause disease. It is to reduce this problem that sewage treatment plants are needed.
2 Poisonous chemicals from waste may get into the air, water or soil. These chemicals can have serious effects on organisms.
3 Nutrients from waste can cause all the oxygen in water to be used up. This results in the death of all the organisms in the water.
4 Waste that is dumped in landfill sites may be smelly, unsightly and attract rats and gulls.
5 Dumping waste at sea may pollute the seas and kill fish stocks.
6 Incinerators burn waste at high temperatures. However, there is a fear that these incinerators may release poisonous gases.

Human effects on the environment

Positive effects

1 By reducing the use of CFCs the ozone layer will begin to rebuild. This will protect the planet from excess ultraviolet rays.
2 In Ireland, the number of plastic bags that are discarded and litter our environment has been reduced. This was achieved by charging a fee for plastic bags in shops.

Negative effects

1 The temperature of the planet is rising due to the increased burning of fossil fuels. The increases in global temperatures will result in flooding and weather changes.
2 By failing to reduce, reuse and recycle materials we contribute to extra waste materials in the environment.

 Key Points

- Ecology is the study of the relationships between plants, animals and their environment.
- A habitat is the area where a plant or animal lives (this chapter highlights a grassland habitat).
- Producers are plants that make their own food.
- Consumers are animals that get their food from plants or other animals.
  - A herbivore is an animal that only eats plants.
  - A carnivore is an animal that only eats other animals.
  - An omnivore is an animal that eats both plants and animals.
- Decomposers are organisms that feed on dead plants and animals.
- A food chain is a list of organisms in which each organism is eaten by the next one in the chain.
- An adaptation is a structure or habit that helps an organism to survive in its habitat.
- Competition takes place when two or more organisms require something that is in short supply.
- A food web consists of two or more interconnected food chains.
- Interdependence means that living things depend on each other for survival.
- The study of a habitat involves the following five steps:
  - Draw a simple map of the habitat (or photograph it if possible).
  - Measure and record the environmental (non-living or abiotic) features in the habitat.
  - Identify and list all the organisms (which are the biotic features) in the habitat.
  - Collect samples of those organisms that cannot be identified in the habitat.
  - Estimate the number of each organism in the habitat.
- Organisms can be named by using a suitable key or by comparing them to drawings or pictures in a book.

BIOLOGY

BIOLOGY

- A pooter, a beating tray, a pitfall trap and a sweep net are devices used to collect animals.

- A quadrat is a square used to estimate the number of plants in a habitat.

- A line transect is a rope marked at regular intervals and laid out across a habitat to estimate the number of plants present.

- Conservation is the protection and wise management of natural resources.

- Pollution is any undesirable addition to the environment.

- Waste management includes composting, recycling, landfill and incineration.

- Effective waste management involves
  - reduce
  - reuse
  - recycle.

- Failure to manage waste properly results in disease, poisoning, damage to water and soil quality, unsightly and unhealthy dumps and fish kills.

? Questions

1 What is meant by:
- (a) the environment?
- (b) ecology?
- (c) ecosystem?
- (d) habitat?

2 (a) Name the type of habitat you have studied.
- (b) Naming living things from your habitat, give examples of:
  - (i) a producer
  - (ii) a consumer
  - (iii) a herbivore
  - (iv) a carnivore
  - (v) an omnivore
  - (vi) decomposers.

3 (a) Separate the following list into producers and consumers:

buttercups, rabbits, slugs, grasses, earthworms, nettles, daisies, thrushes, caterpillars, primroses, butterflies, woodlice, sparrowhawks.

- (b) Which of the organisms in the list above contain chlorophyll?

4 (a) Name a habitat you have studied.
- (b) Name one plant found in the habitat.
- (c) Name one animal found in the habitat.
- (d) Name one animal in your habitat that eats other animals.

5 Define the following words and, in each case, name one example from the habitat you have studied.
- (a) producer
- (b) consumer
- (c) herbivore

- (d) carnivore
- (e) omnivore
- (f) primary consumer
- (g) secondary consumer
- (h) decomposer.

6 (a) Write down a food chain using the following organisms:

thrush, fox, snail, lettuce

- (b) From your food chain name the:
  - (i) producer
  - (ii) secondary consumer
  - (iii) herbivore.

7 Grass ⟶ Rabbit ⟶ Fox

In the food chain shown above suggest one possible result in each of the following cases:
- (a) If all the foxes died out.
- (b) If all the grass died.
- (c) If there were too many rabbits.
- (d) If there were too many foxes.
- (e) If there were no rabbits.

8 Arrange each of the following into a food chain:
- (a) blue tit, rose leaves, greenfly, spider
- (b) slug, cabbage, thrush, hawk
- (d) owl, oak tree, caterpillar, robin
- (e) butterfly, hawk, thrush, dandelion.

? Questions

9 In the habitat you have studied:
   (a) State one way in which a named animal is adapted to the habitat.
   (b) State one way in which a named plant is adapted to the habitat.
   (c) Give one example of competition between plants in the habitat. Name the resource (or feature) for which the plants compete.
   (d) Give one example of competition between animals in the habitat. Name the resource (or feature) for which the animals compete.

10 State one benefit for each of the following adaptations:
   (a) Plants in a woodland or hedge may grow early in the year.
   (b) Rabbits have large ears.
   (c) Butterflies have tube-shaped mouthparts.
   (d) Hawks have excellent eyesight.

11 (a) What is meant by (i) a food chain? (ii) a food web?
   (b) Give one reason why a food web gives a more accurate account of how organisms feed in a habitat.

12 Figure 15.12 represents a food web.

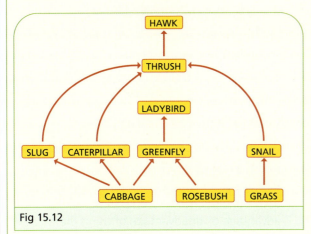

Fig 15.12

   (a) Write out two different food chains based on the diagram.
   (b) From the food web name
      (i) a producer
      (ii) a primary consumer
      (iii) a secondary consumer
      (iv) a tertiary consumer
      (v) one example of competition between animals
      (vi) any one adaptation shown by a named animal.

13 Figure 15.13 represents a food web.

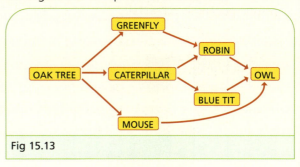

Fig 15.13

   (a) Name the producer in the food web.
   (b) Write out two food chains from this web.
   (c) State one way in which a named plant depends on a named animal in this habitat.
   (d) State one way in which a named animal depends on a named plant in this habitat.
   (e) Name one way in which robins and blue tits might compete.

14 (a) Name the basic steps that you would take to investigate a habitat.
   (b) Name three pieces of equipment that you might use when investigating a habitat.

15 Figure 15.14 shows two pieces of equipment used in a habitat study.

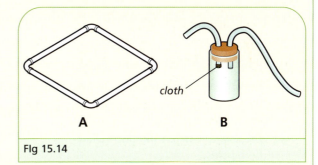

Fig 15.14

   (a) Name apparatus A and B.
   (b) Give one use for each piece of apparatus.
   (c) Explain how each piece of apparatus is used.
   (d) Name and draw a diagram of any other piece of apparatus used to study a habitat.

BIOLOGY

? Questions

16 Figure 15.15 shows the results of a study of four plants growing in a habitat.

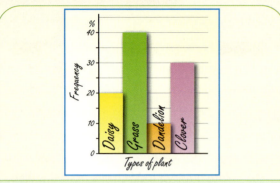

Fig 15.15

(a) Name two pieces of apparatus that might be used to obtain these results.

(b) Which type of plant occurred most often?

(c) Which type of plant occurred least often?

(d) If a quadrat is thrown 10 times in the habitat:
  (i) Which of the four plants is most likely to be found in the quadrats?
  (ii) How many of the quadrats would be expected to contain clover?

17 (a) Draw a diagram of a quadrat.

(b) Describe how you would use a quadrat to study plants in a habitat.

18 Give a reason for each of the following:

(a) All food chains start with a producer.

(b) Decomposers are important in a habitat.

(c) Studying a habitat instead of an ecosystem.

(d) Producers must have chlorophyll.

(e) Primary consumers are also called herbivores.

(f) Caterpillars and butterflies have different types of mouthparts.

(g) Using a key when studying a habitat.

(h) Using a beating tray.

19 What is meant by:

(a) conservation?

(b) pollution?

(c) pollutant?

20 (a) Name two ecosystems that are being destroyed by human activity.

(b) Give one reason why each ecosystem is being destroyed.

(c) Give a reason why ecosystem (and habitat) destruction is harmful.

21 (a) What organism is responsible for conservation?

(b) Suggest three reasons why conservation is important.

22 Name two ways in which human activity results in (a) air pollution, (b) water pollution.

23 Name a pollutant responsible for each of the following:

(a) Killing trees

(b) Destroying the ozone layer

(c) Causing changes in the weather

(d) Harming our lungs

(e) Damaging beaches and seabirds.

24 Give a reason for each of the following:

(a) Using lead-free petrol

(b) The need for treating sewage

(c) Buildings in some cities become blackened

(d) Planting more trees and forests

(e) Not spreading too much slurry or fertiliser on the land

(f) Recycling or reusing as many products as possible

(g) Using CFC-free sprays.

25 Name two problems that result from poor waste management procedures.

26 (a) Name two ways in which human activity harms our environment.

(b) Name two ways in which human activity benefits our environment.

Examination Questions

27 Waste management includes: composting, incineration, landfill and recycling.

Pick one of the methods of managing waste above and say how it works and give one advantage or disadvantage of using the method you have selected.

(JC, HL, 2006)

❓ Questions

28 A number of identical small trees were planted in the same way at different distances from a very big tree. After a few years it was noticed that the trees close to the big tree did not grow as quickly as those further away. The diagram summarises the observations made.

Trees when first planted

Trees a few years after planting

Fig 15.16

A horticultural advisor said that the poor growth of the trees closer to the big tree was due to competition.

(a) List two things for which the trees must compete with each other.

(b) Explain why the trees nearest to the big tree are so small.

(c) (i) In ecology what is meant by conservation?

(ii) Give an example of an Irish animal or plant species that is on the threatened list.

(iii) Many species of plant are protected in national parks. The manager of one of these parks is asked to measure the frequency with which a protected species occurs in a habitat within the park.

Describe how this might be carried out. Include a diagram of any equipment that might be used.
(Adapted from JC, OL, Sample Paper)

29 (a) The piece of equipment drawn below is used in ecology.

Fig 15.17

(i) Name the piece of equipment

(ii) Give one use of this piece of equipment.

(b) In ecology micro-organisms play a major role in recycling nutrients.
Name one decomposer from a habitat you have studied.
(JC, OL, 2006)

30 A number of identical small trees were planted in the same way at different distances from a very big tree. After a few years it was noticed that the trees close to the big tree did not grow as quickly as those further away. All the trees planted were 1 m high to begin with.

The graph summarises the observations made after 5 years.

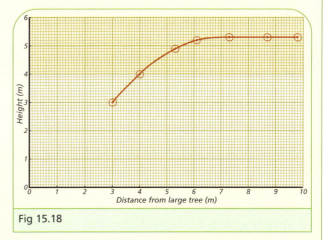

Fig 15.18

(a) How many small trees were planted?

(b) At what minimum distance from the big tree did the planted trees appear to stop being influenced by their proximity to it (i.e. how close they are to it)?

(c) If a small tree was planted 5.5 m from the large tree what height would you expect it to be when the trees were measured for this investigation?

(d) What gain in height did the tree planted at 4 m show?
(Adapted from JC, HL, Sample Paper)

Chapter 16

MICROBIOLOGY AND BIOTECHNOLOGY

Introduction

Micro-organisms (also called microbes) are small living things. Most of them are too small to be seen with the naked eye and can only be seen using a microscope. The study of micro-organisms is called microbiology.

There are three types of micro-organisms: viruses, bacteria and fungi. Viruses and bacteria are tiny but some fungi are large enough to be seen by the eye.

Micro-organisms are found almost everywhere. For example, they are present in the air, water and soil, on our skin, and inside plants, animals and humans.

Micro-organisms are found in huge numbers. For example, there are more bacteria in our intestines than there are human cells in the human body. One gram of soil can contain 100 million bacteria while a few drops of fresh milk may contain more than 3 000 million bacteria.

Viruses

Viruses are the smallest micro-organisms. Up to one million viruses may fit across the thickness of a thumbnail (1 mm).

Viruses cannot reproduce by themselves. For this reason it can be argued that they are not living things.

Viruses increase in numbers by invading other cells where they cause the other cell to form new viruses. For this reason all viruses are said to be parasites (i.e. they live on or in another living thing).

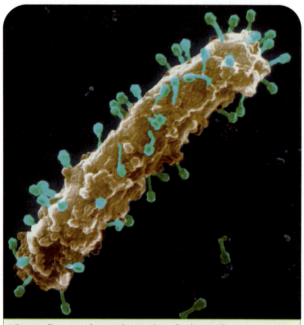

Viruses (in green) on a bacterium (in brown)

Viral diseases

Viruses cause disease to plants, humans and other animals. Examples of human diseases caused by viruses include measles, mumps, chicken pox, polio, colds, flu (influenza), cold sores and AIDS.

There are very few chemicals or medicines that will kill viruses. For example, antibiotics prevent bacterial infections but have no effect on viruses.

Our bodies fight off most viral infections when our white blood cells produce chemicals called antibodies.

HIV (human immunodeficiency virus) is a virus that prevents white blood cells from making antibodies. This results in a condition known as AIDS (acquired immunodeficiency syndrome).

As a result of AIDS the victims may die because they cannot fight off other infections (such as pneumonia).

Bacteria

Bacteria (which is the plural of bacterium) can only be seen using a microscope (although they are larger than viruses). They are very simple organisms; for example they do not have a proper nucleus.

In order to grow bacteria need the following:

1 Food
2 Water
3 A suitable temperature
4 A suitable pH.

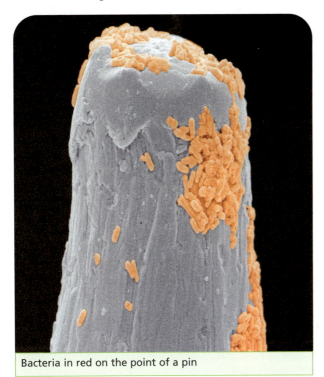

Bacteria in red on the point of a pin

Aerobic bacteria require oxygen (for respiration). Anaerobic bacteria do not require oxygen.

Under ideal conditions bacteria can reproduce very rapidly. Bacteria reproduce asexually. Many bacteria can double their numbers every 20 minutes.

Most bacteria live on dead material. Bacteria cause the material to rot or decay (i.e. they are decomposers).

Decay is vital as it returns materials to the soil to support new growth. In addition, decaying material improves the structure of the soil and allows plants to grow much better.

The main effects of bacteria are shown in Table 16.1.

Table 16.1 Effects of bacteria	
Useful	**Harmful**
Decay dead plants and animals	Cause diseases
Make foods such as cheese, butter, yoghurt	Destroy foods, e.g. they cause milk to turn sour
Used in biotechnology	Destroy crops in the fields

Bacterial diseases

Bacteria cause disease to plants, humans and other animals. Examples of human bacterial diseases are tetanus (lockjaw), tuberculosis (TB), pneumonia, sore throats, boils, tooth and gum decay, food poisoning, cholera and anthrax.

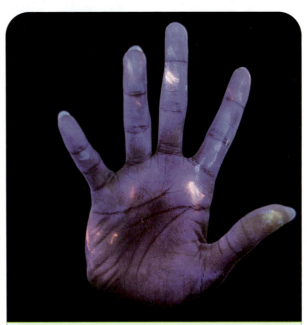

Bacteria (shown using ultraviolet radiation) on the surface of a hand

Antibiotics are chemicals made by bacteria and fungi which kill or prevent the reproduction of other bacteria.

Penicillin is an example of an antibiotic.

Fungi

Fungi (which is the plural of fungus) are simple plants that do not contain chlorophyll. For this reason they are not green and cannot make their own food.

Different fungi are parasites on plants, on animals and even on humans. Many fungi feed on dead material and act as decomposers.

Some fungi are single celled (e.g. yeast), while many form long threads. Very often these threads are underground and only come to the surface as reproductive structures such as mushrooms.

Fungi: edible field mushrooms

The main effects of fungi are shown in the following table:

Table 16.2 Effects of fungi	
Useful	Harmful
Decay dead plants and animals	Cause human diseases, e.g. athlete's foot
Some can be eaten, e.g. mushrooms	Some are poisonous
Used to make alcohol, e.g. yeast	Cause food to rot, e.g. bread mould

Biotechnology

Biotechnology is the use of living things or parts of living things to produce useful products.

The organisms used in the biotechnology industry include plants, animals and micro-organisms. In some cases, enzymes are used instead of entire organisms.

In particular, bacteria and fungi such as yeast are used to produce a vast range of useful products such as foods, drugs, alcohols, hormones and enzymes.

Old examples of biotechnology include the use of yeast to make alcohol or to rise the dough in bread-making. In the last century fungi and bacteria were discovered and are now used to produce antibiotics.

In recent years micro-organisms are being altered (by adding genes) to allow them to produce a huge range of new products.

Biotechnology: the steel tanks contain yeast which has been altered so that it produces the human hormone insulin

Industrial uses of biotechnology

1 Yeast is used in breweries to produce alcohol.
2 Bacteria are used to produce specialised enzymes which are added to washing powders and stain removers. These products are often called biological detergents.

    These enzymes help to remove difficult stains caused by materials such as fat, inks and blood.

Medical uses of biotechnology

1 Bacteria and fungi are used to make antibiotics.
2 Altered organisms (such as bacteria and yeast) are used to produce medically important products such as:
    – hormones (e.g. insulin which is used to treat diabetes)
    – antibodies (which are used to treat infections)
    – chemicals needed to clot blood (these chemicals are used by haemophiliacs whose blood will not naturally clot).

Biotechnology allows these products to be produced cheaply, in huge quantities and without any dangerous by-products.

MANDATORY ACTIVITY 10

To investigate the presence of micro-organisms in air and soil

Method

1 Get three sterile petri dishes containing nutrient agar. (Sterile means that nothing is living in the dishes. The nutrient agar is a solid, jelly-like material which contains the foodstuff that micro-organisms need to grow.)

2 Remove the lid from one petri dish and expose the agar to the air for about five minutes (Dish A).

3 Briefly open the second dish and sprinkle a small amount of soil over the surface of the agar in this dish (Dish B).

4 Do not open the third dish (Dish C). This acts as a control (or comparison) for the experiment.

5 Label the undersurface of each dish with a marker. This allows you to see clearly through the lid any growth that might be present on the agar.

6 Leave the dishes in a warm room or in an incubator at 25°C.

7 Examine the dishes each day without opening them.

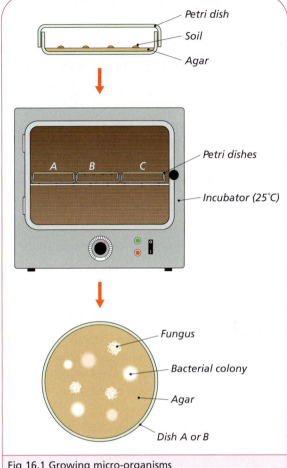

Fig 16.1 Growing micro-organisms

Results

After a few days the following results should be observed.

1 Dish A (exposed to the air) may show a number of shiny, round growths. Each of these is a colony consisting of a huge number of bacteria which grew from one initial bacterium.

The dish may also contain a number of patches of fluffy or mouldy growth. These are patches of fungi each of which grow from a single spore which lands on the agar.

2 Dish B (to which soil was added) shows a greater number of bacterial colonies and fungus growths.

3 Dish C shows no growth at all (showing that the micro-organisms in the other dishes must have come from the air or the soil and were not present in the dish).

Conclusion

This shows that bacteria and fungi are present in the air and in soil.

BIOLOGY

 Key Points

- Micro-organisms (microbes) are small living things.

- Microbiology is the study of small living things.

- The three types of micro-organisms are viruses, bacteria and fungi.

- Viruses are:
  - the smallest type of micro-organism
  - unable to reproduce themselves
  - parasites (get food from a living thing)
  - the cause of diseases such as measles, mumps, chickenpox, polio, colds, flu, cold sores and AIDS
  - not killed by antibiotics
  - controlled by antibodies produced by white blood cells.

- Bacteria:
  - are larger than viruses (but can only be seen with a microscope)
  - require food, water and a suitable temperature and pH in order to grow
  - are divided into aerobes (need oxygen) and anaerobes (do not need oxygen)
  - reproduce asexually
  - may feed on live things or dead things
  - may be useful (for decay, producing foods or in biotechnology)
  - may be harmful (cause diseases, destroy food and crops)
  - cause diseases such as tetanus, tuberculosis (TB), pneumonia, sore throats, boils, tooth and gum decay, food poisoning, cholera and anthrax
  - are controlled by using antibiotics.

- Antibiotics are chemicals made by micro-organisms which kill or prevent the growth of bacteria.

- Fungi:
  - do not contain chlorophyll
  - grow as single cells or threads
  - may be useful (cause decay, may be eaten, make alcohol and used in bread-making)
  - may be harmful (cause diseases, may be poisonous, rot food).

- Biotechnology is the use of living things or their parts to produce useful products.

- Biotechnology uses plants, animals, micro-organisms and enzymes to manufacture many different products.

- Industrial uses of biotechnology include:
  - using yeast to produce alcohol
  - using bacteria to produce enzymes for removing stains.

- Medical uses of biotechnology include:
  - using bacteria and fungi to produce antibiotics
  - using micro-organisms to produce medical products such as hormones, antibodies and blood-clotting chemicals.

- To investigate the presence of micro-organisms in the air and in soil:
  - Expose a petri dish containing sterile nutrient agar to the air.
  - Sprinkle soil on a second petri dish containing sterile nutrient agar.
  - Do not open a third petri dish containing sterile nutrient agar.
  - Leave the dishes in an incubator (or a warm place) for a few days.
  - Micro-organisms grow on the first two dishes but not on the third dish.

? Questions

1 (a) What is microbiology?
  (b) What is a micro-organism?
  (c) Name three types of micro-organisms.
  (d) Which of the three types of micro-organisms are the smallest?
  (e) Which of the three types of micro-organisms may be seen without using a microscope?

2 (a) What is a parasite?
  (b) Why must viruses always be parasites?

3 (a) Name three common human illnesses caused by viruses.
  (b) Why are antibiotics of no use in treating these illnesses?
  (c) Name the chemicals produced in the human body that control virus infections.

? Questions

4 (a) Name three conditions needed for bacteria to grow properly.

(b) Give two ways in which bacteria are of benefit to humans.

(c) Give two ways in which bacteria are a disadvantage to humans.

5 (a) Name three common human illnesses caused by bacteria.

(b) Give one use for antibiotics.

(c) Name an antibiotic.

6 (a) Name one way in which fungi are different to green plants.

(b) Why can fungi never act as producers in a food chain?

(c) Give two examples of fungi.

7 Write out and complete the following:

(a) The three types of micro-organisms are _____, _____ and _____. The smallest of these are _____ while the largest are _____. Viruses can only _____ inside other living cells. For this reason they are called _____. Diseases caused by viruses are _____ and _____.

(b) The conditions needed by bacteria in order to grow are _____, _____ and the correct _____ and _____. Bacteria reproduce by _____ reproduction. Antibiotics are chemicals made by _____ which are used to control infections caused by _____. Antibiotics have no effect on _____ infections.

(c) Fungi, such as _____ and _____, do not contain the green chemical _____. This means they cannot _____ their own _____.

8 Give a reason for each of the following:

(a) Bacteria are studied using microscopes.

(b) Bacteria do not grow well in a fridge.

(c) Antibiotics are not given to people with colds and flu.

(d) Eating wild mushrooms can be dangerous.

(e) Viruses may not be living things.

(f) People with AIDS find it difficult to control infections.

(g) Decay may be a good or a bad thing.

(h) Yeast is used in biotechnology.

(i) Some washing powders have bacteria-made enzymes added to them.

(j) Micro-organisms are often grown in incubators.

(k) Nutrient agar is used to grow micro-organisms.

9 Copy out and complete the following table and say whether each illness is caused by a virus or a bacterium.

Illness	Caused by
Common cold	
Food poisoning	
Tuberculosis	
Pneumonia	
Measles	
AIDS	
Tetanus	

10 (a) What is biotechnology?

(b) Name two types of organisms used in biotechnology.

(c) Name two uses of biotechnology in (i) industry, (ii) medicine.

11 When investigating if micro-organisms are present in the air or in soil, explain each of the following:

(a) Why the original petri dishes must be sterile.

(b) Why the lid is left off one of the dishes.

(c) Why the petri dishes are labelled on the underneath surface.

(d) Why the dishes are left in a warm place.

(e) How you would know the difference between bacteria and fungi growing on the agar.

(f) Why there was no growth on the control dish.

12 Milk turns sour due to bacterial action. Design an experiment to investigate acidity in milk as it turns sour.

13 Design an investigation to find out the effect of temperature on the rate at which milk turns sour.

Examination Question

14 (a) Micro-organisms are used widely in biotechnology. Give one use of biotechnology in industry.

(b) Micro-organisms can be found growing in a variety of locations.

(i) Describe how the presence of micro-organisms in a sample of soil might be investigated.

(ii) Include a diagram of any equipment that might be used.

(JC, OL, 2006)

BIOLOGY

125

Chemistry

Chapter 17

STATES OF MATTER

Introduction

Most of us know about solids, liquids and gases. We know, for example, that rock is a solid, water is a liquid and oxygen is a gas. All substances are either solids, liquids or gases. These are known as the three states of matter.

Matter is anything that occupies space and has mass.

Gold is a very unreactive metal and is found in nature as a pure element. It is widely used in jewellery as it does not tarnish and has an attractive appearance

The three states of matter: solid (ice), liquid (water) and gas (steam)

Properties of the states of matter

The following table outlines the different properties of the three states of matter.

Table 17.1 Properties of solids, liquids and gases		
Solids	**Liquids**	**Gases**
Have a definite shape	No definite shape Take shape of their container	No definite shape Take shape of their container
Have a definite volume	Have a definite volume	No definite volume
Cannot be compressed	Cannot be compressed	Can be compressed
Cannot flow	Can flow	Can flow

ACTIVITY 17.1

To explain solids, liquids and gases (Try this at home!)

(a) Solid

Method

1 Place marbles in a tray as shown.

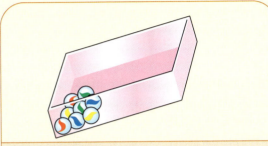

Fig 17.1 To illustrate solids

2 Tilt the tray and all the marbles gather into one corner. How are the marbles arranged?

Result

The marbles cannot move freely without tilting the box. They can only vibrate.

Conclusion

When particles in a solid are tightly packed together they are not free to move. They can only vibrate.

(b) Liquid

Method

1 Place marbles in a tray as shown.

2 Place the tray horizontally and shake it gently. Can you see the marbles sliding past each other?

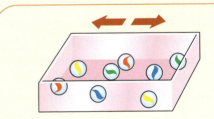

Fig 17.2 To illustrate liquids

Result

Particles are far enough apart to slip and slide over each other.

Conclusion

In a liquid, particles can slide past each other.

(c) Gases

Method

1 Place marbles in a tray as shown.

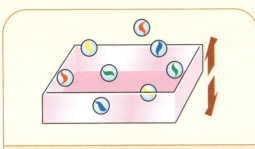

Fig 17.3 To illustrate gases

2 Shake the tray quickly. What happens to the distance between the marbles?

Result

The marbles are free to move in all directions.

Conclusion

The particles in a gas can move in all directions independently of each other.

Changes of state

Most substances can be changed into solids, liquids and gases by heating or cooling them.

An easy example is to examine the three states of water.

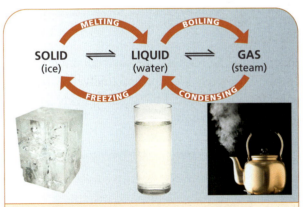

Fig 17.4 The three states of water

● The temperature at which a solid turns into a liquid is called its melting point.
● The temperature at which a solid turns from a liquid to a gas is called its boiling point.

These changes are physical changes. No new substance is being formed.

CHEMISTRY

Evaporation

When water is heated to 100°C it will boil and change from the liquid state to the gaseous state. However, liquids do not have to be boiled in order to change into gases. Rain puddles quickly disappear when the sun comes out. This is due to evaporation.

Evaporation occurs at the surface of a liquid. Some of the water molecules at the surface gain enough energy and escape as gases. Obviously evaporation is a much slower process than boiling, but the overall result is the same.

Steam rising from a boiling kettle hits a cold ladle and as a results condenses back to water

Diffusion

At one time or another we have all smelt a stink bomb. However, we don't actually see the gas particles in the air. The smell quickly spreads throughout the room.

Tea dissolving in water. Chemicals from the tea leaves diffuse throughout the water forming a solution

Diffusion is the name used to describe the way particles in gases and liquids spread throughout the space in which they are placed.

▶▶▶ **ACTIVITY 17.2**

To demonstrate diffusion in a liquid

Method

1 Using a forceps gently place a few potassium permanganate crystals at the bottom of a beaker of water.

2 Observe what happens for a few minutes.

3 Leave your beaker until your next science class. Explain what has happened to the purple particles.

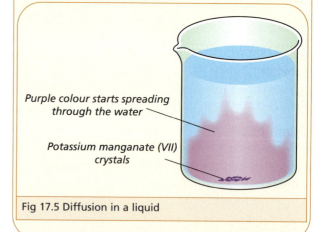

Purple colour starts spreading through the water

Potassium manganate (VII) crystals

Fig 17.5 Diffusion in a liquid

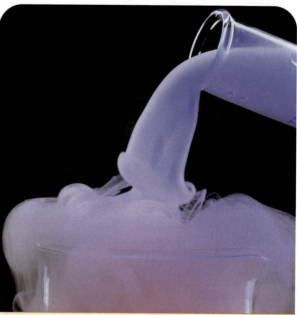

Carbon dioxide only exists in the gaseous state or the solid state. When cooled below −78°C, CO_2 turns from a gas into a solid (dry ice). When water is added to dry ice thick clouds of fog are produced

CHEMISTRY

DEMONSTRATION

To demonstrate diffusion through a gas

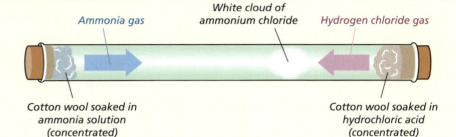

Ammonia gas
White cloud of ammonium chloride
Hydrogen chloride gas

Cotton wool soaked in ammonia solution (concentrated)
Cotton wool soaked in hydrochloric acid (concentrated)

Fig 17.6 Diffusion through a gas. Which gas diffuses faster, ammonia or hydrogen chloride?

Watch what happens when your teacher brings a bottle of ammonia and a bottle of hydrochloric acid near each other. The vapours produced by the two solutions react in the air above the containers to form ammonium chloride fog

Brownian motion

Robert Brown was a Scottish botanist. He observed that, when viewed under a microscope, tiny pollen grains were constantly moving in an erratic way. He tried other tiny particles such as coal dust and found they moved similarly. He thought the movement was due to the particles themselves.

Since then scientists have discovered that the movement is caused by these particles being struck by moving water molecules and air molecules.

ACTIVITY 17.3

To demonstrate Brownian motion

Method

1 Put a drop of milk under a cover slip on a microscope slide and examine under a microscope.

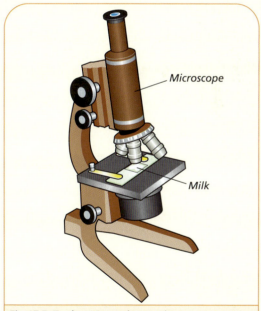

Microscope

Milk

Fig 17.7 To show Brownian motion

2 Look for tiny blobs of fat moving erratically.

Result

You will find that there is random movement. This is called Brownian motion.

Conclusion

Very small particles in a liquid or gas move when they collide with the molecules of the liquid or gas.

Key Points

- Matter is anything that occupies space and has mass.

- The three states of matter are: solids, liquids and gases.

- The particles in a solid are packed tightly together, therefore giving them a fixed shape.

- In a liquid the particles have some space between them and can slide past each other which means they don't have a fixed shape.

- In a gas the particles have lots of space between them. Gases have no fixed shape or volume.

- Substances can be changed from one state to another by heating or cooling them.

- Diffusion is the spreading out of the particles of a liquid or gas throughout the space they are in.

- Brownian motion is the movement of very small particles in a liquid or gas which is caused by their collisions with the molecules of the liquid or gas.

Salt pans are shallow pools of seawater which are allowed to evaporate leaving salt behind

Questions

1 Write out and complete the following:
  (a) Matter is anything that occupies _____ and has _____. There are three states of matter: _____, _____, _____.

  (b) In a solid the particles are packed _____ together. Therefore, solids have a _____ shape. In a gas, particles move around very _____ and there are large _____ between them. In a _____, the particles can slide past each other. As a result, they have no _____.

2 From the following list select the correct word:
  freezing, melting, boiling, condensation

Solid	⟶	Liquid	=
Liquid	⟶	Gas	=
Gas	⟶	Liquid	=
Liquid	⟶	Solid	=

3 The temperature at which a solid turns into a liquid is called its _____ point.

4 The temperature at which a liquid turns into a gas is called its _____ point.

5 What is condensation? Give an example.

6 Examine this cooling curve for X as it changes from a gaseous state to a solid state.
  (a) What is the freezing point of X?
  (b) What is the melting point of X?
  (c) What is the boiling point of X?

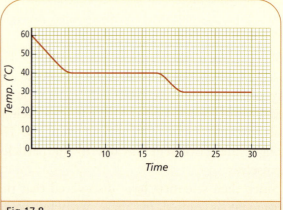

Fig 17.8

? Questions

7 Describe what happens to the water particles as ice is heated to water and eventually to steam.

8 (a) Why does condensation form on the inside of single-glazed windows in winter?

(b) How could it be reduced?

9 Explain the following statements in terms of particles:

(a) When a beam of sunlight shines into a room, dust particles can be seen moving in all directions in the air.

(b) A balloon will deflate slightly over a period of time.

(c) You can smell the dinner cooking in the kitchen from upstairs.

(d) On a cold wet day the windows of a car get 'steamed up'.

10 Place a ✓ or ✗ in each column to indicate whether each of the following statements is true or false.

Statement	Solids	Liquids	Gases
It has a definite volume			
It expands when heated			
It will change its shape to that of the container in which it is placed			
Its particles will spread out to fill the container in which it is placed			

Examination Questions

11 Study the diagram carefully. It shows the ways that particles of gases and solids occupy space.

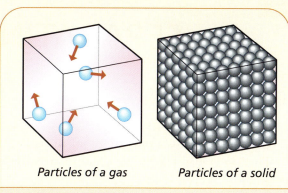

Particles of a gas Particles of a solid

Fig 17.9 States of matter

The particles of gas have lots of space and move randomly at high speeds in three dimensions and collide with each other and with their container.

The arrows represent the velocities of the gas particles.

The particles of a solid are packed closely together and cannot move around but they can vibrate.

Give one property of a gas and one property of a solid, that you have observed, and is consistent with (matches) this micro-view of these states of matter.
(JC, HL, 2006)

12 In each case choose one state of matter from the list on the right, which matches the characteristics in the table below.

Characteristics	State of matter
• Has definite shape	
• Has definite volume	
• It is not easily compressed	
• Has no definite shape	
• Has no definite volume	
• Is easily compressed	

Solid

Liquid

Gas

(JC, OL, Sample Paper)

CHEMISTRY

Chapter 18

SEPARATING MIXTURES

Introduction

Many everyday substances are mixtures.

- Air is a mixture of gases.
- Emulsion paint is a mixture of oil and water.
- The sea is water with lots of substances such as salt dissolved in it.
- Crude oil, which is taken from the ground, is a mixture of hundreds of substances such as home heating oil, petrol and tar used for roads. Unless chemists can separate these components from the oil it is useless.

> **A mixture** is made up of two or more substances mingled together but not chemically combined.

There are many ways to separate mixtures. The method used depends on the physical differences between the substances to be separated.

Let's examine some of the techniques.

Filtration

Filtration is used to separate a liquid and an insoluble solid, e.g. soil and water or chalk and water. You have all seen a coffee filter machine in action. The coffee solution is separated from the ground coffee by passing it through filter paper.

Gold panning involves swirling gravel found in a river in a shallow container filled with water. The lighter stones are lost while the heavier metals such as gold remain

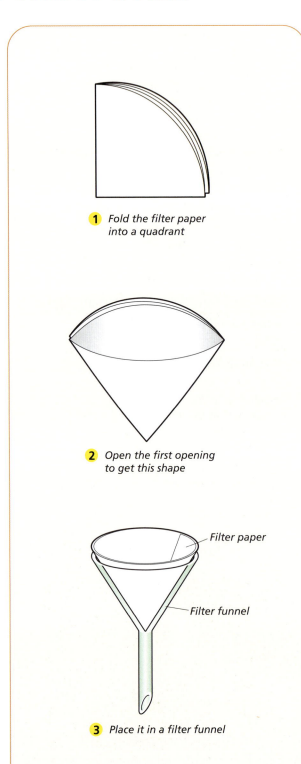

1 Fold the filter paper into a quadrant

2 Open the first opening to get this shape

Filter paper

Filter funnel

3 Place it in a filter funnel

Fig 18.1 Preparing for filtration

MANDATORY ACTIVITY 11A

To separate soil and water

Method

1 Slowly pour the mixture into the filter paper as shown.

2 Allow time for the water to separate from the soil.

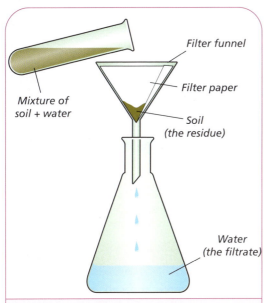

Fig 18.2 To separate soil and water

Result

- The insoluble solid (soil in this case) is trapped in the filter paper. This is called the residue.
- Clean water passes through the paper and is collected in the conical flask. This is called the filtrate.

Conclusion

Soil can be separated from water by filtration.

Decanting

Decanting is another technique used to separate insoluble solids and liquids, e.g. sand and water. Simply allow the solids to settle to the bottom and gently pour off the liquid.

▶▶▶ **ACTIVITY 18.1**

To separate a solid and a liquid

Method

1 Place the solid and the liquid in a beaker.

2 Allow the solid to settle.

3 Carefully pour the liquid off into another beaker.

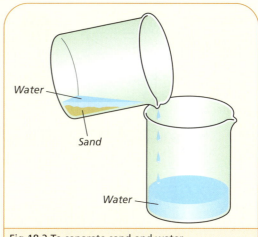

Fig 18.3 To separate sand and water

Result

The solid is left behind. However, not all the liquid can be separated as the last of the water being poured may contain sand.

Conclusion

A solid can be separated from a liquid by decanting.

CHEMISTRY

Gas masks like these could be worn during chemical or biological warfare. The respirators filter the contaminated air during an attack

135

Evaporation

This method is used to separate a soluble solid and a liquid, e.g. salt and water. The technique involves heating the solution until the liquid vapourises away, leaving the solid behind.

A micrograph of the cut end of a piece of filter paper showing its cellulose fibres

⭐ **MANDATORY ACTIVITY 11B**

To separate salt and water through evaporation

Method

1 Pour the solution into an evaporating dish.

2 Set up the water bath as shown. The dish is heated by the steam produced from the water in the beaker.

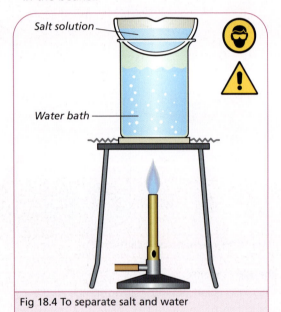

Fig 18.4 To separate salt and water

3 When almost all of the solvent (water) has evaporated, turn off the Bunsen burner and allow it to stand for a while. The dish will be very hot so allow time for it to cool before handling.

Result

The salt crystallises as the water evaporates.

Conclusion

Salt can be separated from water by evaporation.

Distillation

The problem with evaporation is that the liquid (solvent) is lost. Distillation is a technique that can be used to separate a soluble solid (solute) from a liquid, or to separate two miscible liquids (liquids that mix), on the basis that they have different boiling points, e.g. water and alcohol.

⭐ **MANDATORY ACTIVITY 11C**

To separate water and salt through distillation

Method

1 Set up the apparatus as shown in Fig 18.5.

2 Turn on the water tap and water will flow through the outer pipe of the Liebeg condenser and out the top back to the sink.

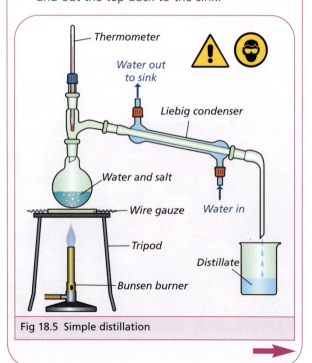

Fig 18.5 Simple distillation

CHEMISTRY

3 Heat the round-bottomed flask gently (with a blue flame) until the solution begins to boil and a steady flow of drops of water distils across. The thermometer should read 100°C.

Result

The water boils at 100°C and the steam then is cooled back to water in the condenser. The salt remains in the round-bottomed flask and is known as the residue.

Conclusion

Water and salt can be separated by distillation.

Simple distillation

Simple distillation can be used to separate miscible liquids (liquids that may be mixed) that have different boiling points, e.g. alcohol and water. Alcohol boils at 78°C and water boils at 100°C. However, some water will evaporate with the alcohol vapour and hence a good separation is difficult.

Fractional distillation

Fractional distillation as shown in Figure 18.6 will greatly improve the quality of separation.

The fractionating column will heat up to the boiling point of the alcohol and the water vapour will condense here and drip back down into the round-bottomed flask.

A distillation apparatus is used to separate substances with different boiling points, e.g. water and alcohol

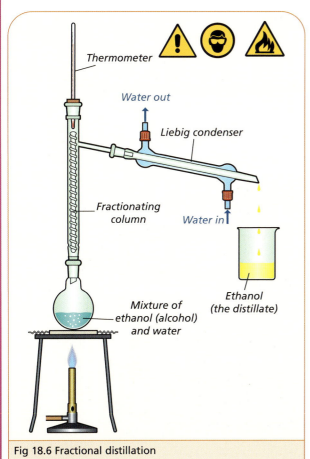

Fig 18.6 Fractional distillation

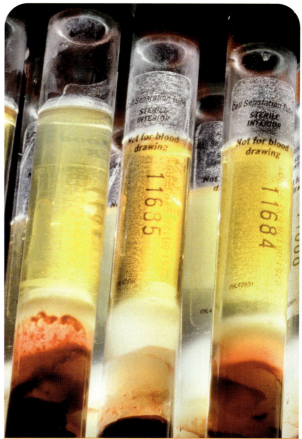

Blood is spun at very high speeds to separate blood into different components; the blood cells fall to the bottom of the tube and the serum (the yellow liquid) remains on top

CHEMISTRY

137

Salt

Salt comes from rock salt which is mined from the ground. Rock salt in its raw state is used to 'grit' the roads and stop them freezing. This is because the salt lowers the freezing point of water.

However, rock salt must be purified before it can be put on your chips! The first step in the purification process is the addition of water. The salt is soluble in water and impurities such as sand and tiny stones are insoluble. These impurities are filtered out and the water is evaporated from the filtrate, leaving salt behind.

CHEMISTRY

▶▶▶ ACTIVITY 18.2

To purify rock salt

Method

1 Crush some rock salt, using a pestle and mortar.

2 Add the crushed rock salt to water in a beaker and stir. (**Hint**: Only use a small volume of water.)

3 Filter the mixture. The filtrate will be salty water and the residue is composed of insoluble impurities.

4 Evaporate the water, using a water bath.

Result
The water evaporates leaving pure salt behind.

Conclusion
Rock salt can be purified, to produce salt (sodium chloride) for food.

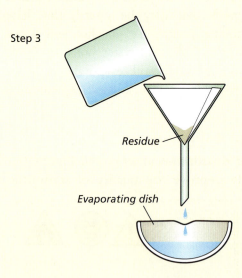

Step 3

Residue

Evaporating dish

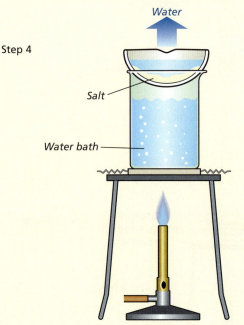

Water

Step 4

Salt

Water bath

Step 1

Step 2

Fig 18.7 To purify rock salt

Chromatography

Chromatography is a technique used to separate a mixture of dyes in ink, for example.

Chromatography can be used to separate a compound into its constituent chemicals

 MANDATORY ACTIVITY 11D

To separate the dyes in water-soluble markers

Method

1 Set up a piece of chromatography paper as shown.

2 Dot the colours as shown. Try to keep the dots small and concentrated.

3 Form a cylinder with paper and allow about 10–15 minutes for the solvent (water in this case) to travel up the paper column by capillary action. (The water level must be below the pencil line!)

Result

The different dyes in each colour are carried different distances by the water and hence are separated out.

Conclusion

Dyes can be separated in water.

4 Allow the paper to dry and stick it in your laboratory book. It is called a chromatogram. If using permanent markers, use methylated spirits instead of water.

Hint

The spots must be above the level of the solvent at the start. Can you think of the reason for this?

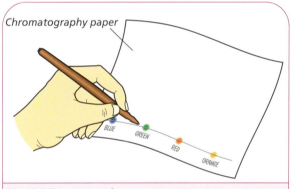

Fig 18.8 To separate dyes

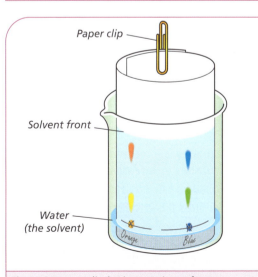

Fig 18.9 Paper cylinder in container of water

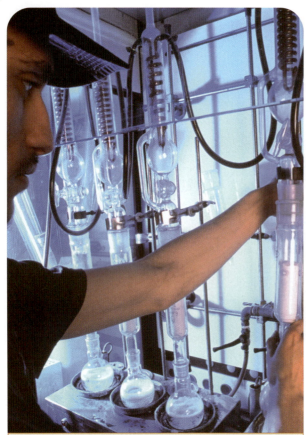

A biofuel production worker analyses samples produced by the distillation of rapeseed oil. This is a biofuel that can be used as an alternative to petrol or diesel

Key Points

- A mixture is composed of two or more substances mingled together but not chemically combined.

- An insoluble solid can be separated from a liquid by either decanting or filtering.

- Evaporation can be used to recover the solute from a solution.

- Distillation is used to separate miscible liquids with different boiling points or to separate a solute and a solvent.

- Chromatography is a technique used to separate substances by carrying them different distances in a solvent along chromatography paper.

? Questions

1 (a) What is a mixture?
   (b) Give five examples.

2 Sugar is soluble in water. What does this mean?

3 Sand is insoluble in water. List two methods that can be used to separate them.

4 The pouring off of a liquid from an insoluble solid is called _____.

5 The clear liquid that flows through filter paper is called _____ and the solids trapped in the paper are known as the _____.

6 Petrol and water are **immiscible**. Explain the highlighted term.

7 Distillation can be used to separate liquids that have different _____ _____.

8 Alcohol boils at __°C, whereas water boils at __°C. Therefore, the distillate will be the _____.

9 What is the purpose of the condenser in the distillation apparatus?

10 Choose from the following list the most suitable method for each of the following separations:
   chromatography, distillation, evaporation, filtration
   (a) to obtain salt from salty water
   (b) to separate a mixture of different coloured inks
   (c) to obtain a sample of pure water from tap water
   (d) to purify whiskey

11 A chemist accidentally pours silver chloride powder into a bottle of glucose powder.
   (a) Describe how the two powders could be separated.
        Hint: Silver chloride is insoluble in water but glucose is soluble.
   (b) Draw and label diagrams for each technique involved.

12 Describe with diagrams how you could separate a mixture of salt, water and sand.

? Questions

13 A, B, C, D and E are five food colourings. The dyes in each of these food colourings were separated by chromatography as shown.

Give the correct answer.

Food colour A could be a mixture of:
(a) B and C
(b) B and D
(c) B and E
(d) C and E

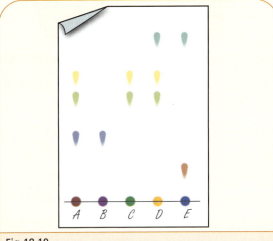

Fig 18.10

Examination Questions

14 A spot of water-soluble ink was put on a piece of chromatography paper and set up as shown in the diagram. The ink used was a mixture of different coloured dyes.

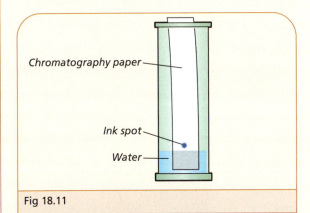

Chromatography paper

Ink spot

Water

Fig 18.11

(a) What happens to the ink spot as the water moves up the paper?
(b) What would happen to a spot of water-soluble ink consisting of a single coloured dye if it were used in the above experiment?
(JC, HL, 2006)

15 Separation techniques are very important in chemistry.

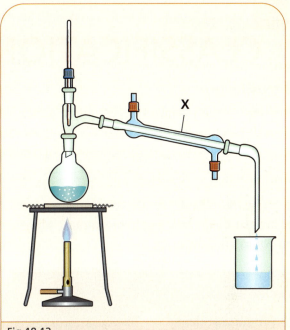

X

Fig 18.12

(a) What is the name given to the separation technique shown in Figure 18.12?
(b) Name two substances which could be separated using this technique?
(c) Name the part of the apparatus labelled X in the diagram.
(d) What is the name given to the separation technique shown below.
(JC, OL, 2006)

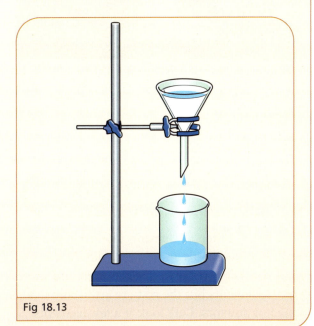

Fig 18.13

Chapter 19 •••

CLASSIFICATION OF SUBSTANCES

Introduction

We have already looked at the three states of matter, i.e. solids, liquids and gases. If we look even closer we can further classify substances. Scientists believe that everything is made up of atoms.

Some substances are made up of atoms that are all the same. Therefore, these substances cannot be broken down into anything simpler. They are known as **elements**.

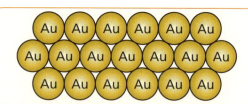

Fig 19.1 Gold (symbol Au) is an element. Note how tightly the atoms are packed together. This is why gold is a solid

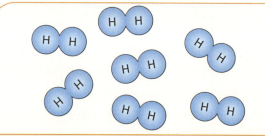

Fig 19.2 Hydrogen (symbol H) is also an element. The hydrogen atoms exist in pairs (H_2). Note also the amount of space between each pair. This is why hydrogen is a gas

Robert Boyle was an Irish scientist; he is known as the Father of Chemistry. He defined an element as a substance that could not be broken down into simpler substances

Copper, mercury and magnesium are elements. Copper (brown) is a good conductor of heat and electricity. Mercury (centre) is a liquid metal at room temperature. Magnesium (silver) is a reactive metal

Elements

> **Elements** are substances made up of only one type of atom.

There are 92 different elements found naturally on Earth. In other words, there are 92 different atoms found naturally on Earth. In fact, there are more than 92 elements but the rest are artificially made in nuclear reactors. Let's look at a few elements.

The elements sulfur (yellow) and phosphorous (which has to be kept under water as it spontaneously reacts in air)

Molecules

We have just looked at hydrogen gas. Its formula is H_2, not just H. This is because the H atoms occur in pairs. These are chemically combined and exist as **molecules**.

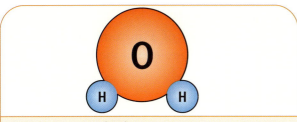

Fig 19.3 A water molecule

> A molecule is made up of two or more atoms chemically combined.

Hydrogen oxide is the chemical name for water. Its chemical formula is H_2O. This means that the basic unit of water consists of 2 hydrogen atoms and 1 oxygen atom all chemically combined.

Fig 19.4 A carbon dioxide molecule

Compounds

There are 92 elements found naturally on Earth. However, there are millions of different substances. So logically these substances must be made up of more than one type of element. These substances are called **compounds**.

> Compounds are made up of two or more different types of atom chemically combined.

Compounds are all around us in everyday life.

- Table salt has the chemical formula NaCl. It is made up of 2 elements. Can you name them? **Hint**: Look up the periodic table!
- Glucose is a type of sugar. It has the chemical formula $C_6H_{12}O_6$. This means each molecule of glucose is made of 6 carbon atoms, 12 hydrogen atoms and 6 oxygen atoms all chemically combined.

Salt crystals are made up in a cubic shape

Mixtures

So far we have looked at elements and compounds. Elements are made up of only one type of atom; compounds are made up of different elements, which are chemically combined. In mixtures, the substances that make up the mixtures are never present in the fixed amounts.

The amounts of the substances in the mixture can vary. A packet of Skittles is a mixture of different coloured sweets. Maltesers would be more like an element because all the sweets are the same.

> A mixture contains two or more different substances mingled together but not chemically combined.

Mixtures are everywhere.

- The sea is a mixture of water and salts.
- The air is a mixture of N_2, O_2 and CO_2 and other gases.
- Soil is a mixture of sand, clay, humus, water, etc.

There are differences between compounds and mixtures.

Compound

1 The elements in a compound are always present in a fixed ratio.
2 There is always a heat change when a compound is formed.
3 It is difficult to separate the elements of a compound.
4 The properties of a compound are different to those of the elements of which it is made.

Mixture

1 Substances that make up a mixture are present in no fixed ratio.
2 Usually no or very little heat can change when a mixture is made.
3 It is usually easy to separate the substances of a mixture.
4 The properties of a mixture are similar to those of the substances in the mixture.

▶▶▶ ACTIVITY 19.1

To compare a mixture of iron and sulfur to the compound iron sulfide (FeS)

Method

1 Weigh out 2 g of iron and 4 g of sulfur on a piece of paper. Describe their appearances.

2 Mix the iron and sulfur together, using a spatula. This is a mixture.

- What does it look like?
- How could you separate the iron and sulfur?
- Can you think of another way? Try it if you have time.

3 Now place the mixture in a test tube and heat it strongly in the fume cupboard until the mixture glows red. (Sulfur dioxide gas is given off – it is poisonous.)

4 Allow the test tube to cool and examine its contents. This compound is iron sulfide.

5 Try bringing a magnet near and notice that it will not be attracted. The iron and sulfur are no longer present as elements. Instead they have formed a compound called iron sulfide.

Result

Notice that the yellow colour of sulfur and grey specs of iron have disappeared.

Conclusion

When elements form compounds they lose some of their properties.

Additional test

This is a teacher demonstration only. This must be done in a fume cupboard.

To the grey/black iron sulfide add 1–2 drops of dilute hydrochloric acid (HCl). The gas given off is hydrogen sulfide (H_2S) and smells of rotten eggs. This is a very toxic gas!

Fig 19.5 Iron and sulfur mixture

Properties of compounds

- The substance produced in Activity 19.1 is clearly a new substance. It has its own properties. These properties differ from those of iron and sulfur, the elements from which the substance is made.

- Water is a compound made up of hydrogen, which is a flammable gas, and oxygen, a colourless gas. Oxygen is a necessary requirement for a fire, whereas water is sometimes used to put out fires. Clearly the properties of the compound differ from those of its elements.

- Magnesium is a shiny, reactive metal. When burned it combines with oxygen very brightly to form a white powder, magnesium oxide, which is quite unreactive.

- Likewise carbon dioxide gas is a colourless gas used in fire extinguishers and exhaled by us all the time. Its formula is CO_2, i.e. it is composed of carbon (a black solid) and oxygen.

These examples show that the properties of a compound differ from those of the elements from which it is made.

Iron reacting with sulfur in a test tube. The compound formed is iron sulfide

Key Points

- Everything is made up of atoms.

- Elements are substances made up of only one type of atom.

- A molecule is made up of two or more atoms chemically combined.

- Compounds are made up of two or more different types of atoms chemically combined.

- A mixture contains two or more different substances mingled together but not chemically combined.

- When a compound is formed during a chemical reaction its properties differ from those of the elements from which it is made.

? Questions

1   (a)   What is an element?
    (b)   Give three examples and their symbols.

2   (a)   What is a molecule?
    (b)   Give three examples and their formulae.

3   Write out and complete the following:
    Compounds are made up of two or more _____ which are chemically _____. A mixture is made up of two or more _____ which are mingled together but not _____ _____.

4   Write out and complete the following:
    The smallest part of an element is called an _____. All the _____ in an element are the same. Therefore, elements cannot be _____ _____ into simpler substances.

5   Write out and complete the following:
    Atoms of the same type which are chemically combined form _____. If a substance is made up of more than one type of atom it is a _____.

6   Match up the elements with their uses.

Table 19.1	
Element	Use
Mercury	In breathing apparatus
Copper	In swimming pools
Oxygen	In thermometers
Chlorine	To make ladders
Aluminium	To make electric wires

7   State two differences between a mixture and a compound.

8   You are given a mixture of sulfur and iron filings.
    (a)   What colour would you expect the mixture to be?
    (b)   How could you separate the sulfur and the iron filings?
    (c)   Can you think of another method of separating the mixtures?

CHEMISTRY

? Questions

9 Figure 19.6 shows a mixture of iron filings and sulfur being heated.

   (a) Name the compound which is formed as a result of the reaction.

   (b) What precautions should always be taken when heating a test tube?

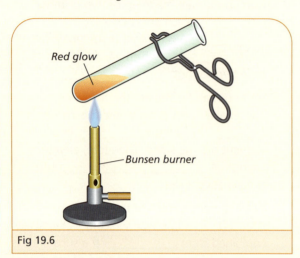

Red glow

Bunsen burner

Fig 19.6

10 Look at the boxes.

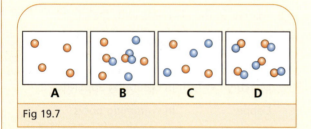

A   B   C   D

Fig 19.7

Which box contains

   (a) a pure compound?

   (b) only one element?

   (c) a mixture of elements and a compound?

   (d) a mixture of elements?

11 Sulfur is a non-metal and iron is a metal. Explain why it is easier to separate iron from a mixture of iron and sulfur than from the compound iron sulfide.

12 The following list contains four elements and four compounds. List which are elements and which are compounds.

	Element or compound?
Carbon dioxide	
Iron sulfide	
Sulfur	
Glucose	
Sodium	
Salt	
Oxygen	
Silicon	

Examination Questions

13 Natural gas is mainly methane (CH_4).

   (a) Name one of the two elements found in methane.

   (b) Name one gas produced when methane is burned in air.

   (JC, OL, 2006)

14 Complete the table below identifying one mixture and one compound from the list on the right.

   (JC, OL, 2006)

Mixture	Compound		Table salt Carbon Air

15 Robert Boyle introduced the word element into the language of chemistry.

Complete the table below identifying each of the substances listed as an element or a compound.

An example is completed in the case of carbon dioxide.

Substance	Element	Compound
Carbon dioxide		✓
Water		
Carbon		

(JC, OL, Sample Paper)

Chapter 20

ATOMIC STRUCTURE

Introduction

The word **atom** comes from a Greek word meaning 'cannot be split'. In the 1800s and 1900s it was believed that nothing smaller than an atom existed. However, as you probably know, atoms can be split! Atoms are extremely small, indeed so small that they have never been seen! So how do we know what they look like?

Many great scientists came up with theories on the shape and structure of the atom. However, later experiments using radioactive particles created a new picture. By 1915 scientists Rutherford and Bohr had developed a model which is still used today.

The atom

> **An atom** is the smallest part of an element, which still has the properties of that element.

The atom is made up of three types of particles. These are:

- protons
- neutrons
- electrons.

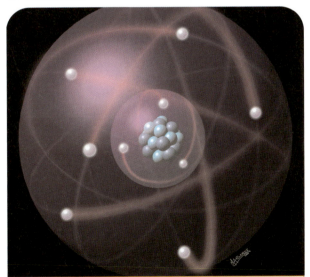

An atom called the Bohr model after the scientist Niels Bohr. In the centre of the atom is the nucleus that is made up of protons and neutrons

The protons and neutrons are found together in the middle of the atom. This is called the nucleus of the atom. It is extremely small and dense. The tiny electrons spin around the nucleus.

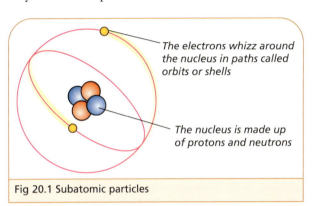

The electrons whizz around the nucleus in paths called orbits or shells

The nucleus is made up of protons and neutrons

Fig 20.1 Subatomic particles

Figure 20.1 shows three particles that make up atoms. They are called subatomic particles. They differ greatly from each other in a number of ways. Table 20.1 summarises the main differences.

Table 20.1 Differences in subatomic particles			
Particle	**Charge**	**Location**	**Mass in atomic mass units**
Proton	+1	In the nucleus	1
Neutron	0	In the nucleus	1
Electron	−1	Orbiting the nucleus	0

We can see from Table 20.1 that protons and neutrons have the same mass (measured in units called atomic mass units). Electrons are so light it would take almost 2 000 of them to weigh as much as one proton or neutron, so we call its mass zero.

An easy way to remember the charges on each:

P for *Proton*　　　P for *Positive*
N for *Neutron*　　N for *Neutral*
So *Electrons* must be *Negative*.

CHEMISTRY

Atoms of each element differ from each other by the numbers of protons, neutrons and electrons which they possess.

Atomic number and mass number

Each element in the periodic table (page 152) has its own atomic number and mass number. The smaller number is the atomic number. It tells us how many protons there are in one atom of that element.

> **Atomic number** is the number of protons in an atom of that element.

All atoms have an overall neutral charge because the number of protons (which are positive) is equal to the number of electrons (which are negative). Therefore, the atomic number also reveals how many electrons an atom of that element has.

The larger number written with an element in the periodic table is called its **relative atomic mass number**.

> **Mass number** is the number of protons **and** neutrons in an atom of that element.

Can you remember why we don't count the electrons when calculating the mass number?

In order to work out the number of neutrons we must subtract the atomic number from the mass number.

Example

To calculate the numbers of protons, neutrons and electrons in an aluminium (Al) atom we look it up in the periodic table.

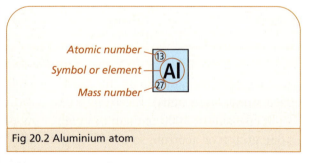

Atomic number — 13
Symbol or element — Al
Mass number — 27

Fig 20.2 Aluminium atom

The atomic number = 13.
So an aluminium atom has 13 protons and 13 electrons.
(Remember: number of protons = number of electrons.)

The mass number = 27.
Therefore 27 = number of protons + neutrons
So number of neutrons
= mass number – atomic number
= mass number – number of protons
= 27–13
= 14

Can you calculate the number of protons, neutrons and electrons in
(a) helium?

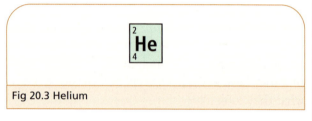

Fig 20.3 Helium

(b) fluorine?

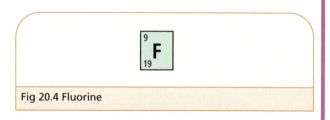

Fig 20.4 Fluorine

Arrangement of electrons

The electrons whizz around the nucleus of an atom. The electrons move around the nucleus in shells or orbits. The arrangement of the electrons in these shells is very important as it helps to explain how and why the element reacts chemically.

- The first shell can hold up to 2 electrons.
- The second shell can hold up to 8 electrons.
- The third shell can hold up to 8 electrons.

Electronic configuration

An electronic configuration is a short-hand way of showing the number of electrons in each shell. It is important to remember that electrons start filling up the shells from the inside outwards.

Niels Bohr, a famous Danish scientist, devised a way to draw the atom in a simple way. This way of representing an atom is called the Bohr model.

Example

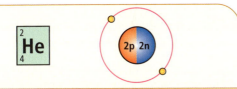

Fig 20.5 Helium
2 protons
2 electrons
2 neutrons (4 – 2)
2 electrons (first shell full)
Electronic configuration (2)

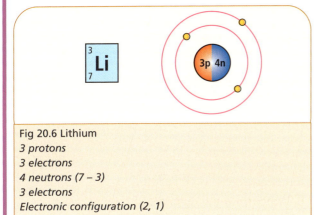

Fig 20.6 Lithium
3 protons
3 electrons
4 neutrons (7 – 3)
3 electrons
Electronic configuration (2, 1)

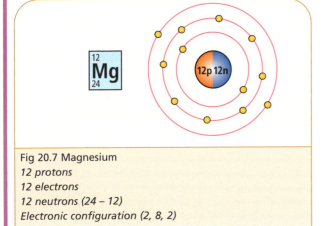

Fig 20.7 Magnesium
12 protons
12 electrons
12 neutrons (24 – 12)
Electronic configuration (2, 8, 2)

Reactivity

Why are some elements very reactive and others quite unreactive? The reason is that chemical reactions involve the electrons in the outer shells of atoms. Every atom 'wants' to have its outer shell completely full of electrons. This situation makes them stable and unreactive (happy!)

Isotopes

Too often we hear stories about nuclear power and the dangers that accompany it. You may have heard the word 'isotope' or the term 'radioactive isotopes' mentioned in the news. But just what is an isotope?

> **Isotopes** are atoms that have the same number of protons but different numbers of neutrons.

In other words, isotopes are atoms that have the same atomic number but different mass numbers.

Many elements exist as isotopes, e.g. carbon and hydrogen.

Example

Figure 20.8 and Figure 20.9 show two isotopes of carbon.

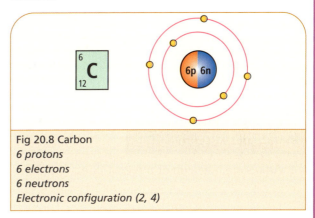

Fig 20.8 Carbon
6 protons
6 electrons
6 neutrons
Electronic configuration (2, 4)

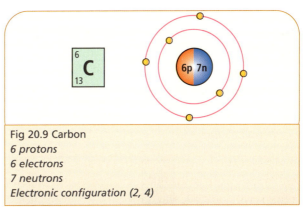

Fig 20.9 Carbon
6 protons
6 electrons
7 neutrons
Electronic configuration (2, 4)

A 'button' of Uranium-235, the radioactive isotope of uranium that is used as a fuel inside nuclear reactors

CHEMISTRY

Key Points

- The atom is made up of three particles: proton, neutron and electron.

- Protons have a positive charge and are found in the nucleus.

- Neutrons are neutral and are also found in the nucleus.

- Electrons are negatively charged and are found orbiting the nucleus shells.

- Protons and neutrons have a mass of 1 atomic mass unit (AMU), whereas the mass of an electron is so small it is taken as O.

- The atomic number of an element is the number of protons in an atom of that element.

- The mass number of an element is the total number of protons and neutrons in an atom of that element.

- Electrons occupy shells or orbits. The first shell can hold up to 2 electrons and the second and third can hold up to 8.

- An atom is stable if its outer shell is full of electrons.

- Isotopes are atoms with the same number of protons but different numbers of neutrons.

? Questions

1 Atoms are made up of three particles: _____ , _____ and _____.

2 The protons and neutrons are found in the _____ and the _____ spin around in orbits or _____.

3 What is the atomic number of an atom?

4 What is the mass number of an atom?

5 Here is the element neon as it appears in the periodic table.

Fig 20.10

(a) What is its atomic number?
(b) What is its mass number?

6 Write down the numbers of protons, electrons and neutrons in each of the following elements.

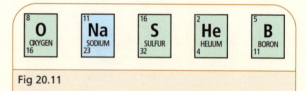

Fig 20.11

7 Draw a simple atomic diagram (Bohr diagram) of:

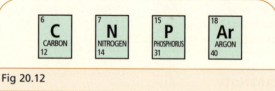

Fig 20.12

8 Copy and complete the following:

Particle	Charge	Location	Mass (AMU)
(i)	+	(ii)	1
(iii)	O	(iv)	1
Electron	(v)	(vi)	0

Questions

9 (a) What are isotopes?

(b) Hydrogen has 3 isotopes. What is the difference between the 3 isotopes?

10 Draw a simple Bohr diagram for each isotope shown in the diagram.

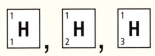

Fig 20.13

11 Look up the periodic table of the elements and name the group of elements that:

(a) have a full outer shell

(b) have 1 electron in their outer shell

(c) have 7 electrons in their outer shell.

12 (a) Identify the atom shown below.

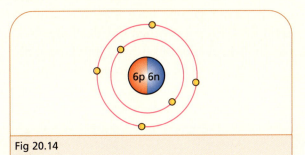

Fig 20.14

(b) Name another atom with the same number of electrons in the outer shell.

13 Copy the table below and fill in the missing numbers.

Examination Questions

14 Define the term 'isotope'.
(JC, HL, 2006)

15 Niels Bohr received the Nobel Prize for physics in 1922 for his model of the electronic structure of the atom. Potassium has an atomic number of 19. Give the arrangement of the electrons in an atom of potassium.
(JC, HL, 2006)

16 Complete the statements below using one of the words from the list on the right in each case.

Protons are_____ charged particles.

Electrons are_____charged particles.
(JC, OL, 2006)

| Negatively |
| Positively |

17 Carbon–12 and carbon–13 are isotopes of carbon.

What does this mean?
(JC, HL, Sample Paper)

Element	Atomic number	Mass number	Number of protons	Number of neutrons	Electronic configuration
Hydrogen	1	1	1		
Boron	5	11			2, 3
Fluorine		19	9		
Aluminium		27	13		
Calcium	20			20	2, 8, 8, 2

CHEMISTRY

Chapter 21

THE PERIODIC TABLE

Introduction

Lots of elements were discovered around 200 years ago. Scientists struggled to find any links between them. Many spent their whole careers trying to link elements and arrange them in some sort of order.

In 1869, a Russian chemist called Dmitri Mendeleev solved the problem, at least partially. He arranged the elements according to their atomic masses. Mendeleev constructed a table and grouped elements that had a similar behaviour.

At that stage only 57 elements were known, so he left blank spaces for elements not yet even discovered! He also predicted the properties of these unknown elements. In 1955 a new element, Mendelevium, was named in his honour.

The modern periodic table

Today there are 92 naturally occurring elements. They are arranged in order of atomic number (the number of protons an atom of the element has) rather than atomic mass. There are no gaps either.

The elements which Mendeleev had left spaces for in his periodic table have all been discovered and fitted in perfectly.

Groups

Each vertical column in the table is called a group. Some groups have a family name. For example, Group 1 is called the alkali metals. Can you find the names of Group 7 and Group 8?

The elements in each group have the same number of electrons in their outer shell and hence behave similarly to each other.

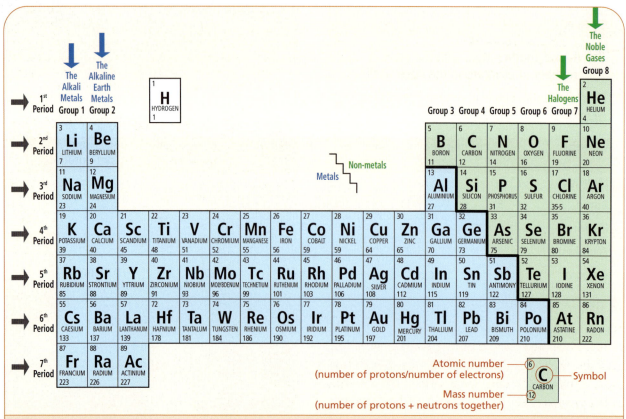

Fig 21.1 Modern periodic table. Metals are to the left of the zig-zag line (in blue). Non-metals are to the right of the line (in green)

Chapter 22 ●●●

CHEMICAL BONDING

Introduction

Water and sodium chloride are compounds, i.e. they are composed of two or more different elements chemically combined. But why do elements react with each other?

When we look at the periodic table we find a group of elements that are very unreactive. They are in Group 8 (or Group O). Can you remember the name of this group?

The members of this group, e.g. helium and neon, are very stable (or happy) because they have a full outer shell of electrons. All other atoms want to have a full outer shell of electrons also and this is why they react with other atoms.

There are two ways to achieve a full outer shell of electrons.

1   One atom loses electrons and another atom gains electrons. This is called ionic bonding.
2   The atoms involved share electrons. This is called covalent bonding.

Ionic bonding

Bonding simply means the way atoms join together.

Ionic bonding occurs when one atom wants to lose electrons and another atom wants to gain electrons in order to become stable, i.e. to end up with an outer shell which is full of electrons. Let's look at an example of this:

Sodium atom

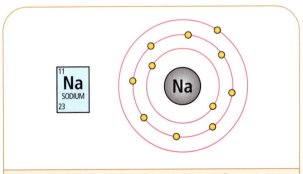

Fig 22.1 Sodium, 11 electrons, electronic configuration (2, 8, 1)

Sodium has 1 electron in its outer shell. It has a choice! It can either gain 7 more electrons to become (2, 8, 8) or else lose 1 electron to become (2, 8). Which would be easier? Obviously to lose 1 electron.

Chlorine atom

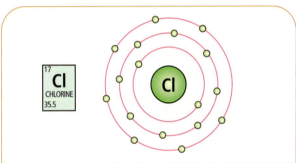

Fig 22.2 Chlorine, 17 electrons, electronic configuration (2, 8, 7)

Chlorine has 7 electrons in its outer shell. The choice is either to gain 1 electron and become (2, 8, 8) or lose 7 electrons and become (2, 8). The obvious thing to do is to gain 1 electron.

When sodium reacts it loses 1 electron. The electron doesn't just disappear; another atom has to be around to accept it. In this case it is a chlorine atom. It accepts the electron and hence becomes negatively charged.

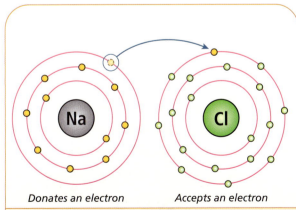

Donates an electron   Accepts an electron

Fig 22.3 An ionic bond is formed by the transfer of electrons between two atoms

Remember that all atoms are neutral because they have the same number of positive protons and negative electrons. However, sodium has just lost 1 electron and chloride has gained 1 electron so the electrons and protons no longer cancel each other out.

Table 22.1			
Sodium		**Chloride**	
11 Protons	= +11	17 Protons	= +17
10 Electrons	= −10	18 Electrons	= −18
	+1		−1

Each of these atoms has become charged. These are now called **ions**.

> An **ion** is a charged atom or a charged group of atoms.

● When an atom loses electrons it becomes positively charged.
● When an atom gains electrons it becomes negatively charged.

The positive sodium ion and the negative chloride ion attract each other (opposite charges attract). The attraction between the ions is very strong and the result is an ionic bond. Because of the strong attraction between these ions, millions of them bond together forming crystals.

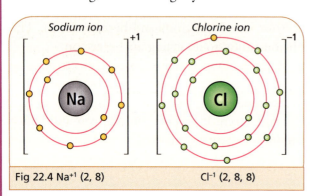

Fig 22.4 Na^{+1} (2, 8)   Cl^{-1} (2, 8, 8)

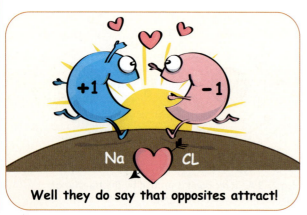

Well they do say that opposites attract!

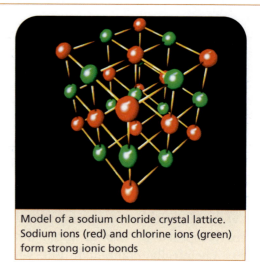

Model of a sodium chloride crystal lattice. Sodium ions (red) and chlorine ions (green) form strong ionic bonds

Ionic bonds form between metals and non-metals. This is because all metals want to lose electrons in order to have a full outer shell, whereas non-metals want to gain electrons to fill their outer shells and become stable.

> An **ionic bond** is the electrical force of attraction between oppositely charged ions in a compound.

Naming ionic compounds

When naming ionic compounds the positive ion (metal) is named first and the negative ion (non-metal) makes up the last part. For example, the ionic compound formed between sodium and chlorine is sodium chloride (NaCl).

● If a name ends in **-ate** it means the compound also contains oxygen.
● If it ends in **-ide** this means it only contains the elements named.

Let's look at another ionic bond.

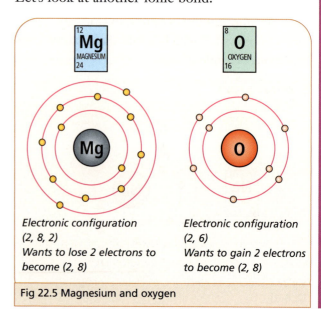

Electronic configuration (2, 8, 2)
Wants to lose 2 electrons to become (2, 8)

Electronic configuration (2, 6)
Wants to gain 2 electrons to become (2, 8)

Fig 22.5 Magnesium and oxygen

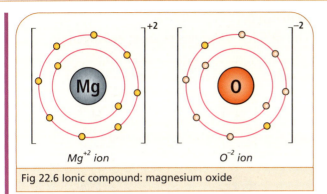

Fig 22.6 Ionic compound: magnesium oxide

The ionic compound formed is called magnesium oxide, formula MgO. Notice that the charges cancel each other out. In other words, all ionic compounds are neutral. Knowing this, we can work out the formula for any ionic compound.

Example 1

If magnesium and chloride were to react the magnesium atom loses 2 electrons and becomes a Mg^{+2} ion. Chlorine wants to gain 1 electron and will become a Cl^{-1} ion. The charges must cancel out so therefore it would take 2 chloride ions to cancel out 1 magnesium ion.

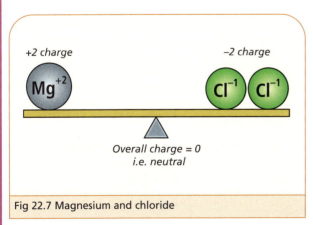

+2 charge   –2 charge

Overall charge = 0
i.e. neutral

Fig 22.7 Magnesium and chloride

The name of the ionic compound formed will be magnesium chloride and its chemical formula is $MgCl_2$.

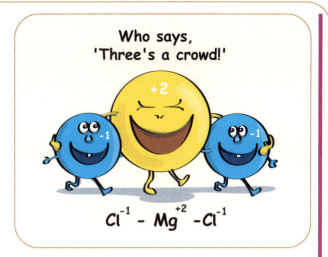

Who says, 'Three's a crowd!'

$Cl^{-1} - Mg^{+2} - Cl^{-1}$

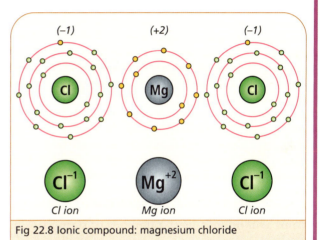

Fig 22.8 Ionic compound: magnesium chloride

Example 2

Aluminium is in Group 3. Thus it has 3 electrons in its outer shell. Will it lose 3 electrons or gain 5 electrons to become stable? Obviously it will lose 3 electrons! So the aluminium ion formed will be Al^{+3}. Chlorine forms a Cl^{-1} ion, so what will be the formula for aluminium chloride?

From this see-saw it would take $3 \times (-1)$ charges to cancel out $1 \times (+3)$ charge, so the formula is $AlCl_3$.

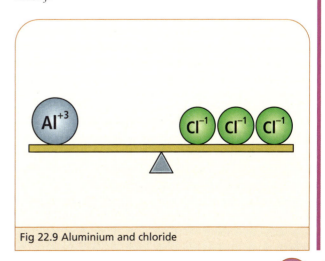

Fig 22.9 Aluminium and chloride

CHEMISTRY

See if you can copy out and complete this table.

Table 22.1				
	Chloride (Cl^{-1})	Fluoride (F^{-1})	Oxide (O^{-2})	Sulfate (SO$_4$$^{-2}$)
Sodium (Na^{+1})	NaCl			
Magnesium (Mg^{2+})		MgF$_2$		
Aluminium (Al^{+3})				
Hydrogen (H^{+1})				

Properties of ionic compounds

Ionic compounds have a number of properties in common:

1   They are crystalline solids with high melting and boiling points (due to strong attraction between molecules).
2   They are usually soluble in water.
3   They conduct electricity when molten (melted down) or in solution, i.e. dissolved in a solvent.
4   Reactions between ionic substances are usually fast.

Covalent bonding

In ionic bonding we saw how metals bond with non-metals. The metals want to lose electrons and the non-metals want to gain electrons. But there are many compounds formed between non-metals only, e.g. H_2O and CO_2. How can they bond or react if they both, by definition, want to gain electrons? The answer is: they share electrons.

> A covalent bond consists of a pair of electrons shared between two non-metal atoms.

Hydrogen

The simplest example occurs in hydrogen (H_2) gas.

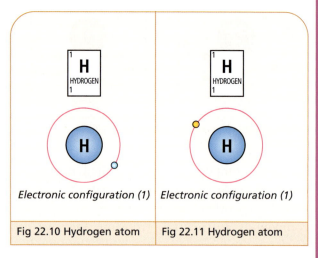

Electronic configuration (1)   Electronic configuration (1)

Fig 22.10 Hydrogen atom   Fig 22.11 Hydrogen atom

Two hydrogen atoms share their single electrons with each other by overlapping their outer shells.

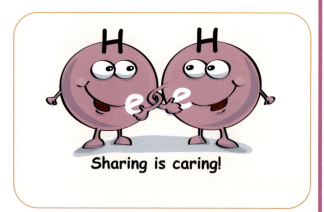

Sharing is caring!

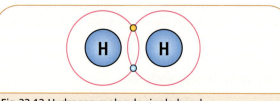

Fig 22.12 Hydrogen molecule single bond

Each hydrogen atom now has 2 electrons in its first and only shell and both are stable. Hence hydrogen exists as H_2 molecules which are stable, and not as H atoms which are unstable. A simplified diagram would be:

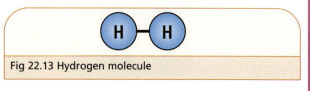

Fig 22.13 Hydrogen molecule

Notice the single bond is because the 2 hydrogen atoms share 1 pair of electrons.

Oxygen

Oxygen gas exists as O_2 molecules.

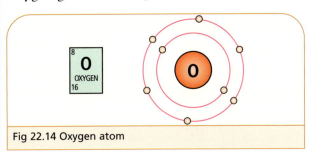

Fig 22.14 Oxygen atom

Each oxygen atom wants to gain 2 electrons and become (2, 8). So 2 oxygen atoms overlap their outer shells and share 2 electrons each, i.e. 2 pairs. This results in a double bond between them.

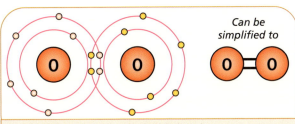

Can be simplified to

Fig 22.15 Oxygen molecule double bond

Nitrogen

Nitrogen exists as N_2 molecules also. See if you can draw a molecule of N_2.

Water

Another example of a covalent molecule is water (H_2O). As the formula suggests, water is made up of 2 hydrogen atoms bonded to 1 oxygen atom. Why?

As we have already seen each hydrogen atom wants to share or gain 1 electron. Oxygen wants to gain or share 2 electrons. So 2 hydrogen atoms each share with a single oxygen atom as follows:

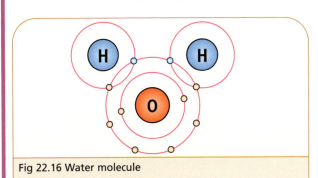

Fig 22.16 Water molecule

Each hydrogen atom shares 1 pair of electrons with the oxygen atom.

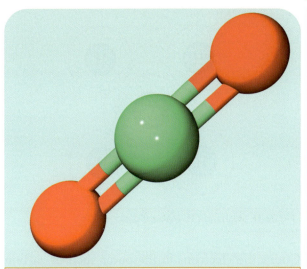

A carbon dioxide molecule (CO_2). The carbon atom (green) is joined to the oxygen atoms by double bonds

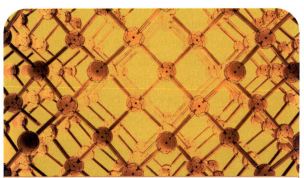

The molecular structure of a diamond, which is a form of carbon. Diamonds are the hardest naturally occurring substance

Methane

Natural gas is mainly methane. Its formula is CH_4, i.e. there are 4 hydrogens bonded to a carbon atom.

Why is it 4 hydrogens and not 2 or 3? To find out why it is 4 hydrogens we need to examine the carbon atom.

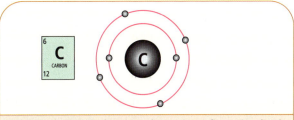

Fig 22.17 Carbon, 6 electrons, electronic configuration (2, 4)

Carbon atoms could lose 4 electrons and have an electronic configuration of (2) or gain 4 to become (2, 8).

In methane the carbon atom 'shares' 4 electrons with 4 hydrogen atoms.

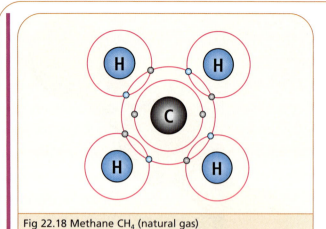

Fig 22.18 Methane CH$_4$ (natural gas)

Count the electrons in the outer shell of carbon and hydrogen. What do you notice?

From these examples of covalent bonding we notice that since no electrons are being transferred there are no changes on the atoms. There is little force or attraction between the molecules so they don't build up into crystals like ionic compounds do. Because of this, covalent compounds usually exist as liquids or gases. They do not conduct electricity and do not dissolve in water.

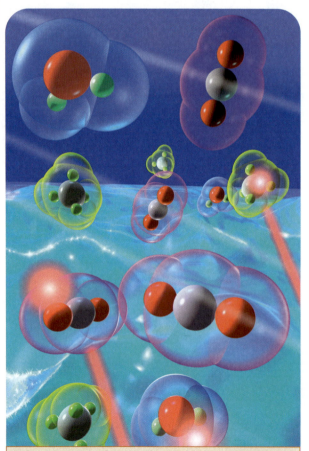

Molecular models of carbon dioxide (CO$_2$) in red/blue, methane (CH$_4$) in blue/green and water (H$_2$O) in red/green. The increase in the levels of these gases leads to a gradual warming of the planet

►►► ACTIVITY 22.1

To investigate the ability of ionic and covalent substances to conduct electricity

Method

1 Set up a circuit as shown.

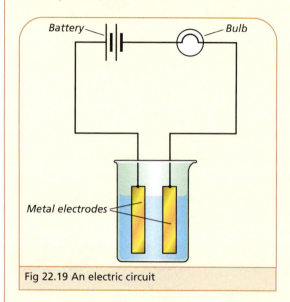

Fig 22.19 An electric circuit

2 Pour distilled water into the beaker. Does the bulb light? What does this tell us about pure water?

3 Add a small amount of table salt (or any salt) to the water. What do you notice? What type of substance is salt? (ionic or covalent)

4 Repeat this experiment with a range of substances, e.g. oil, dilute acid, sugar solution, copper sulfate.

Result

The bulb will light when ionic substances are placed in the beaker, but will not light when covalent substances are used.

Conclusion

Ionic compounds conduct electricity. Covalent compounds do not.

CHEMISTRY

Properties of ionic and covalent compounds

Table 22.2	
Ionic compounds	**Covalent compounds**
1 Usually solids	1 Usually liquids or gases
2 High melting and boiling points	2 Low melting and boiling points
3 Soluble in water	3 Insoluble in water
4 Conduct electricity when molten or dissolved in water	4 Do not conduct electricity usually
5 Undergo fast reactions	5 Undergo slow reactions

Valency

> **The valency** of an element is the number of electrons an atom of that element wants to gain, lose or share in order to have a full outer shell (to be chemically stable).

For example, sodium has an electron configuration of (2, 8, 1). It would like to lose its outer electron to become Na^{+1} (2, 8). Thus the valency of sodium is 1.

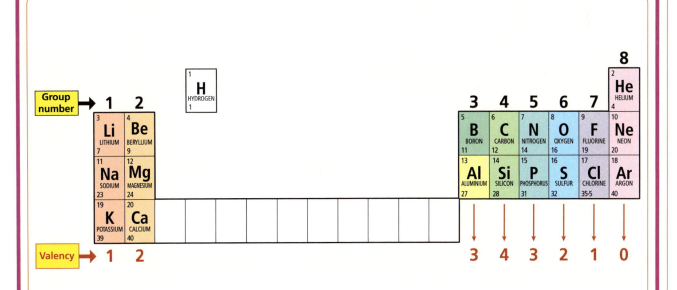

Fig 22.20 A section of the periodic table showing how the valency of an element can be calculated

? Questions

1 Why do most elements react?

2 Noble gases do not react because they have a _____ outer _____.

3 There are two types of bonds: _____ and _____.

4 Write out and complete the following: _____ bonds form between metals and non-metals. The metal _____ electrons and becomes _____ charged and the non-metal _____ electrons and becomes _____ charged. These oppositely charged ions are attracted to each other resulting in _____ bonds. Hence these compounds tend to be _____ with _____ melting and boiling points.

5 Write out and complete the following: A covalent bond is formed when electrons are _____. These bonds are weak and as a result covalent compounds are usually _____ or _____ with _____ melting and boiling points.

6 What is an ion?

7 A negative ion is an atom that has _____ electrons.

8 A positive ion is an atom that has _____ electrons.

9 By referring to the periodic table decide whether the following compounds are ionic or covalent: calcium oxide, sodium fluoride, hydrogen oxide, sulfur dioxide, aluminium oxide, potassium bromide, carbon dioxide.

10 What is meant by the valency of an element?

11 Describe how sodium forms an ionic bond with chlorine.

12 Name an element that generally forms ions with a charge of +2. Explain why this happens.

13 Name an element that generally forms ions with a charge of −1. Explain why this happens.

14 State the type of bond that is formed between fluorine atoms. Show with the aid of a diagram how the bond is formed.

15 Using neon as an example, explain why noble gases are unreactive.

16 Name an element that generally forms ions with a charge of:
(a) +2
(b) +1
(c) +3
(d) −1

Explain why.

Examination Questions

17 The bond in a molecule of hydrogen gas is formed by a shared pair of electrons.

(a) Name the type of bond found in hydrogen gas.

The bonds in sodium chloride are formed by sodium atoms losing electrons and chlorine atoms gaining electrons.

(b) Name the type of bond found in a sodium chloride crystal.
(JC, OL, 2006)

18 Sodium chloride, NaCl, is common salt.
(a) Draw Bohr structure diagrams showing the arrangement of electrons in a sodium and in a chlorine atom.
(b) Describe how a sodium atom and a chlorine atom combine to produce sodium chloride. You may use a diagram if it helps.
(c) What word is used to describe the type of bond formed between sodium and chlorine in sodium chloride?
(JC, HL, Sample Paper)

Chapter 23

METALS AND NON-METALS

Introduction

There are 92 naturally occurring elements. Over three-quarters of these elements are metals. The rest are non-metals. A few elements such as silicon have metallic and non-metallic properties. They are known as semi-metals.

Metals

We are surrounded in our everyday lives with metals. They have many uses. We use particular metals for particular products depending on their properties. Aluminium is quite light yet strong, so we use it in the manufacture of aircraft. Copper is a brownish metal, which can be stretched and pulled. We use it in electrical wiring.

Chemically, metals generally want to lose electrons from their atoms in order to become more stable.

As we have seen, most of the elements from the periodic table are metals. Their uses vary enormously from building construction to jewellery. So just how is an element classified as a metal?

Let's look at some typical properties of metals.

▶▶▶ ACTIVITY 23.1

Investigation of physical properties of metals

Method

1 Heat the metal, e.g. a nail, strongly in a Bunsen burner. Does it melt? What can we say about the melting point of metals?

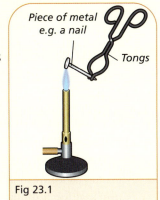
Piece of metal e.g. a nail
Tongs
Fig 23.1

Result
The nail does not melt.

Conclusion
Metals have high melting points.

2 Rub a metal, e.g. a magnesium ribbon, with a piece of sandpaper. What do you observe? Try another metal or two.

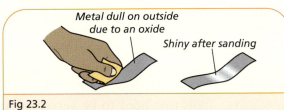
Metal dull on outside due to an oxide
Shiny after sanding
Fig 23.2

Result
The metal is lustrous (shiny) after sanding.

Conclusion
Metals are shiny, but many turn dull due to oxidation in air.

3 Set up the apparatus as shown in Figure 23.3. What do you see happening? How can you tell which is the best conductor of heat?

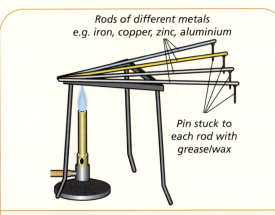
Rods of different metals e.g. iron, copper, zinc, aluminium
Pin stuck to each rod with grease/wax
Fig 23.3

Result
The pins will fall off the metal rods at different times as the wax melts due to heat travelling by conduction.

Conclusion
Metals conduct heat.

CHEMISTRY

4 Set up a circuit like the one shown in Figure 23.4. Try various metals across the gap. What happens each time?

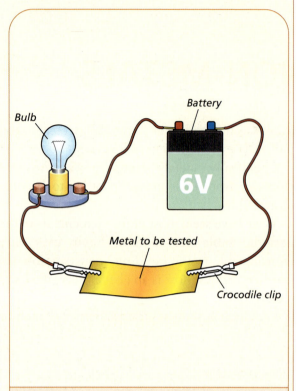

Fig 23.4

Result

The bulb will light when any metal is tested. The bulb will not light when non-metals are tested.

Conclusion

Metals conduct electricity, non-metals do not.

5 Strike any two metals against each other. What type of noise do they make?

Result

Metals give a sonorous (ringing) sound when struck.

Conclusion

Metals are sonorous.

Each of these short investigations shows some of the physical properties of metals. In addition, metals:

(a) are **malleable**, that is, they can be hammered into different shapes. For example, aluminium is flattened into thin sheets and sold as 'tinfoil'.

(b) are **ductile**, that is, they can be pulled or drawn out into wire. For example, copper is the conductor in electric wires.

(c) are generally **hard** and **strong** and have high densities.

Figure 23.5 outlines the physical properties of metals.

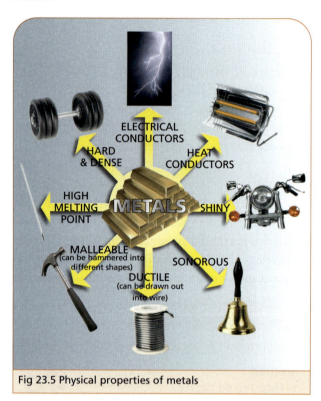

Fig 23.5 Physical properties of metals

Exceptions

There are a few exceptions to those just listed. For example:

● The alkali metals (lithium, sodium potassium, etc.) are soft metals with low melting points and low densities.
● Mercury is a liquid at room temperature; this is unusual for a metal.

Hardness of metals

A simple way to compare the hardness of metals is to try and scratch or score one metal with another.

A softer metal will not scratch a harder one. For example, try scratching a piece of steel with a piece of lead. What do you notice?

Tungsten and diamond are very hard substances. Tools that need to be very hard are 'tipped' with tungsten and other hard metals.

Physical properties of non-metals

In Activity 23.1 the physical properties of metals were investigated. The physical properties of non-metals are generally the opposite to those of metals, for example, non-metals do not conduct electricity or heat. Non-metallic elements are found on the right-hand side of the periodic table. Many are liquids or gases, for example, chlorine is a yellow gas and oxygen is a colourless gas. The following diagram summarises the physical properties of non-metals.

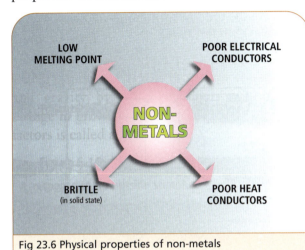

LOW MELTING POINT

POOR ELECTRICAL CONDUCTORS

NON-METALS

BRITTLE (in solid state)

POOR HEAT CONDUCTORS

Fig 23.6 Physical properties of non-metals

Exceptions

There are exceptions. For example:

● Carbon in the form of graphite is a good conductor of electricity.
● Carbon in the form of diamond is extremely hard, with a very high melting point.

Alloys

Aeroplane bodies are mainly made of aluminium. Aluminium is used because it is quite light. However, pure aluminium would not be strong enough for the body of a plane.

If small amounts of other metals are added to aluminium it increases its strength greatly.

The metals are heated until molten and mixed together. They form a mixture that has very different properties from the metals on their own. This is how alloys are formed.

> An **alloy** is a mixture of metals.

Memory metal glasses are a type of alloy that can be easily deformed but quickly regain their shape once released

Aluminium cans are recycled, which cuts costs and reduces environmental damage

Table 23.1 Alloys, their uses and composition		
Alloy	**Composition**	**Use**
Brass	Copper and zinc	Musical instruments Ornaments
Bronze	Copper and tin	Statues
Solder	Lead and tin	Soldering
Mild steel	Iron and carbon	Building reinforcement
Stainless steel	Iron, chromium and nickel	Knives Sinks
Alnico	Aluminium, nickel and cobalt	Powerful magnets

CHEMISTRY

Corrosion of metals

Many metals such as gold and silver do not corrode. They remain shiny and are durable. This makes them ideal for use in jewellery. Many other metals corrode. For example, copper becomes covered with a green-coloured substance on exposure to air.

> **Corrosion** is an undesirable process whereby a metal changes to its oxide or some other compound by combining with oxygen from air.

Rusting

Rusting is the corrosion of iron. It costs us millions of euro every year. Rust forms on the surface of iron (or steel). Unfortunately, it is a soft crumbly substance which flakes off and exposes the next layer of iron to rust.

A rusting ship found along the Dingle peninsula

Rusting experiment

Rusting occurs when iron is exposed to air and water. The iron turns into hydrated iron oxide, a brown flaky substance.

★ MANDATORY ACTIVITY 12

To investigate the conditions necessary for rusting to occur

Method

1 Set up four test tubes as shown in Figure 23.7.

2 Leave the test tubes for a week. What do you see in each test tube?

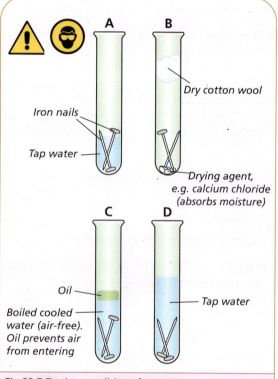

Fig 23.7 Testing conditions for rusting

Result

The nails in test tubes A and D are rusted. Those in test tubes B and C will not rust.

Conclusion

In order for rusting to occur both air and water must be present.

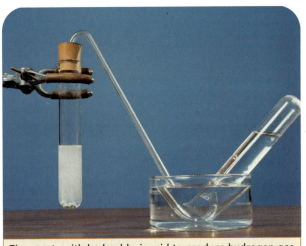

Zinc reacts with hydrochloric acid to produce hydrogen gas

Give rust a shock!

Concrete is reinforced with steel rods and is the most widely used building material. However, some buildings and bridges are weakening because the steel rods are rusting away! Rust needs more space than the steel and so the concrete cracks. Luckily, scientists have discovered that by passing a small electric current through the steel it is protected from rusting.

Rust prevention

When iron (or steel) rusts it weakens the metal and hence the metal has to be replaced. As we have seen earlier in this chapter iron must be exposed to both oxygen and water for rusting to occur. Rust prevention is usually achieved by coating the metal with a material that prevents water and oxygen coming into contact with the metal.

Fig 23.8 Methods of rust prevention

The activity series

All metals have similar properties but they do not react at the same rate. Some metals are more reactive than others. For example, sodium reacts vigorously with water, whereas an iron bar will react slowly with water, thus forming rust.

The activity (or reactivity) series is like a 'league table' for metals. The most reactive metals are at the top and the least reactive ones are at the bottom. One way to compare metals is to examine how they react with substances such as water and acid.

▶▶▶ ACTIVITY 23.2

To investigate the reactivity of a number of metals with water

Use of alkali metals in this experiment should be teacher demonstrated only.

Method

1 Set up the apparatus as shown in Figure 23.9.

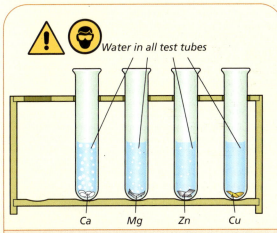

Fig 23.9 Reactivity of metals with water

2 What do you see happening?

3 Which fizzes and which do not fizz?

From this simple test you can arrange metals in order of reactivity.
(**Hint**: It is a good idea to sand each metal with sandpaper before the experiment. Why?)

Result

The 'fizzing' is caused by hydrogen gas being produced in a chemical reaction between the metals and water.

Conclusion

From this experiment we can see that some metals react more vigorously than others.

ACTIVITY 23.3

To investigate the reactivity of a number of metals with dilute acid

1 Set up the apparatus as shown in Figure 23.10.

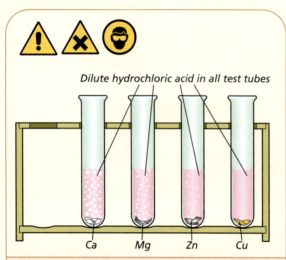

Dilute hydrochloric acid in all test tubes

Ca   Mg   Zn   Cu

Fig 23.10 Reactivity of metals with dilute acid

2 What do you see happening?

Result
The amount of effervescence or fizzing is increased in comparison to Activity 23.2.

Conclusion
Metals react more strongly with dilute acid than with water.

3 Try as many metals as are available. **Alkali metals cannot be used in this experiment as they react explosively with acids.**

4 Observe closely what is happening in each test tube. What happens to the magnesium and calcium metals?

5 To test the gas being produced add a small piece of magnesium to the first test tube again, then place a cork (or your thumb) on top of the test tube. When you feel the pressure building up quickly remove the cork and place a lighted splint over the mouth of the test tube. What happens?

The word equation for the reaction that occurred is:

zinc + hydrochloric acid ➝ zinc chloride + hydrogen

The chemical equation is:

$$Zn(s) + 2HCl(aq) \longrightarrow ZnCl_2(aq) + H_2(g)$$

From the results of the two experiments list the metals tested in decreasing order of activity.

The following list places some important metals in decreasing order of activity. There is a mnemonic to help remember the order.

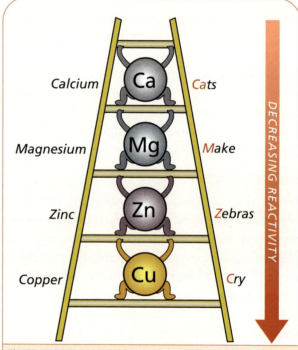

Calcium   Ca   Cats

Magnesium   Mg   Make

Zinc   Zn   Zebras

Copper   Cu   Cry

DECREASING REACTIVITY

Fig 23.11 Decreasing reactivity of metals

The alkali metals and other reactive metals are never found free (on their own) in nature. They react with other elements and water and air to form compounds such as oxides. On the other hand, unreactive metals such as silver and gold are found in nature in their elemental state, as they do not react with their surroundings.

Key Points

- Metals are materials that are usually shiny, hard, ductile, malleable and sonorous.

- Metals are good conductors of electricity and heat.

- Metals usually have high melting points.

- The properties of non-metals are usually the reverse of metals.

- An alloy is a mixture of metals.

- Corrosion is a reaction whereby a metal oxidises to an oxide or other compound.

- The activity series lists metals in order of reactivity.

- Very reactive metals such as potassium and sodium are never found as elements in nature but are found as compounds.

- The reverse is true for unreactive metals like gold.

? Questions

1 Metals are good conductors of _____ and _____.

2 Gold is a malleable metal. What does this mean?

3 Bells are always made of metal. What property of metals makes them ideal for this purpose?

4 Copper can be drawn out into thin wire. This means it is a _____ metal.

Table 23.2		
Property	Typical metal	Typical non-metal
Conducts electricity	Yes	
Conducts heat		No
Shiny		
Malleable		
Ductile		
Low melting point		
Sonorous		

5 What is unusual about the metal mercury?

6 Copy and complete this table:

7 (a) What is an alloy?
  (b) Give two examples.

8 Machines used to drill through rock are diamond-tipped. Why?

9 Give the name and symbol for a metal which:
  (a) is used in hot water pipes
  (b) is a liquid at room temperature
  (c) is the main element in steel
  (d) is used in expensive jewellery
  (e) is used in 'tinfoil'
  (f) is used to galvanise iron.

10 Describe with the aid of a diagram an experiment to show that both water and air are necessary for iron nails to rust.

11 Suggest a reason for the following:
  (a) Aeroplanes are made of aluminium not steel.
  (b) Tinfoil is made of aluminium not steel.
  (c) Bicycles are painted.
  (d) 'Silver' coins are not pure silver.
  (e) Gold rings are not pure gold.
  (f) Copper is used in electrical wires.

Micrograph showing silver crystals. Silver is a lustrous, white precious metal, which is an excellent conductor of electricity and heat

CHEMISTRY

? Questions

12 Write out and complete the following:
The activity series puts _____ in order of _____. The alkali metals are found at the _____ of the series and gold and platinum, which are unreactive, are found at the _____.

13 The following is a list of metals in decreasing order of activity:
calcium, magnesium, zinc, iron, copper
(a) Which metal will react most vigorously with dilute acid?
(b) Which metal will not react with dilute acid?
(c) What gas is produced when zinc reacts with dilute hydrochloric acid?

14 Draw three test tubes showing what happens when (a) magnesium, (b) zinc, (c) copper are added to dilute hydrochloric acid.

15 Thomas and Petra set up an experiment to investigate the conditions for rusting to occur.

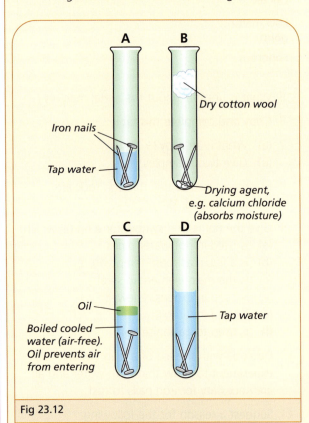

Fig 23.12

(a) What results should they expect after a few days? Why?
(b) If they had used galvanised nails what would the result be? Why?

16 Separate the following list of elements into metals and non-metals and answer the following questions:
iron, carbon, copper, zinc, mercury, sodium, chlorine, helium
(a) Which metal and non-metal form steel?
(b) Which non-metal is used in swimming pools?
(c) Which two metals form the alloy brass?
(d) Which is a metal element that is a liquid at room temperature?
(e) Which non-metal forms the 'lead' in a pencil?

Examination Questions

17 Reactivity tests were carried out on calcium, copper, magnesium and zinc in four test tubes containing an acid. The test carried out using magnesium is shown.
(a) List one thing you would do to make the tests fair.
(b) List the four metals in order of reactivity with the acid, starting with the most reactive.
(JC, HL, 2006)

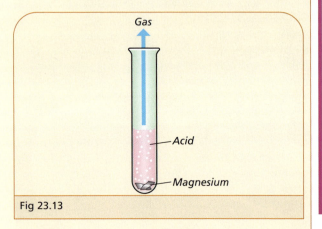

Fig 23.13

❓ Questions

18 The diagram shows three experiments, which were set up to investigate rusting. Study the diagram and answer the questions below.

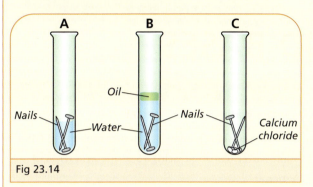

Fig 23.14

(a) In which test tube A, B or C will the nails rust?

(b) What is the function of the calcium chloride in the test tube C?
(JC, HL, 2006)

19 The Millennium Spire, in Dublin, is made from steel. Iron and steel can suffer from corrosion. Iron and steel show visible signs of corrosion. Give one visible sign of corrosion.

Oxygen and water together are necessary for the corrosion of iron or steel.

Describe, with the aid of labelled diagrams, experiments to show that:

(a) Oxygen alone will not lead to the corrosion of iron (or steel).

(b) Water alone will not lead to the corrosion of iron (or steel).
(JC, HL, 2006)

CHEMISTRY

Chapter 24

ACIDS AND BASES

Introduction

Most people think of acids as dangerous fuming liquids that are to be avoided. Some acids do indeed fit this description, but many are quite harmless. Many of us like a little acid on our chips!

Table 24.1 Examples of acids

Everyday acids	Lab acids
Lemon juice (citric acid)	Hydrochloric acid (HCl)
Rain water (carbonic acid)	Sulfuric acid (H_2SO_4)
Vinegar	

Examples of everyday acids and sulfuric acid, which is used in car batteries

Tomato ketchup?

Just sodium chloride and ethanoic acid, please!

Bases

A base is the opposite of an acid. Many bases are very corrosive and equally as dangerous to handle as acids. Most bases are the oxides (or hydroxides) of metals. Household bases include items such as bicarbonate of soda, bleach and toothpaste.

Burns caused by sodium hydroxide (caustic soda), this is a strong alkali

Acids

The word 'acid' is Latin for 'sour'. Acids have a sour, sharp taste. Here are a few examples of acids, which we need to know the names of and/or their formulae.

Vinegar (acid) poured into a beaker of washing soda (base) creates a neutral solution

Bases which we will meet in the lab and which we need to know the names and formulae of are:

- Sodium hydroxide (NaOH)
- Calcium carbonate or limestone ($CaCO_3$)
- Calcium hydroxide or limewater ($Ca(OH)_2$).

Some bases are soluble in water and are called alkalis. All alkalis are bases but not all bases are alkalis.

Lime is a base; it is spread on fields to reduce soil acidity and increase crop yield

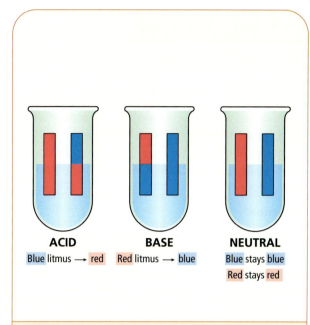

ACID	BASE	NEUTRAL
Blue litmus → red	Red litmus → blue	Blue stays blue Red stays red

Fig 24.2 Demonstrating litmus indicator

▶▶▶ ACTIVITY 24.1

To test a variety of substances with litmus

Method

1 Place a selection of household and laboratory chemicals in different test tubes. If the substance is a solid, then dissolve a spatula full of it in a test tube half-filled with water.

   Take care when handling these substances as some will be corrosive.

2 Into each test tube dip a piece of blue litmus and red litmus paper.

3 Remove the papers and observe any colour changes.

4 Fill in your results as shown in Table 24.2.

Table 24.2			
Substances being tested	**Acidic**	**Basic**	**Neutral**
Vinegar	✓		
Water			✓

Fig 24.1 Acids, bases and alkalis

Indicators

Indicators are substances (usually plant extracts) which change colour depending on whether they are in an acidic solution or a basic solution.

Litmus indicator turns red in acids and blue in bases (or alkalis).

ACTIVITY 24.2

To extract and use red cabbage indicator

Red cabbage contains a dye (an indicator) which changes colour at various pHs (pH measures the strength of an acid or a base). In this activity we will extract the pigment from the leaves of red cabbage and add it to three different solutions, one acidic, one basic and one neutral, and note the colour change in each.

Method

1 Tear up a few leaves of red cabbage and add to a beaker containing about a 100 cm³ of water.

2 Heat the water to boiling point and allow it to simmer for 2–3 minutes.

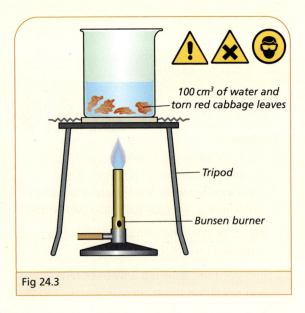

100 cm³ of water and torn red cabbage leaves

Tripod

Bunsen burner

Fig 24.3

3 Allow to cool. Then decant into another beaker.

4 Add acidic, neutral and alkaline solutions to three different test tubes.

5 Using a dropper, add several drops of the red cabbage indicator to each test tube. Stopper and shake each.

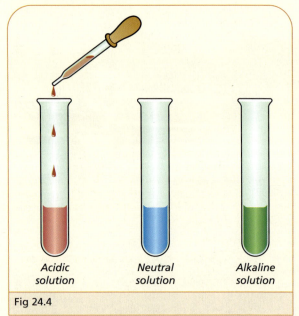

Acidic solution   Neutral solution   Alkaline solution

Fig 24.4

Result

The indicator turns red in the acidic solution, blue in neutral and green in the alkaline.

Conclusion

Red cabbage contains a chemical that acts as an indicator.

The pH scale

From Activity 24.1 you may have noticed that all acids turn blue litmus paper red. But not all acids are equal. Some, such as the vinegar we have on chips, are safe, whereas others such as car battery acid are very dangerous indeed.

> The pH scale is a scale, which runs from 0 to 14, which indicates the level of acidity or basicity of a solution.

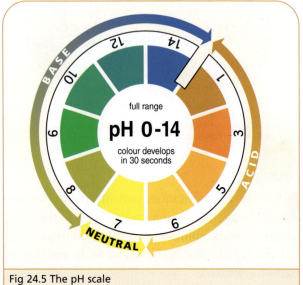

full range

pH 0-14

colour develops in 30 seconds

BASE

ACID

NEUTRAL

Fig 24.5 The pH scale

A solution with pH of 7 is said to be neutral. Below 7 indicates that the substance is acidic. The lower the pH the more acidic the substance is. Bases have pH values. The higher the pH the more alkaline (basic) the substance is.

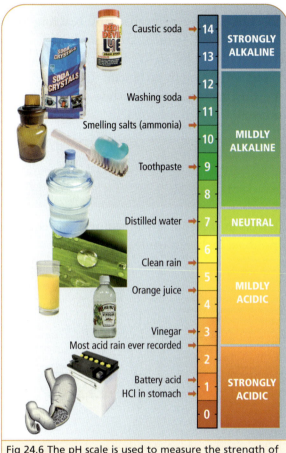

Fig 24.6 The pH scale is used to measure the strength of acids and bases

Universal indicator

Universal indicator is a mixture of indicators. It will change to a variety of different colours depending on the pH of the indicators.

Universal indicator tells us whether the substance being tested is a strong or weak acid or base, or if it is neutral.

MANDATORY ACTIVITY 13

To investigate the pH of a variety of substances using the pH scale

Method

1 Place each substance to be tested in a test tube (again if it is a solid, dissolve a spatula full of it in a test tube half-filled with water).

2 Dip a piece of universal paper into the test tube, observe the colour change and, using a colour chart, find its pH.

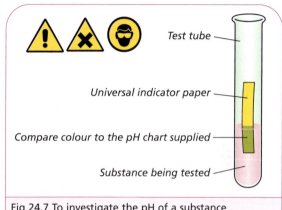

Fig 24.7 To investigate the pH of a substance

3 Fill in your results in a table:

Table 24.3		
Substance	**pH**	**Nature of substance**
Lemon juice	5	Weak acid

Note 1
Universal indicator is also available in liquid form. Simply add a few drops of the indicator to each test tube in this case.

Note 2
An electronic pH meter could also be used to accurately determine pH.

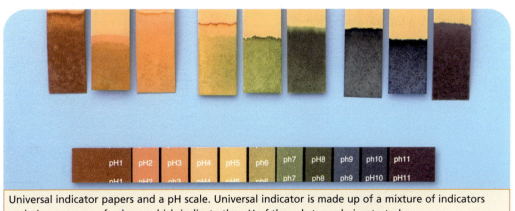

Universal indicator papers and a pH scale. Universal indicator is made up of a mixture of indicators and gives a range of colours which indicate the pH of the substance being tested

Neutralisation

We have already said that a base is the chemical opposite of an acid. So what would happen if an acid and a base were mixed together? They react together to cancel each other out and form a salt and water.

> acid + base ———————→ salt + water

1 You will remember from biology that there is hydrochloric acid in your stomach. When people suffer from indigestion the 'burning' feeling is due to an excess of acid being produced. The problem can be remedied by taking an antacid (anti-acid) tablet like Alka Seltzer, which is a base.

2 A bee sting is acidic and can be cured by rubbing a mild base, e.g. baking soda on it.

3 A wasp's sting is alkaline. Can you think of a household substance to rub on it to ease the pain?

Acids can also be neutralised by reacting them with carbonates (which are also bases). Some lakes become acidic, so to counteract this they are sprayed with calcium carbonate (limestone).

In this neutralisation carbon dioxide gas is produced.

> an acid + a carbonate ———→ a salt + water + carbon dioxide

Later on in the course you will use this equation to produce carbon dioxide.

▶▶▶ **ACTIVITY 24.3**

To demonstrate a neutralisation reaction

Method

1 Place 20 cm³ or dilute hydrochloric acid (HCl) in a conical flask.

2 Add in 3–4 drops of universal indicator solution. What colour is it? What is its approximate pH?

3 Using a dropper slowly add dilute sodium hydroxide (NaOH) to the acid and observe any colour changes. (Gently swirl the flask to allow the acid and base to react.)

4 See if you can achieve a neutral solution. What colour are you trying to get? Check by looking at the pH scale.

5 If you add too much sodium hydroxide you need not start again. Why? Because by adding some more acid you can neutralise, i.e. cancel out, the excess base. It's a bit like a see-saw!

Conclusion

A base is used to neutralise an acid and an acid will neutralise a base.

Conical flask

20 cm³ of dilute hydrochloric acid + universal Indicator

Dropper containing dilute sodium hydroxide solution

Fig 24.8 Neutralising an acid

Titration

Activity 24.3 can be done in a very accurate way using specific glassware. The process is called a titration.

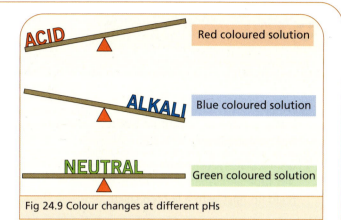

Fig 24.9 Colour changes at different pHs

 MANDATORY ACTIVITY 14

To titrate HCl against NaOH and form sodium chloride

Introduction

There are two parts to this experiment. In the first part you will neutralise the base (NaOH) with the acid (HCl). The products will be a salt (NaCl) and water (H_2O).

In the second part you will use a separation technique to isolate salt.

HCl	+ NaOH	$\longrightarrow$	NaCl	+	H_2O
hydrochloric	+ *sodium*	$\longrightarrow$	*sodium*	+	*water*
acid	*hydroxide*		*chloride*		

Method

1 Using a pipette, measure 25 cm³ of sodium hydroxide into a clean conical flask.

2 Add a few drops of indicator, e.g. litmus.

3 Using a funnel, fill the burette with dilute hydrochloric acid.

4 Slowly, while continuously swirling the conical flask, add the acid from the burette until the indicator changes colour (if litmus indicator is being used the colour change will be from blue to red). Note the volume of acid added.

5 Repeat the whole experiment two more times, taking great care to add the acid drop by drop near the end.

6 If the second and third volumes of acid are not the same take an average of the two. (The first titration was only to act as a guide.)

7 Now repeat this experiment without using any indicator, adding the exact volume of acid that is needed to neutralise the base.

8 The conical flask now contains a solution of salt and water. Pour into an evaporating dish and heat until the water has evaporated leaving white sodium chloride (table salt) behind.

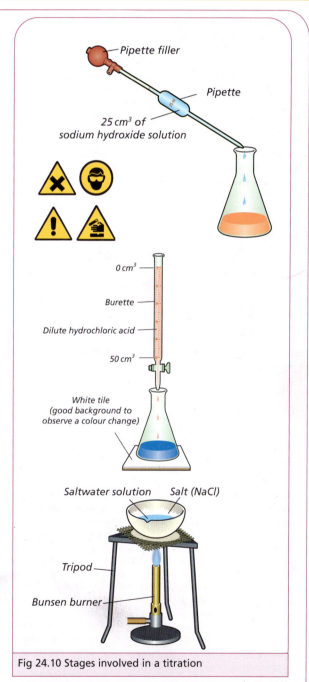

Fig 24.10 Stages involved in a titration

Conclusion

An acid and a base react to form a salt and water.

CHEMISTRY

CHEMISTRY

To investigate the reaction between zinc and hydrochloric acid

Method

1 Lower a piece of zinc into a test tube of dilute hydrochloric acid.

2 Gently place your thumb over the top of the test tube and when you feel the pressure build up, test with a lighted splint.

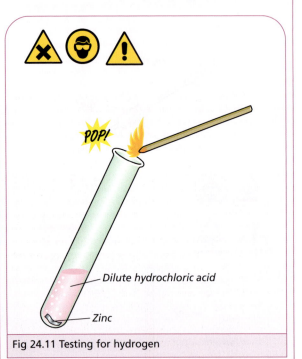

Fig 24.11 Testing for hydrogen

Result

The splint burns with a 'popping' sound.

Conclusion

Hydrogen gas burns with a 'pop'. This is the conclusive test for hydrogen gas.

Reaction of metals with acids

Many metals react with acids and hydrogen gas is produced. The hydrogen gas burns explosively in air.

The word equation for the reaction in Figure 24.11 is:

zinc + hydrochloric acid ⟶ zinc chloride + hydrogen

The chemical equation for this reaction is:

$$Zn + 2HCl \longrightarrow ZnCl_2 + H_2$$

In general:

an acid + metal ⟶ a salt + hydrogen

Note: Only metals above hydrogen in the activity series will react with acids to produce hydrogen.

🔑 Key Points

- Acids have a sour taste.

- Acids turn blue litmus red and have a pH below 7.

- The lab acids are hydrochloric acid and sulfuric acids.

- Bases turn red litmus blue and have a pH greater than 7.

- An alkali is a base which is soluble in water.

- The pH scale on which the strength of acids and bases is expressed runs from 0–14. A pH of 7 indicates a neutral substance.

- An acid and a base react to form a salt and water.

- An acid and a metal react to form a salt and hydrogen.

- An acid and a carbonate react to form a salt and water and carbon dioxide.

? Questions

1 Write out and complete the following:
   Acidic solutions turn blue litmus _____.
   Their pH is always _____ than 7 and they
   have a _____ taste.

2 Name and give the formula of two acids and
   two bases used in the lab.

3 What are alkalis?

4 What effect do acids have on
   (a) blue litmus paper?
   (b) red litmus paper?

5 What advantage does universal indictor have
   over litmus indicator?

6 What is the pH scale?

7 Helen tested some solutions with universal
   indicator. She wrote down their pHs but forgot
   to fill out the table below. Help her match the
   pHs to the correct solutions.

   pHs 1, 5, 7, 10, 14

Solution tested	pH
Distilled water	
Car battery acid	
Vinegar	
Baking soda	
Sodium hydroxide	

8 Give an explanation for the following:
   (a) Taking Milk of Magnesia for indigestion
   (b) Rubbing vinegar on a wasp sting
   (c) Rubbing baking soda on a nettle sting
   (d) A mixture of vinegar and baking soda
        'fizzing' when mixed.

9 (a) Name an indicator.
   (b) What colour will this indicator be in
        sulfuric acid?

10 Describe an experiment you would carry out
    in the lab to measure the pH of a solution.

11 (a) You are given two test tubes as shown in
         Figure 24.12, one with dilute sulfuric acid
         and the other with dilute potassium
         hydroxide. State the pH value you would
         expect to obtain for each.

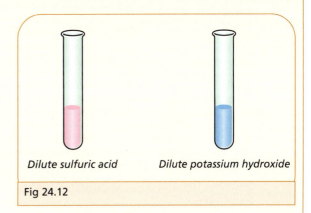

Dilute sulfuric acid   Dilute potassium hydroxide

Fig 24.12

   (b) What term is used to describe the reaction
        that takes place when the contents of the
        two tubes are mixed together?
   (c) What are the products of this reaction?

12 Olivia titrated hydrochloric acid against sodium
    hydroxide solution and found that 18.6 cm³ of
    the acid neutralised the base.
    (a) Draw a fully labelled diagram of the
         apparatus she would have used.
    (b) What piece of equipment would she have
         used to accurately measure 25 cm³ of
         sodium hydroxide solution?
    (c) Name an indicator she could have used
         and state the colour change observed.
    (d) Which was more concentrated, the acid
         or the base?
    (e) Name the salt formed during the
         experiment.
    (f) What else was produced during the
         experiment?
    (g) Name the technique used to separate the
         two products.

? Questions

Examination Questions

13 The pieces of laboratory equipment shown, in Figure 24.13 together with some other items, were used to prepare a sample of sodium chloride.

(a) Name item A or item B.

(b) There were 25 cm³ volumes of base used in this experiment. Describe how the piece of equipment A was used to measure the volume of acid required to neutralise this amount of base.

(c) Note a suitable acid and name a suitable base for the preparation of sodium chloride by this method.

(d) Write a chemical equation for the reaction between the acid and the base that you have named.

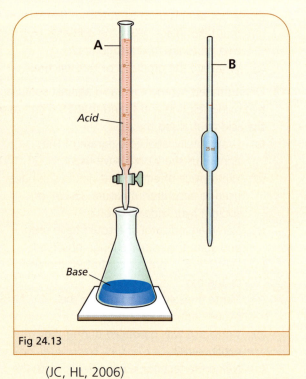

Fig 24.13

(JC, HL, 2006)

14 Many substances found in the home are acids and bases.

Complete the table below identifying **one acid** and **one base** from the list on the right.

Acid	Base

Vinegar
Water
Oven cleaner

(JC, OL, 2006)

Chapter 25 ○○○

CHEMICAL REACTIONS

Introduction

Our study of chemistry leads us to a better understanding of our world and the processes by which materials can change and be changed. These changes can be of two types:

1 physical changes
2 chemical changes.

Physical changes

Physical changes differ from chemical changes in a number of ways. Physical changes often just involve changing the state of a substance. Butter melting in a pan, the windows of a car becoming fogged up, or sugar dissolving in tea, are all examples of physical changes.

It is usually quite easy to reverse a physical change, as nothing new or different has been created.

> **A physical change** is a change in which no new substance is formed.

Examples of physical changes include:

● melting ice
● chopping wood
● blowing up a balloon.

A hotel made entirely of ice and snow. Drinks are served in glasses made of ice! The building melts every May

Chemical changes

Chemical changes or chemical reactions occur all around us every day. Indeed, inside our bodies chemical reactions are taking place which are vital to our wellbeing. They generate heat energy to keep our bodies warm.

The clothes you are wearing are made from artificial fibres that were produced in chemical reactions.

The pens in your schoolbag are probably made of plastic that was chemically produced from crude oil.

In a chemical change a reaction always takes place and a new substance is produced. It is very difficult, often impossible, to reverse these changes.

> **A chemical change** is a change (reaction) in which a new substance is formed.

Example of chemical changes include:

● burning a piece of magnesium ribbon
● a banana ripening
● an acid reacting with a base. (Can you remember the two new substances produced in this reaction?)

The main differences between physical and chemical changes are summarised in Table 25.1.

Table 25.1	
Physical change	**Chemical change**
1 No new substance formed	1 One or more new substances formed
2 Can be easily reversed	2 Very difficult or impossible to reverse
3 Often no heat change occurs	3 Heat is always needed or produced in the change
4 No change in mass	4 A change in mass

Ice crystals have formed on a mountain signpost in the direction of the prevailing wind which was recorded at −36°C

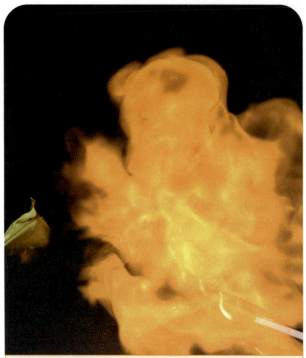

A chemical reaction: a hydrogen balloon explodes after ignited by a candle. Hydrogen reacts with oxygen producing water

Fehling's reaction: Fehling's solution is blue but forms a brick red precipitate in the presence of simple sugars

A fireball and 'mushroom cloud' formation caused by the detonation of the first atomic bomb in New Mexico, 1945

Burning fuels

> **A fuel** is a substance that burns in oxygen and produces heat.

Fuels store chemical energy. When they are burned this energy is released. This is a chemical change or, more accurately, a chemical reaction. In order to burn, the fuel must

- combine with oxygen and
- have sufficient heat present.

So, the three conditions needed for a fire to happen are: fuel, oxygen and heat.

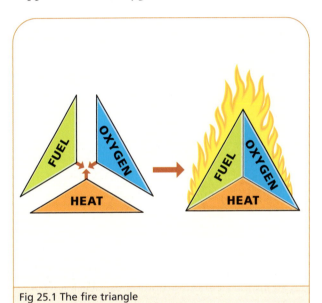

Fig 25.1 The fire triangle

Mouldy foods caused by fungal growth, which is a chemical reaction

Explosions and steam are caused as a flow of molten rock (lava) enters the sea. The dramatic colours are caused by metals in the lava reacting with seawater

Putting out a fire

A fire can be quenched by removing any one of these conditions, i.e. fuel, oxygen or heat. For example, in a forest fire an area of trees in the fire's path can be removed. The fire then runs out of fuel.

In a chip pan fire, a fire blanket or a lid is placed on the pan. Which side of the fire triangle is being removed?

Fire extinguishers

Fire can be extinguished by removing one or more sides of the fire triangle. The type of fire extinguisher to be used depends on the fire involved.

Firebreaks, strips of land cleared of trees, prevent fires from spreading, as the fire runs out of fuel

CHEMISTRY

Key Points

- A physical change is a change in which no new substance is formed.

- A chemical change is a change in which a new substance is formed.

- A fuel is a substance that burns in oxygen to produce heat.

- The fire triangle is made up of the three conditions necessary for a fire, i.e. fuel, heat and oxygen.

- Fires can be extinguished by removing any one of the three sides of the fire triangle.

Questions

1 Write out and complete the following:
A _____ change is one in which no new substance is formed. It is usually _____ to reverse, whereas in a _____ change a new substance is produced and it is usually _____ to reverse. Boiling water is an example of a _____ change.

2 Describe an experiment you have done which involved (a) a chemical change, (b) a physical change. Explain how you decided whether the change involved was chemical or physical.

3 What type of change or changes take(s) place when a candle is burning?

4 State whether each of the following changes is physical or chemical:
 (a) a candle burning
 (b) a candle melting
 (c) a nail rusting
 (d) switching on a torch
 (e) squeezing oranges
 (f) an orange rotting
 (g) milk going sour
 (h) inflating a balloon
 (i) blue litmus turning red
 (j) freezing food.

5 The three conditions necessary for a fire can be shown in the _____ _____. Copy and complete Figure 25.2.

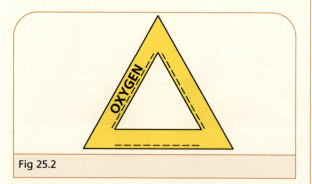

Fig 25.2

6 Suggest a method of putting out the following fires:
 (a) a waste-paper bin fire
 (b) a chip pan fire
 (c) an out-of-control bonfire
 (d) a coat on fire
 (e) a television on fire.

7 When new forests are planted, large channels are left with no trees. Why is this a good idea?

8 What type of fires should you never use water on?

9 State two of the requirements for a fire to continue to burn.

10 When you turn off a Bunsen burner which part of the fire triangle are you removing?

11 List three precautions that should be taken to prevent accidental fires in homes.

Chapter 26

THE ATMOSPHERE

Introduction

We live surrounded by a mixture of gases called air. The air makes up the Earth's atmosphere and is essential for life. Not only do we need air to breathe, we also need it to burn substances and keep the planet warm.

Gases in the air

Nitrogen and oxygen between them account for about 99 per cent of air! Do you know which other gases make up the remainder of air? Look at the table:

Table 26.1 Composition of air	
Gas	**% of air**
Nitrogen	78%
Oxygen	21%
Carbon dioxide	
Water vapour	1% approximately
Noble gases	
Other gases	

These percentages are not exact. The amount of water vapour in the air (humidity) varies a lot. As you ascend a mountain the level of oxygen decreases and eventually you may need breathing apparatus to climb to the top!

Nitrogen

Nitrogen gas makes up almost four-fifths of the air. It is an unreactive gas. When food is packed, nitrogen is used inside the sealed packaging. The oxygen is kept out and this helps the food stay fresh longer. Why?

Liquid nitrogen is very cold. It is almost –200°C. Therefore, it is used to freeze things very quickly.

Oxygen

We need oxygen to respire. Oxygen is used up when substances burn. Only one-fifth of the air is oxygen. In a day, one person will breathe in about 15 000 litres of air. About 3 000 litres of this is oxygen. Our lungs are able to trap about a quarter of this, i.e. 750 litres of oxygen. The amount of oxygen we need depends on how active we are.

The aurora borealis, or northern lights, are caused by energy charged particles from the sun colliding with gas molecules in the upper atmosphere

An experiment to measure the oxygen content of air. The candle burns until the oxygen in the jar is used up. When the candle goes out water rises up to replace the used-up oxygen

CHEMISTRY

187

CHEMISTRY

⭐ MANDATORY ACTIVITY 16A

To show that approximately one-fifth of air is oxygen (O_2)

Method

1 Pack some steel wool into the bottom of a graduated cylinder or test tube.

2 Invert the cylinder or test tube and place in a basin containing water.

3 Leave for about one week.

4 Measure the height that the water rises up the cylinder or test tube compared to the total height.

Result

The steel wool rusts using up the oxygen in the air in the cylinder in the process. Water rises up the cylinder to replace the oxygen used up.

Conclusion

The water rises approximately one-fifth of the height of the graduated cylinder to replace the oxygen used up as the steel wool rusts.

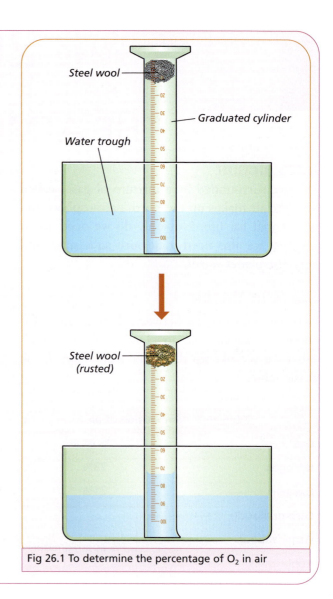

Fig 26.1 To determine the percentage of O_2 in air

⭐ MANDATORY ACTIVITY 16B

To show that air contains carbon dioxide

Method

1 Set up the apparatus as shown.

Result

Air is bubbled through the clear limewater. Eventually the limewater will turn cloudy or milky in colour. This is the definitive test for carbon dioxide gas.

Conclusion

Air contains carbon dioxide.

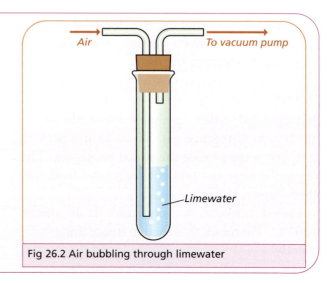

Fig 26.2 Air bubbling through limewater

MANDATORY ACTIVITY 16C

To show there is water vapour in air

Method

1 Fill a test tube with ice and stopper it.

2 Dry the outside of the test tube.

3 After a short time a liquid is seen to form on the outside of the test tube.

4 Place a piece of cobalt chloride paper against the side of the test tube.

Result

The cobalt chloride paper changes colour from blue to pink. This is the definitive test for the presence of water.

Conclusion

Air contains water vapour.

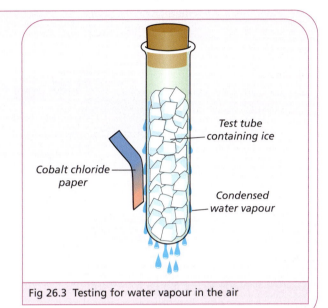

Fig 26.3 Testing for water vapour in the air

Test tube containing ice

Cobalt chloride paper

Condensed water vapour

ACTIVITY 26.1

An alternative method to measure the percentage of oxygen in air

Method

1 Set up the apparatus as shown in Figure 26.4.

(b) What happens to the copper turnings?

(c) Why wait until the apparatus cools before taking the final readings on the syringe?

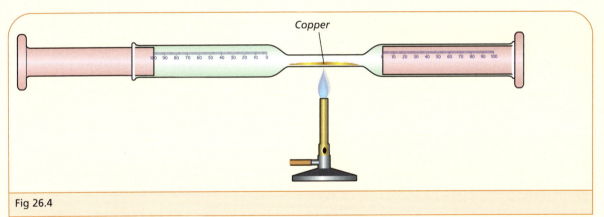

Copper

Fig 26.4

2 Before you begin heating the copper metal turnings, make sure that one syringe contains 100 cm³ of air and the other syringe is completely closed.

3 Heat the copper turning and slowly pass the 100 cm³ of air from one syringe to the other.

4 Pass the air from syringe to syringe several times or until the volume does not change.

5 Let the apparatus cool down and read the final volume of air left.

(a) What can you conclude about the amount of oxygen in air?

Result

The volume of air in the syringe will go down to approximately 80 cm³ and the copper turnings will turn grey-black in colour. This is because copper reacts with the oxygen present forming copper oxide.

Conclusion

Approximately 20 per cent of air is oxygen.

An oxygenation barge pumps oxygen into a river during periods of poor water quality to help sustain fish life

 MANDATORY ACTIVITY 17

To prepare and examine oxygen gas

Oxygen is made in the lab by the decomposition of hydrogen peroxide. The process is very slow so a chemical called manganese dioxide is used to speed it up. Manganese dioxide acts as a catalyst in the reaction.

> **Catalysts** are substances that alter the rate of a chemical reaction but are not used up themselves.

The word equation for the preparation of oxygen is:

$$\text{hydrogen peroxide} \xrightarrow{\text{manganese dioxide}} \text{oxygen} + \text{water}$$

The chemical equation for the preparation of oxygen is:

$$2H_2O_2 \xrightarrow{MnO_2} O_2 + 2H_2O$$

Method

1 Set up the apparatus as shown in Figure 26.5.

2 Slowly allow a few cm³ of the hydrogen peroxide to fall onto the black manganese dioxide powder.

3 A gas will be seen bubbling up through the water in the gas jar. When the jar is empty of water, i.e. full of gas, place a lid on it and quickly put another gas jar of water in its place.

4 Collect 4–5 jars of the gas. The first jar collected will contain air from the conical flask and will be quite impure.

5 Carry out the following tests:

(a) **Note the colour and smell of oxygen. Test its pH using universal indicator paper.**

Result

Oxygen has no colour or smell. The universal indicator paper turns green in colour.

Conclusion

Oxygen is a colourless, odourless gas. It has a pH of 7, i.e. it is a neutral gas.

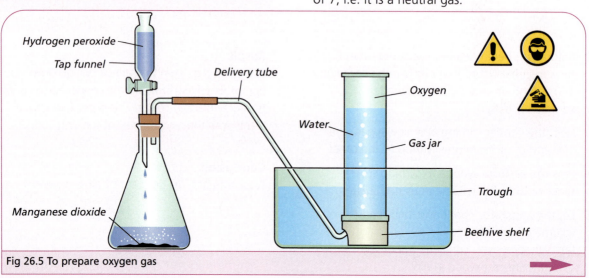

Fig 26.5 To prepare oxygen gas

(b) Place a glowing splint into a jar of oxygen.

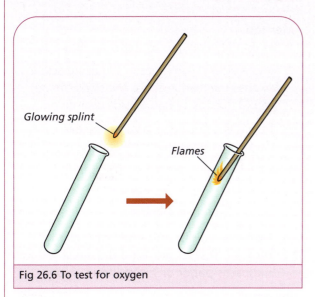

Fig 26.6 To test for oxygen

Result
It will relight.

Conclusion
Oxygen will relight a glowing splint.

(c) Heat a piece of charcoal (carbon) in a combustion spoon over a Bunsen burner until it glows. Quickly insert it into a jar of oxygen (Figure 26.7).

Result
The charcoal will catch fire in the jar of oxygen.

Conclusion
Oxygen aids combustion.

When the flames have gone out remove the piece of charcoal and place pieces of moist red

and blue litmus papers in the jar. What colour do they turn?

Now add a small volume of limewater to the jar. Stopper and shake the jar. What do you notice?

Result
The litmus papers both turn red. The limewater turns milky in colour.

Conclusion
When charcoal burns carbon dioxide is produced. Carbon dioxide is an acidic gas. The oxides of non-metals tend to be acidic.

(d) Hold a piece of magnesium ribbon in a metal tongs over a Bunsen burner until it ignites with a bright white flame. Quickly place in a jar of oxygen gas.

Result
The magnesium burns with an extremely bright flame. Do not look directly at this flame as it could damage your eyes.

Conclusion
Oxygen aids combustion.

(e) Add pieces of moist red and blue litmus papers to the jar when the flames have gone out.

Result
The litmus papers both turn blue in colour.

Conclusion
Magnesium oxide is a base. All metal oxides tend to be bases.

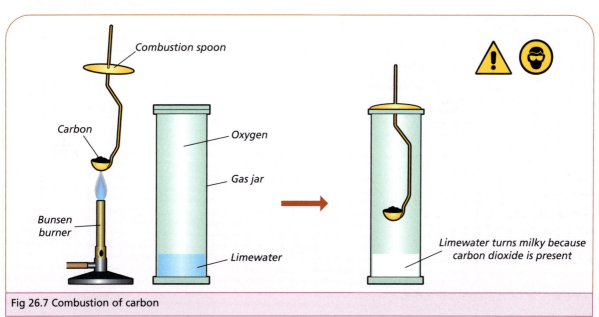

Fig 26.7 Combustion of carbon

CHEMISTRY

MANDATORY ACTIVITY 18

To prepare and examine carbon dioxide gas

Carbon dioxide can be made by reacting an acid with a carbonate. The other products produced are a salt and water.

The word equation for the preparation of carbon dioxide is:

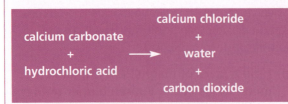

calcium carbonate
+
hydrochloric acid
⟶
calcium chloride
+
water
+
carbon dioxide

The chemical equation for the preparation of carbon dioxide is:

$$CaCO_3 + 2HCl \longrightarrow CaCl_2 + H_2O + CO_2$$

Carbon dioxide can be collected by
(a) the upward displacement of air or
(b) the downward displacement of water.

Some of the carbon dioxide dissolves in the water when it is collected over water.

Method

1 Set up the apparatus as shown in Figure 26.8.

2 Slowly allow a few drops of the dilute acid to fall onto the marble chips. The chips can be seen to 'fizz' as carbon dioxide is given off.

3 The gas is collected in glass jars by the downward displacement of water.

4 Collect four jars of the gas. Discard the first as it contains air from the reaction flask.

5 Carry out the following tests:

(a) Note the colour and smell of carbon dioxide. Test its pH using universal indicator paper or moist litmus paper.

Result

Carbon dioxide has no colour or smell. It will turn moist blue litmus paper red and universal indicator paper red/orange.

Conclusion

Carbon dioxide is a colourless, odourless, acidic gas.

(b) Pour a small volume of limewater into a jar of carbon dioxide. Stopper and shake the jar.

Result

The limewater turns from clear to milky in colour.

Conclusion

Carbon dioxide turns limewater milky.

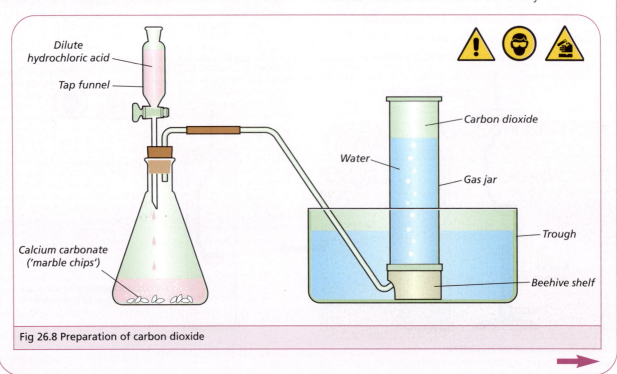

Fig 26.8 Preparation of carbon dioxide

CHEMISTRY

(c) Light a wooden splint and place in a jar of carbon dioxide.

Result

The lighted splint will go out.

Conclusion

Carbon dioxide does not support combustion.

(d) Invert a jar of carbon dioxide and hold it above a lighted candle.

Result

The candle will go out.

Conclusion

Carbon dioxide is more dense than air.

Test for presence of carbon dioxide

If carbon dioxide comes into contact with limewater (which is calcium hydroxide) they react together to form calcium carbonate. This is insoluble and hence makes the solution turn 'milky' in colour. Carbon dioxide is the only gas that turns limewater milky. So this is the test to identify the gas.

The word equation for the reaction between limewater and carbon dioxide is:

$$\text{calcium hydroxide} + \text{carbon dioxide} \rightarrow \text{calcium carbonate} + \text{water}$$

The chemical equation for the reaction between limewater and carbon dioxide is:

$$Ca(OH)_2 + CO_2 \longrightarrow CaCO_3 + H_2O$$

Uses of carbon dioxide

● **Fizzy drinks**: Carbon dioxide is slightly soluble in water. It forms a weakly acidic solution which puts the 'fizz' in drinks.

$$H_2O + CO_2 \longrightarrow H_2CO_3$$

$$\text{water} + \text{carbon dioxide} \longrightarrow \text{carbonic acid}$$

This equation explains why rainwater is slightly acid.

● **Dry ice**: If cooled to below −78°C, carbon dioxide sublimes into a solid called dry ice. This makes a very useful refrigerant as it does not leave a 'mess' of water when it thaws.

Carbonated drinks are caused by carbon dioxide being injected under pressure

● **Fire extinguishers**: Carbon dioxide does not support combustion and is denser than air. This makes it ideal as a fire extinguisher as it can safely be used on most fires.

Steel wool burning in a jet stream of oxygen gas

● **Stage effects**: When water is added to dry ice it causes it to sublime and produce thick heavy smoke, which is harmless. As it is denser than air it stays near the floor. This effect is used in stage shows, videos and films.

CHEMISTRY

Other components of air

Water

The amount of water vapour in the air varies a lot. For example, in the desert there is very little moisture in the air, whereas in a rainforest there can be up to 4 per cent water vapour in the air. Humidity is a measure of the water vapour present in the air.

Argon

Argon is present in air at about 0.8 per cent. It is a noble gas and as expected does not chemically combine with other substances. Household light bulbs are filled with argon, as it prevents the filament from burning and melting.

Other gases

Air pollutants such as sulfur dioxide and carbon monoxide can be present in air. They have a damaging effect on air quality. They are caused by the burning of fossil fuels. This will be discussed further in Chapter 30.

Air pollution over China: the poor air quality is due to the increased use of vehicles, fires and coal-fired power stations

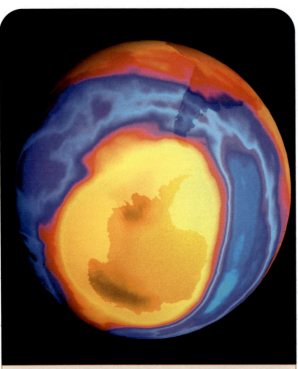

A satellite map shows the hole in the ozone layer over the Antarctic

🔑 Key Points

- Air contains approximately 78 per cent nitrogen and 21 per cent oxygen; 1 per cent is a mixture of argon, water vapour, carbon dioxide and pollutants.

- Humidity is a measure of the amount of water vapour in the air.

- Oxygen can be prepared by reacting hydrogen peroxide with manganese dioxide (which acts as a catalyst).

- Oxygen gas will relight a glowing splint.

- Carbon dioxide gas can be prepared by reacting dilute hydrochloric acid with calcium carbonate.

- The conclusive test for carbon dioxide is that it turns limewater (calcium hydroxide) milky when bubbled through it.

- Carbon dioxide is used in fizzy drinks and fire extinguishers.

CHEMISTRY

 **Questions**

1 Write out and complete the following:
The Earth's atmosphere is about four-fifths
_____ and one-fifth _____.
_____ is present in small amounts and
used to fill light bulbs.

2 Air is a mixture of nitrogen, oxygen, carbon
dioxide, water vapour and noble gases.
 (a) What is a mixture?
 (b) Name two compounds present in air.
 (c) Describe a simple experiment to show
that each of these two compounds is
present in air.

3 State what you would expect to see happening
in each flask in Figure 26.9.

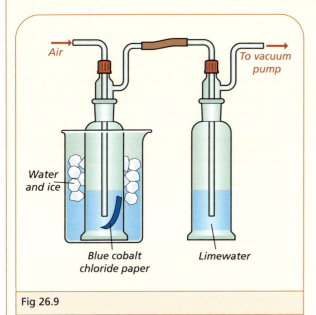

Water
and ice

Blue cobalt
chloride paper

Air

To vacuum
pump

Limewater

Fig 26.9

4 Identify the following:
 (a) Turns limewater milky: _____.
 (b) Makes up 78 per cent of air: _____.
 (c) Used as a catalyst to prepare oxygen:
_____.
 (d) Cobalt chloride paper is used to test for
presence of this substance: _____.
 (e) A noble gas found in air: _____.
 (f) Carbon dioxide will turn this substance
red: _____.
 (g) Is used in fizzy drinks: _____.
 (h) Is used in fire extinguishers:
_____.
 (i) Will relight a glowing splint: _____.
 (j) Will cause magnesium to burn much
brighter: _____.

5 (a) Why do sealed food packages contain
nitrogen but not oxygen?
 (b) Why is oxygen said to be the active part
of air?
 (c) What advantages has 'dry ice' over
ordinary ice?
 (d) Why does 'smoke' from 'dry ice' stay close
to the floor?

6 Write out and complete the following:
To prepare oxygen liquid _____ _____
is dropped onto a black powder called
_____ _____. This substance speeds up
the reaction and is called a _____.

7 Write out and complete the following:
In the preparation of _____ _____,
hydrochloric acid is dropped onto calcium
_____. The gas produced will _____
a lighted splint and turn _____ milky.

8 Draw a labelled diagram of the apparatus used
to prepare oxygen gas.

9 You are given three jars of gas and told they
contain oxygen, nitrogen and carbon dioxide.
How could you identify which is which?

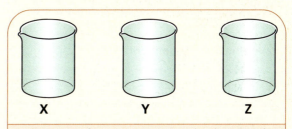

X   Y   Z

Fig 26.10 Jars of oxygen, nitrogen and carbon dioxide
in unknown order

10 An element (A) was burned in a jar of a gas
(B) and the gas produced, (C), turned
limewater from clear to milky. Identify A, B
and C.

CHEMISTRY

? Questions

11 Tomas did an experiment to find out how much oxygen is in the air. He started with 100 cm³ of air and passed it back and forth over the hot copper (Fig 26.11). He saw the copper turn to black copper oxide.

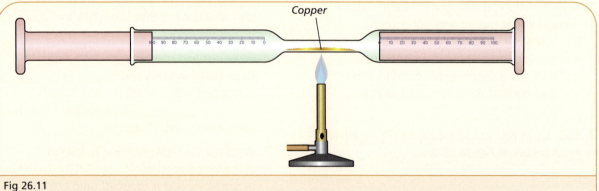

Copper

Fig 26.11

(a) Give a word equation for what is happening.

(b) When the reaction was complete what volume of air would you expect to be left in the syringe?

(c) Thomas let the air cool down before getting his final reading. Why?

(d) The gas left in the syringe is mainly

_____.

12 (a) Why does the amount of water vapour in the air vary?

(b) Oxygen accounts for approximately one-fifth of the air. Describe an experiment you could use to demonstrate this.

Examination Questions

13 In 1774 Joseph Priestley, an English chemist, discovered oxygen.

(a) Name the two chemicals that you reacted together to prepare oxygen in the school laboratory. One of the chemicals acted as a catalyst.

(b) Which one of the two chemicals used was a catalyst?
(JC, HL, 2006)

14 Carbon dioxide turns limewater milky. Complete the chemical equation for the reaction of carbon dioxide with limewater.

$$Ca(OH)_2 + CO_2 \longrightarrow$$

(JC, HL, 2006)

15 A gas jar of oxygen was prepared by decomposing hydrogen peroxide using a suitable catalyst. This preparation may be described as follows:

$$H_2O_2 \longrightarrow H_2O + O_2$$

(a) Balance the above equation.

A piece of magnesium burns very brightly in a gas jar of oxygen and produces a white powder.

(b) What is observed when this white powder is added to water and litmus paper added?

(c) What conclusion can be drawn from this observation?
(JC, HL, Sample Paper)

CHEMISTRY

? Questions

16 The composition of air can be investigated in different ways. Two experiments are shown in the diagram.

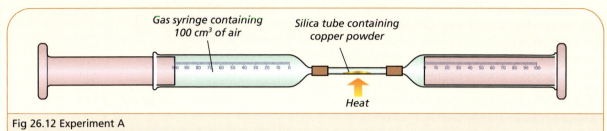

Gas syringe containing 100 cm³ of air

Silica tube containing copper powder

Heat

Fig 26.12 Experiment A

In experiment A the air was pushed repeatedly over the heated copper powder and only 79 cm³ of gas remained at the end of the experiment.

(a) Why is it necessary to let the apparatus cool down before measuring the volume of the remaining gas?

(b) Why did the volume of gas decrease and then remain steady?

(c) What is the remaining gas mainly composed of?

(d) Experiment B is less accurate than Experiment A.

Give a reason why this is so.

(JC, HL, 2006)

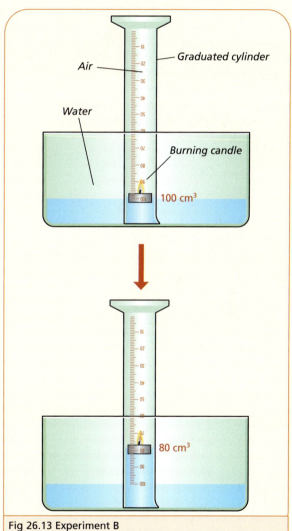

Air

Water

Graduated cylinder

Burning candle

100 cm³

80 cm³

Fig 26.13 Experiment B

? **Questions**

17 The diagram shows a gas jar of carbon dioxide gas being poured onto a lighting candle.
   (a) What happens to the lighting candle when the carbon dioxide gas is poured over it?
   (b) What does this tell us about carbon dioxide gas?
   (c) Name the chemical that turns milky white if carbon dioxide is bubbled through it.
   (JC, OL, 2006)

18 Oxygen gas can be prepared in a school laboratory using the apparatus in Figure 26.15.
   (a) Identify a liquid X and a solid Y that can be used in this preparation.
   (b) Solid Y speeds up the breakdown of liquid X. What name is given to this type of chemical?
   (c) What happens when a 'glowing splint' (very hot piece of wood) is placed in a gas jar of oxygen?
   (d) Give one property of oxygen that this demonstrates.
   (JC, OL, 2006)

Fig 26.14

Fig 26.15

Chapter 27

WATER

Introduction

In this chapter we will look in detail at the most abundant and essential compound on Earth – water. The availability of water has been one of the factors that determined which parts of the world were populated and which were not.

We use water for drinking, cleaning, washing, cooking, cooling and for recreational activities such as swimming and sailing.

Electrolysis of water

The chemical name for water is hydrogen oxide, which obviously means it is a compound. In earlier times it was believed that water was an element.

A scientist called Henry Cavendish proved this was not the case, by actually splitting water into its elements, hydrogen and oxygen. He did so by passing electricity through water using a Hoffmann voltameter.

Electrolysis of water is an example of the chemical effect of an electric current, which you will meet later in the physics section.

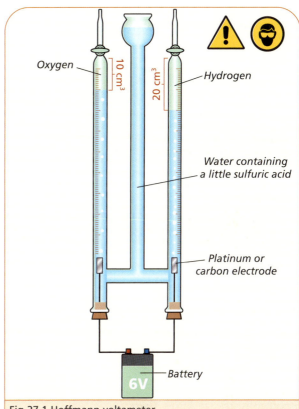

Fig 27.1 Hoffmann voltameter

Electrolysis of water using a Hoffmann apparatus. Oxygen gas is produced at the left electrode and hydrogen gas at the right electrode

Pure water is a poor electrical conductor so a small amount of dilute sulfuric acid is added to improve its conductivity.

- Hydrogen forms at the negative electrode (cathode).
- Oxygen forms at the positive electrode (anode).

The volume of gases produced is in the ratio of 2:1. Every molecule of water splits to produce 2 hydrogen atoms for every 1 oxygen atom. Hence, twice as much hydrogen gas is produced as oxygen gas.

> **Electrolysis** is the splitting up of a compound by passing electricity through it.

water ⟶ hydrogen + oxygen

CHEMISTRY

>>> **ACTIVITY 27.1**

To electrolyse water

Method

1 Set up the apparatus as shown.

2 Connect the electrodes to the battery and note the bubbles of gas being given off from each electrode. At which electrode are more bubbles forming? Which gas do you think this is?

3 Collect a test tube of each gas and test the gases as follows:
   (a) To test for hydrogen, place a lighted splint over the mouth of the test tube.
   (b) To test for oxygen, place a glowing splint into the mouth of the test tube.

Result

● Hydrogen gas burns with a 'pop'.
● Oxygen gas will relight a glowing splint.

Conclusion

Water is split into its elements, hydrogen and oxygen, in the ratio 2:1 respectively.

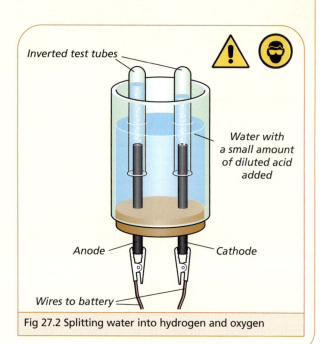

Fig 27.2 Splitting water into hydrogen and oxygen

Expansion of water

As a general rule all substances expand when heated and contract when cooled. Water behaves as expected until it is cooled to 4°C. On further cooling it actually expands!

From physics we know that:

$$density = \frac{mass}{volume}$$

Therefore, if the volume increases, i.e. by expansion, the density will decrease. At 0°C water freezes to ice, which has a larger volume and a smaller density. As a result, water freezes from the top downwards. This property is vital to aquatic life. Can you think why?

Air below 0°C

Ice below 0°C insulates the water under it

0°C Water close to
1°C freezing point
2°C
3°C
4°C Water at 4°C is denser so it sinks to the bottom

Fig 27.3 Fish can survive at the bottom of ice-bound waters

Hot water causes the metal cap to expand

Testing for water purity

CHEMISTRY

▶▶▶ **ACTIVITY 27.2**

To demonstrate that water expands when frozen

Method

1 Fill a glass bottle to the very top with water and screw the top on tightly.

2 Place the sealed bottle in a plastic bag and leave overnight in the freezer.

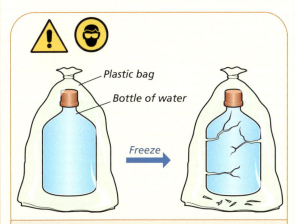

Plastic bag

Bottle of water

Freeze

Fig 27.4 Expansion of ice

3 What happens?
  ● Why is this unusual?
  ● Can you explain why waterpipes sometimes burst in winter?

Result
The bottle will break.

Conclusion
Water expands when it freezes.

Meniscus

The surface of water is not flat. It is curved. Water clings to the sides of the container it is placed in. Hence, the surface of water, called the meniscus, is concave in shape.

> **The meniscus** is the curved surface of a liquid in a vessel.

Some liquids, e.g. mercury, have a convex meniscus.

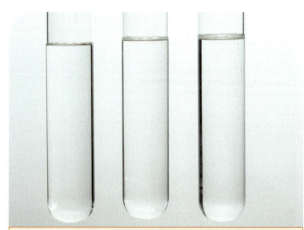

The test tubes contain from left, distilled water, tap water and seawater. The height of the meniscus in each tube is different due to the amount of dissolved chemicals present

Properties of water

1 Water freezes at 0°C. Ice will melt to form water at 0°C. Thus 0°C is the freezing point of water and the melting point of ice.

2 Water is an excellent solvent. A huge variety of substances dissolves in it.

3 Water is a colourless, odourless liquid.

4 The boiling point of water is 100°C.

To test for water

To test for the presence of water, either of the following tests can be used:

(a) Water will turn anhydrous (without water) copper sulfate from white to blue.

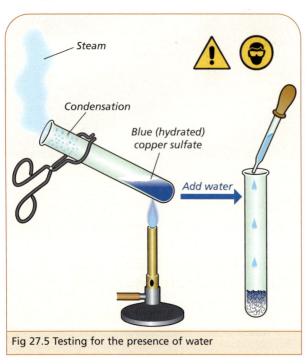

Fig 27.5 Testing for the presence of water

(b) Cobalt chloride paper is blue. Water will change its colour from blue to pink.

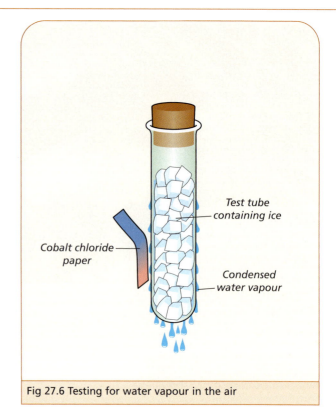

Fig 27.6 Testing for water vapour in the air

The water cycle

The amount of water on Earth remains constant. However, it is constantly moving and changing state. The reason it does not run out is because it is being recycled constantly. Figure 27.7 summarises this cycle.

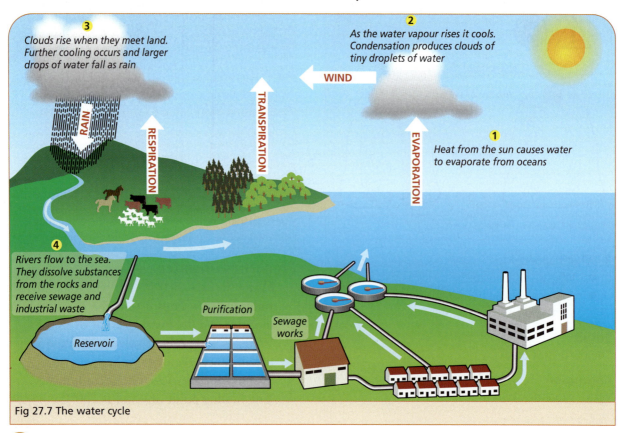

3 Clouds rise when they meet land. Further cooling occurs and larger drops of water fall as rain

2 As the water vapour rises it cools. Condensation produces clouds of tiny droplets of water

WIND

RAIN

RESPIRATION

TRANSPIRATION

EVAPORATION

1 Heat from the sun causes water to evaporate from oceans

4 Rivers flow to the sea. They dissolve substances from the rocks and receive sewage and industrial waste

Reservoir

Purification

Sewage works

Fig 27.7 The water cycle

CHEMISTRY

Water treatment

Unclean water is responsible for millions of deaths each year. Diseases such as typhoid and cholera are widespread in countries where water is not treated.

In Ireland water supplies are taken from rivers and lakes. Before it is fit to drink it must undergo a number of stages of treatment.

Stages in water treatment

1 Screening

The water passes through screens or meshes which trap any floating material, e.g. twigs and leaves.

2 Settling

Water is stored in large reservoirs where heavy particles settle at the bottom. A flocculant such as aluminium sulfate is added. This causes small particles in the water to 'flock' together into larger lumps and settle to the bottom.

3 Filtration

The water is next passed up through thick beds of sand. Any remaining insoluble particles are removed from the water at this stage. Now the water will be clear in colour. However, it may contain bacteria.

4 Chlorination

Chlorine is now added to kill any bacteria that may be present. The amount of chlorine added is carefully monitored. Too little and it may not kill all the bacteria, too much and it can be smelled and tasted!

In some countries where the water is not chlorinated it is recommended that water be boiled before drinking it.

5 Fluoridation

In many countries fluorine is added to the water to help prevent tooth decay.

Finally, the pH is checked. If it is too high some dilute acid is added. If it is too low lime is added.

The stages can be easily remembered in the correct order using a simple mnemonic.

Silly	**Sean**	**Follows**	**Chelsea**	**Football**
c	e	i	h	l
r	t	l	l	u
e	t	t	o	o
e	l	r	r	r
n	i	a	i	i
i	n	t	n	d
n	g	i	a	a
g		o	t	t
		n	i	i
			o	o
			n	n

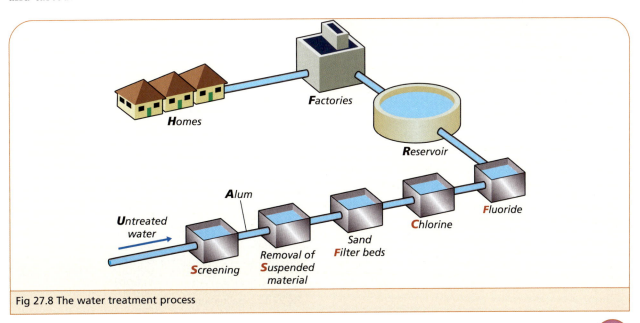

Factories

Reservoir

Homes

Alum

Untreated water

Screening

Removal of **S**uspended material

Filter beds

Sand

Chlorine

Fluoride

Fig 27.8 The water treatment process

203

CHEMISTRY

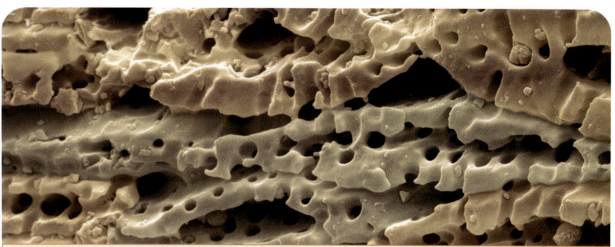

A micrograph of a water purifier made from charcoal, which removes odours and gases from the water passed through it

▶▶▶ ACTIVITY 27.3

To make a model of a filtration bed

Method

1 Cut the base of an empty soft drink bottle.

2 Arrange the various layers as shown in the diagram.

3 Pour some muddied water, e.g. soil and water, into the top.

- What happens to some of the larger heavy particles?
- What is this process called?
- If the filtrate is not clear, what could you do to further purify it?

Result

Clear water emerges from the base of the filter bed.

Conclusion

The layers in the filtration bed remove the insoluble particles.

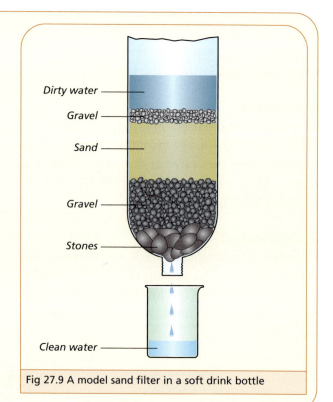

Fig 27.9 A model sand filter in a soft drink bottle

Desalination

Approximately 97 per cent of all water is saltwater. In countries where fresh water is scarce, saltwater is desalinised, i.e. has the salt removed from it.

The process is a separation technique comprising evaporation followed by condensation.

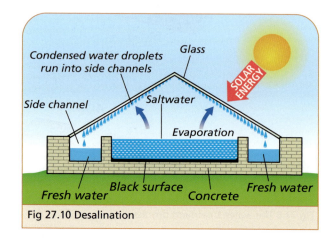

Fig 27.10 Desalination

A desalination plant separates water from saltwater

▶▶▶ **ACTIVITY 27.4**

To obtain a sample of fresh water from seawater by distillation

Distillation is an effective method of purifying water. The main drawback is that it is not cost-effective as large amounts of heat and energy would be required to produce fresh water on a large scale.

Method

1 Set up the apparatus as shown in the diagram.

2 Approximately half fill the flask with seawater.

3 Turn on the water through the Liebig condenser. There should be a steady flow of water through it.

4 Heat the flask gently so that it boils steadily. Anti-bumping granules should be added for smooth boiling.

Results

● The thermometer will read 100°C as the water distils in the condenser. The liquid formed is the distillate.

● The residue will be the salt that can be seen on the sides of the flask.

Conclusion

Saltwater can be desalinised by distillation.

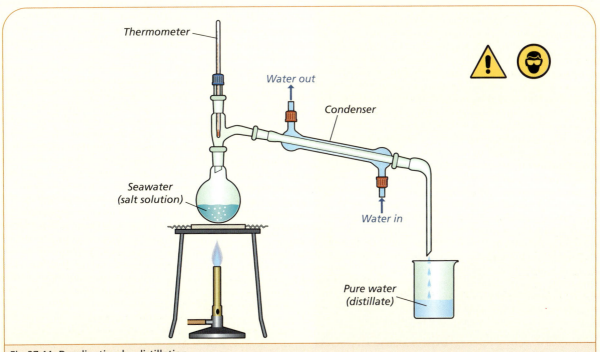

Fig 27.11 Desalination by distillation

Key Points

- Water is a compound of hydrogen and oxygen.

- Upon electrolysis, water is split into its elements, hydrogen and oxygen, in the ratio 2:1, respectively.

- A Hoffmann voltameter is used to electrolyse water.

- Pure water freezes at 0°C and boils at 100°C.

- As water freezes it expands. Hence, ice is less dense than water.

- Water turns white anhydrous copper sulfate blue.

- Water turns cobalt chloride from blue to pink.

- The five stages of water treatment are:
  – screening
  – settling
  – filtration
  – chlorination
  – fluoridation.

- Water can be purified by distillation, i.e. evaporation followed by condensation.

- Desalination is used in many parts of the world to obtain freshwater from saltwater.

A water test: the addition of water to copper sulfate causes a colour change from white to blue

Questions

1 Water is made up of two elements. Name them.

2 Write out and complete the following:
A Hoffmann voltameter can be used to _____ water. The two gases produced are _____ and _____, in the ratio of __ : __.

3 Why is a small amount of dilute acid added to the water before it is electrolysed?

4 Describe how you could test each gas produced during the electrolysis of water to identify them.

5 Write out and complete the following:
(a) The freezing point of water is ___ °C and ____0°C is its boiling point.
(b) The _____ is the curved surface of a liquid in a vessel. It is usually _____ in shape.
(c) As water freezes it _____. Hence ice is _____ _____ water.
(d) Water will turn cobalt chloride from _____ to _____.
(e) Water will turn _____copper sulfate _____.

6 Water supplies are usually treated in a water treatment plant before domestic use.
(a) Name three stages in the treatment of water.
(b) State what happens to the water during each of these stages.
(c) Why is it recommended in some countries that water should be boiled before drinking?

7 What is involved in the desalination of water?

? Questions

8 (a) Name the method of separation of mixtures shown in Figure 27.12.

(b) Name the part labelled A.

(c) How does the design of A enable it to carry out is function?

(d) Identify B.

(e) What is purpose of the thermometer?

(f) Name two changes of state that take place in this apparatus.

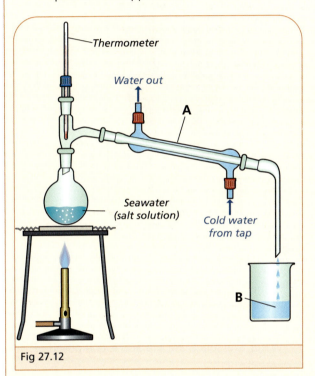

Fig 27.12

9 Describe with the aid of a labelled diagram a simple experiment to show that water contains dissolved solids.

10 What are the following used for in the treatment of water?

(a) beds of sand

(b) fluoride

(c) screens

(d) chlorine

(e) aluminium sulfate.

11 Figure 27.13 shows the apparatus used in the electrolysis of water.

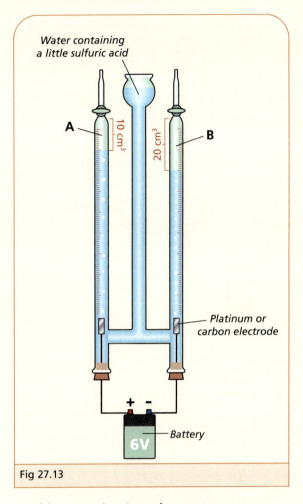

Fig 27.13

(a) Name the piece of apparatus.

(b) Name the two gases produced at A and B.

(c) What information does this experiment give about the composition of water?

(d) What substance must be added to the water at the beginning? Why is it added?

❓ Questions

Examination Questions

12 The diagram shows the electrolysis of water.

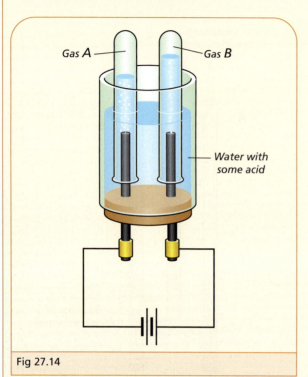

Gas *A* — Gas *B*

— Water with some acid

Fig 27.14

(a) Why is some acid added to the water?
(b) Give a test for gas A.
(c) The volume of gas A is twice that of gas B. What does this tell us about the composition of water?
(JC, HL, 2006)

13 The diagram shows an arrangement of apparatus suitable for the electrolysis of acidified water.

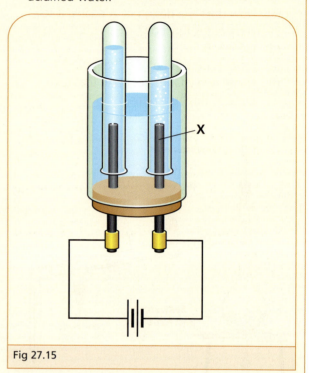

— X

Fig 27.15

Name the gas produced at the electrode X and state a test for this gas.
(JC, OL, 2006)

14 Water is essential for life and is composed of two elements.
(a) Name one of the elements that make up water.
(b) Name a chemical that can be used to test for the presence of water.
(JC, OL, 2006)

CHEMISTRY

Chapter 28

SOLUTIONS AND CRYSTALLISATION

Introduction

Solutions are a type of mixture. We usually think of a solution as a solid dissolved in a liquid. This is true of most solutions. For example, seawater is a solution of salts dissolved in water.

Not all solutions are made up of solids dissolved in liquids.

- Bronze consists of solid tin dissolved in solid copper.
- Marshmallows are examples of air dissolved in a sugary paste.
- Oxygen dissolves in water, forming a solution which fish need in order to 'breathe'.

Solutes, solvents, solutions

1   All solutions are composed of a substance that dissolves. This is called a **solute**.
2   The substance in which they dissolve is called a **solvent**.
3   A solute and a solvent together form a **solution**. For example, a cup of coffee is a solution made up of coffee (solute) dissolved

> A solute dissolves in a solvent to form a solution.

If one substance will not dissolve in another it is said to be **insoluble**.

When a colourless solution of potassium iodide is added to a colourless solution of lead nitrate, an insoluble yellow substance is formed

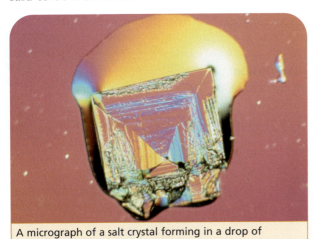

A micrograph of a salt crystal forming in a drop of seawater as the water evaporates

Frost crystals on a leaf

Concentrated or dilute

If there is a large amount of solute and a small amount of solvent then the solution formed is said to be **concentrated**.

If on the other hand, there is a small amount of solute and a large amount of solvent the solution is said to be **dilute**.

A solution can be made more concentrated by removing some solvent, e.g. by evaporation, or by adding more solute.

A solution can be made more dilute by adding more solvent or removing some solute.

Solubility

How many spoons of sugar will dissolve in a cup of tea? The amount depends on two factors: (a) the amount of solvent (water) and (b) its temperature. If too much sugar is added, the tea will become saturated with sugar (solute) at a given temperature.

> A **saturated solution** is a solution that contains as much dissolved solute as possible **at that temperature**.

We can measure how soluble a solute is by seeing how much of it dissolves in 100 g of water.

▶▶▶ ACTIVITY 28.1

To investigate the solubility of a variety of substances in water

A variety of substances will be tested to discover if they are soluble (dissolve) or insoluble (do not dissolve) in water.

Method

1 Half fill a test tube with water.

2 Add a half spatula full of the substance to be tested to the water.

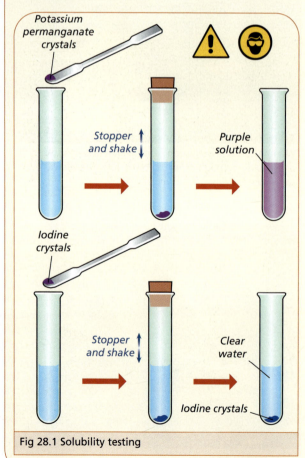

Fig 28.1 Solubility testing

3 Stopper the test tube and shake for 15–20 seconds.

4 Examine the contents of the test tube and decide whether the substance being tested is soluble or insoluble. How will you know?

5 Rinse out the test tube and repeat the experiment with the next substance to be tested.

6 Try as many substances as are available and record your results as shown in the table.

Substance being tested	Soluble in water	Insoluble in water
Sodium chloride (salt)	Yes	
Sulfur		
Wax		
Sand		
Copper sulfate		

Conclusion

If the substance being tested disappears (dissolves) then it is soluble in water. If it does not disappear then it is said to be insoluble.

For example, the solubility of copper sulfate is 19 g in 100 g of water at 15°C. Why is the temperature mentioned, do you think? Would the figure be higher or lower at 25°C?

The following is a graph of the solubility of copper sulfate in water at various temperatures. A solubility curve shows how solubility changes with temperature. In this case, we see that more copper sulfate dissolves at higher temperature.

As you can see the temperature of the solvent has quite an effect on the solubility of the solute. For example, only 17 g of $CuSO_4$ will dissolve in 100 g of water at 10°C, whereas if the water was heated to 60°C it will dissolve 40 g of the salt.

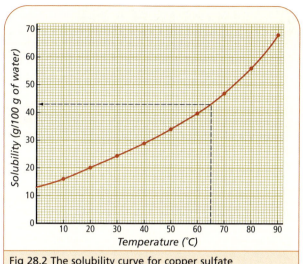

Fig 28.2 The solubility curve for copper sulfate

▶▶▶ ACTIVITY 28.2

To investigate the effect of temperature on solubility

Method

1 Place approximately 100 cm³ of water in a beaker and note its temperature.

2 Using a pestle and mortar, grind up a sample of copper sulfate crystals.

3 Slowly and with constant stirring add the powdered copper sulfate to the water. Continue until the powder will no longer dissolve but instead settles to the bottom.

4 Heat the beaker to 50°C and note what happens to the undissolved copper sulfate.

5 Add more copper sulfate until no more will dissolve.

6 Heat the beaker to near boiling (80–90°C) and again try to dissolve more solute. What do you notice about how temperature affects solubility?

7 Allow the solution to cool. This can be quickened by holding the beaker under running tap water. What do you notice?

Result

More copper sulfate dissolves in the water at higher temperatures.

Conclusion

The amount of solute that will dissolve in a solvent depends on the temperature of the solvent.

Copper sulfate crystals

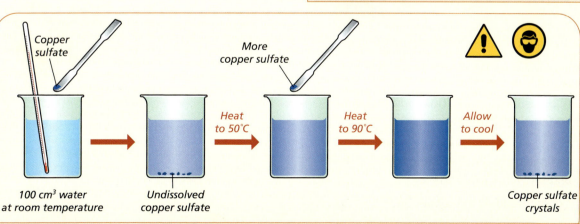

Fig 28.3 Investigating the effect of temperature on solubility

CHEMISTRY

MANDATORY ACTIVITY 19

To examine the preparation and formation of copper sulfate crystals (or aluminium sulfate crystals)

In this experiment copper sulfate crystals can be used directly or they can be produced by reacting copper (II) oxide with dilute sulfuric acid. This is also a neutralisation reaction.

copper oxide (black)	copper sulfate (blue)
+	+
sulfuric acid	water

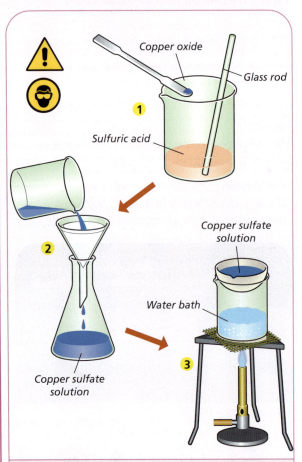

Fig 28.4 The preparation and formation of copper sulfate crystals

Method

1 Add approximately 40 cm³ of dilute sulfuric acid to a small beaker and heat to near boiling point.

2 Add the black copper oxide a little at a time and stir with a glass rod.

3 Continue adding the copper oxide until no more reacts. How will you know when this point has been reached?

4 Filter the mixture.
   ● What is the residue?
   ● What is the filtrate?

5 Pour the filtrate into an evaporating dish and heat on a water bath as shown.

6 Heat until some crystals form at the edge of the solution.

7 Quickly pour about half the hot solution into a small beaker and cool it rapidly by standing it in iced water or by allowing running water to pour over the beaker.

8 Leave the remaining half in the evaporating dish and allow to cool slowly.

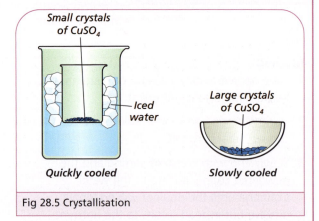

Fig 28.5 Crystallisation

Result
● Small crystals of copper sulfate will form in the beaker that was cooled rapidly.
● Large crystals will form in the evaporating dish, as this solution was cooled slowly.

Conclusion
A rapidly cooled saturated solution results in the formation of small crystals, whereas slow cooling produces large crystals.

Other solvents

- Water is the most commonly used solvent, but there are many others. Detergents are added to water to increase the range of substances it will dissolve.
- Alcohol is a solvent used in biology to remove chlorophyll from plant leaves.
- Propanone is used to remove nail varnish.
- 'Dry cleaning' of clothes involves using a liquid called trichloroethane instead of water to 'wash' the clothes. So it isn't really 'dry cleaning'!

Can you name some household substances that contain solvents used to remove stains?

Many solvents are harmful to inhale. Their vapours can cause damage to the brain, liver and lungs.

Many solvents are very flammable and must be kept well away from naked flames.

Household products that contain non-water-based solvents. These solvents evaporate quickly which speed up the drying of paint, the setting of glue and the diffusion of perfume

🔑 Key Points

- A solute is a substance that dissolves.

- A solvent is a liquid in which a solute dissolves.

- A solution is a mixture of a solute and a solvent.

- A solute dissolves in a solvent to form a solution.

- If a substance dissolves it is said to be soluble. If it does not dissolve it is said to be insoluble.

- A liquid that has a lot of solute dissolved in it is said to form a concentrated solution. A liquid that has only a small amount of solute in it is said to be dilute.

- A solution is said to be saturated when no more solute will dissolve in it at that temperature.

- Crystallisation is the formation of crystals by cooling a saturated solution.

- The solubility of a solvent increases with temperature.

❓ Questions

1 Write out and complete the following:
Most solutions are made up of a substance called the _____ which dissolves in the liquid called the _____. Substances that dissolve are said to be _____ while those that don't are _____. A cup of coffee is a _____; the hot water is the _____ and the coffee granules are the _____.

2 Sugar is _____ in water, whereas sand is said to be _____.

3 What is a concentrated solution?

4 What is a saturated solution?

5 Describe two ways by which a dilute solution could be made more concentrated.

6 Name a solvent other than water and a substance that dissolves in it.

7 Give an everyday example of
 (a) a gas dissolved in a liquid
 (b) a liquid dissolved in a liquid
 (c) a solid dissolved in a liquid
 (d) a gas dissolved in a solid.

8 Describe an experiment you could carry out to test whether a substance is soluble or insoluble in water.

? Questions

9 Give two examples of substances containing solvents that produce harmful vapours.

10 (a) Explain the terms (i) solution and (ii) saturated solution.
   (b) Describe how you would make a saturated solution of copper sulfate in water.

11 If a hot concentrated solution is cooled some of the solute comes out of the solution.
   (a) What is this process called?
   (b) How would you cool the solution if you wanted (i) large crystals, (ii) small crystals?

12 Below is a table showing the mass of copper sulfate (g) that will dissolve in 100 g of water at different temperatures.

Table 28.1 How the solubility of copper sulphate changes with temperature	
Temperature °C	Mass of copper sulfate (g) that will dissolve in 100 g of water
0	14
10	17
20	20
30	25
40	29
50	34
60	40
70	47
80	56
90	68

   (a) From this information graph a solubility curve putting temperature on the *x*-axis.
   (b) From your graph estimate:
      (i)   The mass of copper sulfate that will dissolve at 25°C.
      (ii) The minimum temperature required to dissolve 36 g of copper sulfate.

Examination Question

13 Describe how you could carry out an experiment to grow crystals using alum or copper sulfate.

Include a diagram of any equipment used. (JC, OL, 2006)

Chapter 29

WATER HARDNESS

Introduction

In this chapter we examine some characteristics of water that affect our everyday lives.

Hard water

We know from earlier experiments that water is an excellent solvent and as a result it almost always has substances dissolved in it. These solutes alter the nature of water. Many bottled waters taste different due to the dissolved solids which the water picks up as it flows over and through different rocks.

Soap is added to water to help it dissolve even more substances like grease and oil. The soap forms a lather on top of the water. However, in some regions a 'scum' will form on the water and more soap must be added to get a good lather. Water from these regions is described as being hard.

Rainwater is slightly acidic due to dissolved carbon dioxide forming a weak solution of carbonic acid. Because of its acidic nature, rainwater has a greater dissolving capacity than ordinary water. Limestone rock is very widespread in Ireland and as a result when rain falls, calcium and magnesium compounds become dissolved.

> **Hard water** is water that does not lather easily with soap.

> **Soft water** is water that lathers easily with soap.

The cause of water hardness is the presence of calcium and/or magnesium ions, Ca^{+2} or Mg^{+2}. Can you remember why these ions have a +2 charge? When these ions react with soap they form an insoluble compound which floats on top of the water as a scum.

For example:

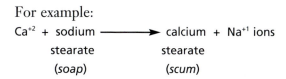

Ca^{+2} + sodium $\longrightarrow$ calcium + Na^{+1} ions
　　　　stearate　　　　　　　stearate
　　　　(soap)　　　　　　　　(scum)

> calcium ions + soap ions $\longrightarrow$ scum

It is only when all of the calcium ions in the water have reacted with the soap that a proper lather will form. As a result, some soap is wasted.

It is worth noting that no scum is formed when using modern synthetic detergents.

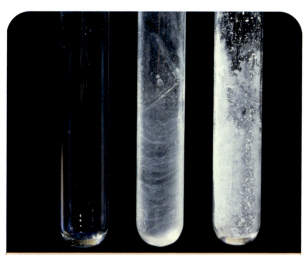

A test to show different types of water hardness; the samples show the deposits left behind from (left) pure water, hard water and very hard water

Problems associated with hard water

One of the main problems of hard water is that when heated, for example in a kettle, a white insoluble solid called **limescale** or **fur** is deposited. This substance binds to heating elements and reduces their efficiency. Washing machines and hot waterpipes can become blocked. There are products available to counteract this problem.

Thick deposits of limescale seen on the interior of a domestic metal pipe

Kettle descalers are basically solutions of weak acids which, when placed in a kettle, react with and dissolve the calcium carbonate.

Because of the problems caused by hard water it is often necessary to 'soften' it before it can be used. It involves removing the calcium and/or magnesium ions.

This can be done in a number of ways:

1 Addition of washing soda
2 Distillation
3 Ion exchanges
4 Deionisers.

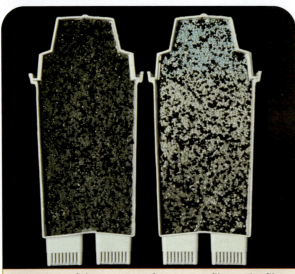

Comparison of the contents of two water filters. The filter on the left is new and the other is six weeks old

1 Addition of washing soda (sodium carbonate)

When washing soda is added to hard water, the calcium ions in the water react with the carbonate ions in the washing soda to form insoluble calcium carbonate. Thus the calcium ions are removed and the water is softened.

Bath salts contain carbonate ions. These soften the water and hence the soap will form lots of suds.

calcium + carbonate $\longrightarrow$ calcium
ions   ions   carbonate
soluble   *soluble*   *insoluble*

Most modern washing powders have their own water softeners added, for the same reason.

An ion exchange water softener removes calcium and magnesium ions and replaces them with sodium ions

2 Distillation

Distilling water removes all soluble and insoluble particles and hence will soften water. However, it is not cost-effective on a large scale due to the high energy costs of boiling large amounts of water.

3 Ion exchangers

An ion exchanger works by simply replacing (exchanging) the ions that cause hardness (i.e. Ca^{+2} and Mg^{+2}) with ions that do not cause hardness. Sodium ions are very often used as the replacement ion because they do not affect hardness. Look at Figure 29.1 overleaf.

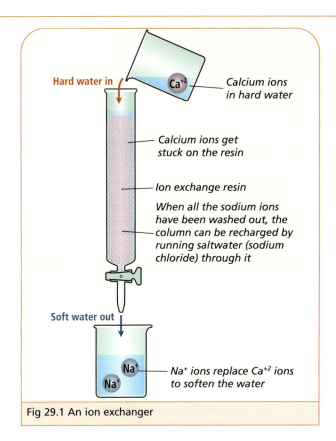

Hard water in

Ca⁺² Calcium ions in hard water

Calcium ions get stuck on the resin

Ion exchange resin

When all the sodium ions have been washed out, the column can be recharged by running saltwater (sodium chloride) through it

Soft water out

Na⁺ Na⁺ ions replace Ca⁺² ions to soften the water
Na⁺

Fig 29.1 An ion exchanger

The column is filled with a resin full of sodium ions Na^{+1}. As the hard water flows through the resin, the calcium ions are deposited and sodium ions replace them. This leaves the water soft.

Eventually, the resin will fill with calcium ions. This can be reversed by adding a sodium salt, e.g. sodium chloride (table salt), to the column and 'washing out' the calcium ions.

4 Deionisers

A deioniser removes all ions from the water, unlike the ion exchanger above, which swaps one type of ion for another. Deionised water is used for topping-up car batteries and many experiments require water that is free from ions. So how is this achieved?

The process is similar to that in an ion exchanger. A column composed of a resin is used. As the water passes through, all the positive ions are replaced with H^+ ions and negative ions are replaced with OH^- ions. These two ions combine to form hydrogen oxide, better known as water.

To investigate hardness in a water sample

Method

1 Half fill a test tube with the water sample to be tested.

2 Place a mark on the test tube 5 cm above the level of water.

3 Using a dropper, add soap solution, 1 cm³ at a time, to the test tube. Stopper and shake the test tube in between.

4 Continue adding the soap solution until a lather is formed, up to the mark on the test tube.

Record the volume of soap added.

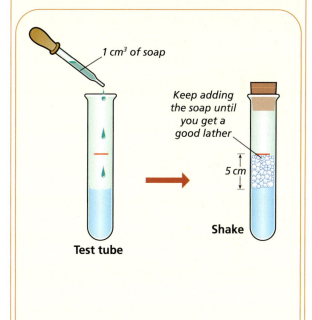

1 cm³ of soap

Keep adding the soap until you get a good lather

5 cm

Shake

Test tube

Fig 29.2 Testing water for hardness

Result

A large volume of soap is required to form a lather in hard water, whereas a small volume of soap is required in soft water.

Conclusion

Hard water does not lather easily with soap, whereas soft water does.

Water softening pellets

Indicator paper being used to test water for hardness

CHEMISTRY

 MANDATORY ACTIVITY 20A

To test for water hardness

In this experiment samples of water from different sources should be tested, e.g. rainwater or river water. An artificially made-up sample of hard water can be made by dissolving calcium sulfate or magnesium sulfate in water.

Method

1 Place 10 cm³ of each of the different water samples in different test tubes.

2 Place a marker on each test tube at a fixed height above the water, e.g. 5 cm.

3 Make a soap solution using natural soap flakes.

4 Using a dropper (or a burette) add the soap solution, one drop at a time, to the first test tube. Stopper and shake the test tube each time.

5 Continue this procedure until a lather forms which is level with or above the marker on the test tube.

6 Record the volume of soap solution required to generate this amount of lather.

   Repeat steps 4–6 on each water sample.

Result

Different water samples will require different volumes of soap solution to produce the required lather.

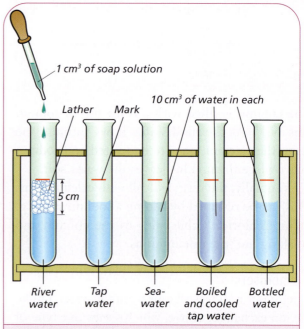

1 cm³ of soap solution

Lather   Mark

10 cm³ of water in each

5 cm

River water | Tap water | Sea-water | Boiled and cooled tap water | Bottled water

Fig 29.3 Comparing hardness in different water samples

Conclusion

● The water sample that required the greatest volume of soap solution to form a sufficient lather is said to be the hardest.
● The water sample that required the least volume of soap is said to be the softest.

Advantages of hard water

● Hard water is a source of calcium, which is vital for bones and teeth.
● Most people argue that hard water tastes nicer than soft water.
● It makes better-tasting beer.
● Some doctors believe it reduces heart disease.

Disadvantages of hard water

● Limescale can block pipes and reduce the efficiency of kettles, washing machines, etc.
● It wastes soap and leaves a scum which can be difficult to remove.

MANDATORY ACTIVITY 20B

To show the presence of dissolved substances in a water sample

Because water is such a good solvent, there will always be dissolved substances present. Water also contains tiny insoluble particles which we firstly remove by filtration.

Method

1 Filter the water sample to remove the undissolved (insoluble) particles as in Figure 29.4.

2 Find the mass of a clean, dry evaporating dish.

3 Half fill a clean evaporating dish with the filtrate.

4 Heat the dish until all the water has evaporated.

5 Carefully examine the dish.

6 Reweigh the dish and there will be an increase in mass.

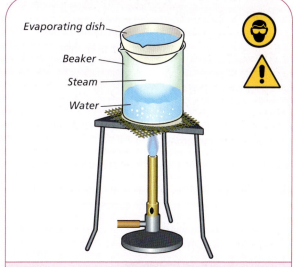

Fig 29.5

Result

There are solids in the dish and they increase the mass of the dish.

Conclusion

The water sample contained dissolved solids.

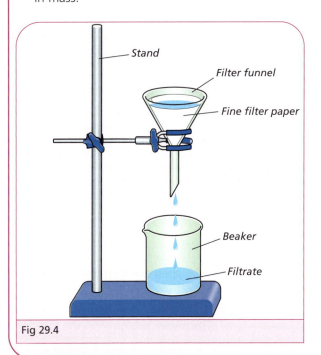

Fig 29.4

 Key Points

● Hard water is water that does not lather easily with soap.

● Soft water is water that lathers easily with soap.

● Water hardness is caused by calcium sulfate and magnesium sulfate dissolved in the water.

● Hardness in water can be removed by either:
  (a) the addition of washing soda (sodium carbonate)
  (b) distillation
  (c) passing the water through an ion exchanger or a deioniser.

● Hard water causes scale in kettles and boilers and it wastes soap. However, it provides calcium for our bones and teeth, makes nicer beer and tastes better than soft water.

? Questions

1 Write out and complete the following:
When compounds of _____ and
_____dissolve in water it becomes
_____. This means it does not
_____ easily with _____.

2 (a) Explain the term 'quick cycle'.
   (b) Why is it very difficult to obtain pure water?
   (c) Why is hard water commonly found in limestone areas?

3 Name three methods of removing hardness.

4 Why is distillation not a viable option to remove water hardness on a large scale?

5 In an ion exchanger the _____ ions are usually replaced with _____ ions, which do not cause hardness.

6 Why do bath salts contain baking soda?

7 List three advantages of hard water.

8 List two disadvantages of hard water.

9 Petra tested three water samples with soap solution. She recorded how much soap was needed to get a good lather, before and after boiling water.

Water sample	Volume of soap added
Sample X	8 cm^3
Sample Y	2 cm^3
Sample T	5 cm^3

   (a) Which sample was the softest water?
   (b) Which sample was the hardest?
   (c) Which was most likely to be distilled water?

10 How would you show that a water sample was soft?

11 In what sort of area is hard water usually found?

12 In an experiment to test for hardness, Geraldine prepared three test tubes as follows:

Test tube	Contents
A	Deionised water
B	Deionised water with calcium hydrogen carbonate
C	Deionised water with calcium sulfate

   (a) Describe how you would test each of the samples for hardness.
   (b) What results would you expect to obtain?
   (c) Explain how you would remove hardness from water.

13 Geraldine was given a sample of water and was told it contained two dissolved compounds, one causing temporary hardness and the other permanent hardness.
   (a) Name a compound that could have caused the temporary hardness. How could it be removed?
   (b) Name a compound that leads to permanent hardness. List three ways by which she could have removed the permanent hardness.

14 (a) Explain how an ion exchanger works.
   (b) Why will it cease to work after a period of time?
   (c) How can it be recharged?

Examination Question

15 A student investigated the hardness of a number of water samples by testing them with soap flakes. In each case the same volume of water was tested. The results are given in the table.

Water sample	Number of soap flakes needed to give a lather
A	1
B	7
C	4

   (a) Which of the three water samples had the softest water?
   (b) State one disadvantage of hard water.
   (c) State one way by which water hardness can be removed.
   (JC, HL, Sample Paper)

CHEMISTRY

Chapter 30 ○○○

FOSSIL FUELS

Introduction

> **Fuels** are substances which burn in oxygen to produce heat energy.

Our bodies use food as a fuel to provide us with energy. Some fuels such as sunflower oil and biomass are renewable, but the fuels we use most commonly are non-renewable – they are fossil fuels.

A domestic waste power station converts domestic waste into electricity

Fossil fuels

Fossil fuels are produced over millions of years. They are formed from the remains of plants and animals.

- Coal was formed from trees and ferns which died and were buried and compressed under swamps.
- Crude oil and natural gas were formed from animals and plants that lived in the sea. Their remains were buried under layers of sand on the seabed. As the pressure increased they eventually formed oil and natural gas.

Fossil fuels are mainly made up of two elements: hydrogen and carbon. Such compounds are called hydrocarbons. When burned, hydrocarbons produce water vapour and carbon dioxide.

> **Hydrocarbons** are compounds made up of hydrogen and carbon.

▶▶▶ ACTIVITY 30.1

To examine the products of combustion of a hydrocarbon

You can use a candle or a natural gas Bunsen burner as a hydrocarbon fuel.

Method

1 Set up the apparatus as shown.

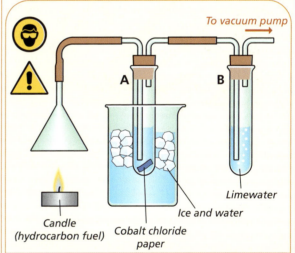

Fig 30.1 Examining the products of the combustion of a hydrocarbon

2 Allow the fuel to burn until a change is observed in each test tube.

3 What do you notice in test tube A? Why is test tube A placed in a beaker of iced water?

Result

The cobalt chloride paper changes from blue to pink in colour, and the limewater turns from clear to milky.

Conclusion

When a hydrocarbon is burned water and carbon dioxide are produced.

The greenhouse effect

A greenhouse heats up in the following way:

1 The glass allows the short wavelength heat rays from the sun to pass through into the glasshouse.
2 It does not allow the longer wavelength heat rays produced by the plants inside to escape from the glasshouse.
3 As a result the temperature inside the glasshouse is increased.

Our atmosphere behaves like the glass in a greenhouse. This effect keeps the average temperature of the Earth relatively steady.

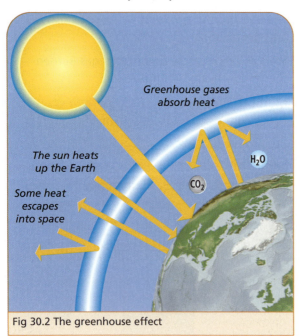

Greenhouse gases absorb heat

The sun heats up the Earth

Some heat escapes into space

H_2O

CO_2

Fig 30.2 The greenhouse effect

Some of the gases in the air are known as greenhouse gases. Without them the Earth would be much colder. The main greenhouse gases are:

- carbon dioxide
- methane
- water vapour.

Nitrogen and oxygen, which between them make up 98–99 per cent of air, absorb very little heat radiation and as a result have little greenhouse effect.

Global warming

The problem of global warming follows from an increase in the greenhouse effect, that is, an increase in the greenhouse gases.

1 The level of carbon dioxide in the air has increased.

- This is mainly due to the burning of fossil fuels.
- Plants absorb carbon dioxide, but we are cutting down huge numbers of trees everyday.
- Cattle graze on the clearing of the trees, but they produce methane gas, and this further adds to the problem.

2 **Methane** is the second most important greenhouse gas. Its chemical formula is CH_4 and it is the simplest hydrocarbon. Levels of methane are rising due to:

- emissions from swamps, bogs and paddy fields
- leaks of natural gas
- emissions from urban waste dumps
- increased cattle population.

These increases in the levels of the main greenhouse gases are causing the Earth to heat up. This could lead to major changes in climate. For example, the melting of the polar ice caps would lead to an expansion of the oceans and a rise in sea level. As a result, lowland areas would become flooded.

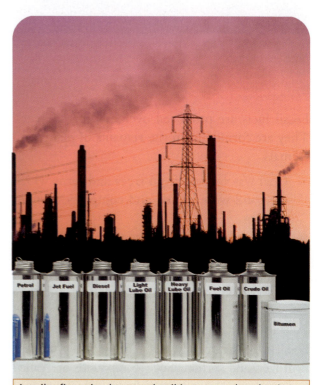

An oil refinery is where crude oil is separated so that it can be used in various oil products

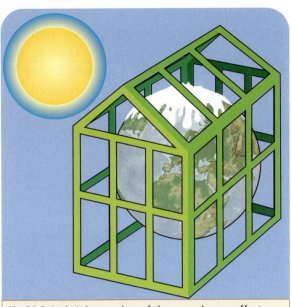

Fig 30.3 Artist's impression of the greenhouse effect

Acid rain

Rain has always been slightly acidic. Water vapour and carbon dioxide react to form carbonic acid. Hence, rain is a very dilute solution of a weak acid.

Rainwater naturally has a pH value of between 5.5 and 6.5. Nowadays rain in parts of the world can have a pH value as low as 3. The reasons for the lowering of the pH are due mainly to the burning of fossil fuels.

- Many fossil fuels, especially coal, contain sulfur as an impurity. When burned the sulfur combines with oxygen, forming sulfur dioxide. This gas dissolves in rainwater forming sulfuric acid which is a stronger acid than carbonic acid.
- Fumes from car exhausts also make our rain acidic. They give off oxides of nitrogen which dissolve in rain to form another strong acid – nitric acid.

Effects of acid rain

1 Lakes and fish

There are hundreds of lakes in Scandinavia that now have no fish. Acid rain causes minerals, such as aluminium and mercury, to be washed out of the soil and into lakes where it poisons fish. Aluminium causes fish gills to become blocked and as a result they cannot absorb the oxygen out of the water.

2 Trees

Acid rain attacks trees and it dissolves many important minerals out of the soil. As a result, trees are damaged or even killed. Many trees are partially defoliated, i.e. lose their leaves.

Acid rain damages trees

3 Buildings and structures

Acid rain attacks buildings and statues, causing them to erode. Limestone buildings are the most easily eroded. Metal objects such as bridges and railways are also corroded more quickly by acid rain.

Acid rain erodes statues and building quicker than normal rain would

Acid soil

If soil is too acidic most crops will not grow well. Farmers often spread powdered limestone or lime on their soil to neutralise it.

Smog

When fossil fuels such as coal and petrol are not fully burned, smoke is produced. Smog is the name used for what results when this smoke and fog combine.

Smog reduces visibility and is associated with many breathing problems. In Dublin only smokeless coal can now be burned and the incidence of smog has dramatically reduced.

▶▶▶ ACTIVITY 30.2

To find the pH of a soil sample

Method

1 Put a small amount of the soil sample in a test tube half-filled with water.

2 Stopper the test tube and shake vigorously.

3 Filter the test tube into a second clean test tube.

4 Test its pH using universal indicator or a pH meter.

Result

Universal indicator paper will remain yellow above the water and turn green below the water.

5 If the soil is acidic, add powdered lime to raise its pH. What could you add if the soil sample had a pH above 7?

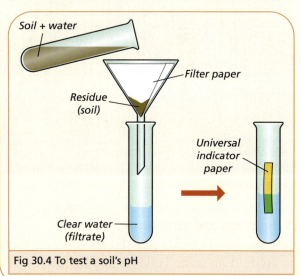

Fig 30.4 To test a soil's pH

▶▶▶ ACTIVITY 30.3

To examine the effect of sulfur dioxide on plants

Sodium metabisulfite gives off sulfur dioxide which is toxic. Therefore, the experiment should be performed in a fume cupboard.

Method

1 Set up the apparatus as shown.

2 Leave for a few days.

Result

The seeds in container B germinate while those in container A do not.

Conclusion

Sulfur dioxide prevents germination.

Note: This experiment can be repeated using seedlings.

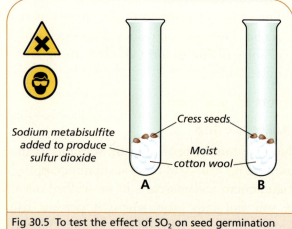

Fig 30.5 To test the effect of SO_2 on seed germination

CHEMISTRY

Reducing the greenhouse effect and the effects of acid rain

The greenhouse effect and the effects of acid rain can be reduced in the following ways:

- The use of fossil fuels as a source of energy should be reduced.
- Scientists are looking at alternative sources of energy such as solar and wind.
- The use of low-sulfur fuels such as natural gas (which is almost entirely methane) instead of coal, turf and peat, which have larger amounts of sulfur, will also help.
- Cars are now fitted with catalytic converters, which convert oxides of nitrogen into harmless nitrogen.
- The planting of trees is essential as they absorb carbon dioxide from the air and produce oxygen.
- Some industries use limestone 'scrubbers' to remove pollutants such as sulfur dioxide from their waste gases before they are emitted into the atmosphere.
- Governments are providing grants to encourage people to switch heating systems in their homes from oil/gas to wood pellets, which are a much cleaner burning fuel.

A hydrogen-fuelled car is filled up at a hydrogen garage

A solar-powered house which is self-sufficient in producing electricity

Key Points

- A fuel is any substance that can be burned to produce heat energy.
- Hydrocarbons are compounds containing hydrogen and carbon.
- Fossil fuels are mainly hydrocarbons formed over millions of years from the remains of animals and plants.
- Hydrocarbons burn to produce carbon dioxide and water.
- Greenhouse gases such as carbon dioxide and methane trap infrared energy and warm the Earth.
- An increase in greenhouse gases is leading to global warming.
- Acid rain is caused by the burning of fossil fuels which produce sulfur dioxide and oxides of nitrogen. These gases react with rainwater, producing sulfuric acid and nitric acid.
- Acid rain causes minerals such as aluminium to be washed into lakes, killing fish.
- Acid rain attacks trees and soil.
- Acid rain attacks limestone buildings and metallic structures causing them to corrode more quickly.

CHEMISTRY

225

? Questions

1 What is a fuel?

2 What is a fossil fuel?

3 Write out and complete:
   (a) Hydrocarbons are compounds made up of
       _____ and _____.
   (b) When burned, hydrocarbons produce
       _____ and _____.

4 Some power stations burn fossil fuels for energy.
   (a) Name two fossil fuels.
   (b) During the burning of fossil fuels, which
       gas is produced that is a major contributor
       to acid rain?
   (c) Name three effects of acid rain on the
       environment.

5 Distinguish between the greenhouse effect and
   the increased greenhouse effect.

6 (a) Rainwater has always been slightly acidic.
       Why?
   (b) Name the two main gases that contribute
       most to acid rain.
   (c) Name the acids that are produced as
       these gases dissolve in rainwater.

7 Thomas and Petra want to test the products that
   are formed when a hydrocarbon fuel is burned.

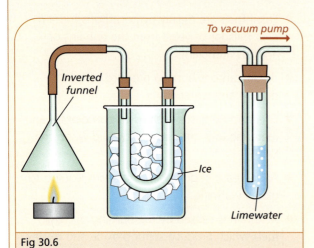

Fig 30.6

   (a) What is the function of the inverted
       funnel in Figure 30.6?
   (b) Why is the U-tube placed in ice?
   (c) How could they determine if the liquid
       formed in the U-tube is water?
   (d) The limewater turns milky in colour. What
       does this indicate?
   (e) What is the function of the vacuum
       pump?

8 Design and draw a poster that explains about
   the increased greenhouse effect.

9 Look at the diagram of a catalytic converter.

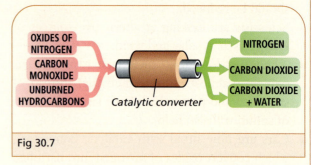

Fig 30.7

   (a) How does a catalytic converter help
       reduce acid rain?
   (b) How does the use of catalytic converters
       affect global warming? Explain.

10 Give a brief explanation for the following:
   (a) Lime is often spread on fields by farmers.
   (b) Catalytic converters are fitted to car
       exhausts.
   (c) Limestone buildings are being eroded
       more quickly now than they were 100
       years ago.
   (d) Many Scandinavian lakes have no fish left
       in them.

11 (a) Describe a simple experiment you would
       do to test the effect of acid rain on
       different types of rock.
   (b) What type of rock would you expect to be
       eroded most?

12 A lot of power stations burn fossil fuels for
   energy. Some of the gases produced lead to
   acid rain.
   (a) Name three fossil fuels.
   (b) Name a gas produced which leads to
       acid rain.
   (c) List three effects of acid rain on the
       environment.
   (d) List three things we can do to reduce the
       problem of acid rain.

Questions

Examination Questions

13 Fossil fuels are burnt to provide energy to generate electricity.
  (a) Give the name or formula of a compound of sulfur formed when a sulfur containing fossil fuel burns in air.
  (b) Acid rain is formed when this sulfur compound dissolves in and reacts with water in the atmosphere. Describe the effect of acid rain on limestone.
  (c) How would you show that water contains dissolved solids?
      (JC, HL, 2006)

14 Sulfur dioxide, SO_2, emissions are environmentally harmful.
  (a) Identify a problem caused to the environment by SO_2 emissions.
  (b) Identify a specific activity which gives rise to SO_2 emissions.
      (JC, HL, Sample Paper)

15 Hydrocarbons are important fuels.
  (a) Give an example of a fuel that is a hydrocarbon.
  (b) Carbon and what other element are always present in hydrocarbons?
      (JC, OL, Sample Paper)

CHEMISTRY

Chapter 31 ⚬⚬⚬

PLASTICS

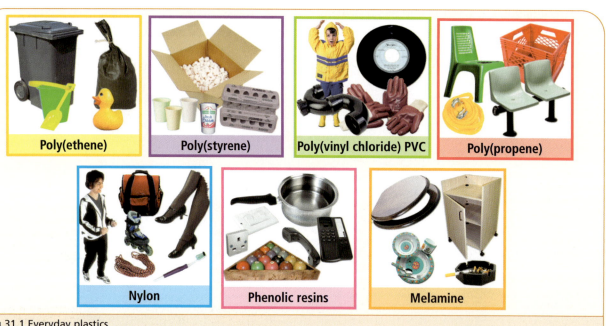

Poly(ethene)   **Poly(styrene)**   **Poly(vinyl chloride) PVC**   **Poly(propene)**

Nylon   **Phenolic resins**   **Melamine**

Fig 31.1 Everyday plastics

Introduction

Plastics are man-made materials. Most are made from crude oil. The world's supply of crude oil will probably run out in your lifetime.

Think of all the materials you use every day that are made out of plastic.

It is hard to believe that plastics have only been made on a large scale since the 1930s. Think of alternative materials you could use to replace the plastics shown in Figure 31.1. What advantages do plastics have over the alternatives?

The origins of plastics

Crude oil is a thick, black, foul-smelling liquid. Oil in this form is not of much use! However, it consists of a mixture of hydrocarbons that are very useful. In an oil refinery, crude oil is separated into groups of hydrocarbons called fractions. Each fraction is separated on the basis of its boiling point. This is done on a large scale in fractionating columns.

The fractionating column is hottest at the bottom and gets gradually cooler towards the top. Larger molecules have higher boiling points and will condense near the bottom of the column, whereas smaller molecules liquify nearer the top. Approximately 4 per cent of these fractions are used to make plastics.

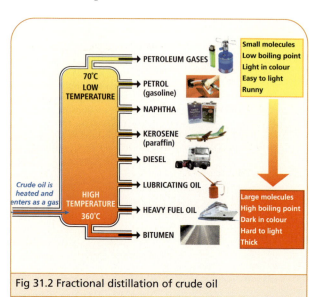

Fig 31.2 Fractional distillation of crude oil

►►► DEMONSTRATION

Distillation of crude oil

Note: This experiment must be performed in a fume cupboard.

Method

1 Set up the apparatus as shown.

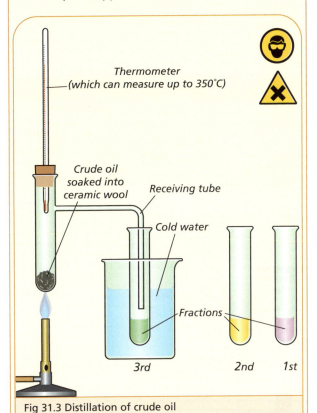

Thermometer (which can measure up to 350°C)

Crude oil soaked into ceramic wool

Receiving tube

Cold water

Fractions

3rd   2nd   1st

Fig 31.3 Distillation of crude oil

2 Heat the test tube gently at first and collect the distillate until the temperature reaches 70°C.

3 Change the receiving tube and heat the crude oil to 150°C. Collect the distillate between 70°C and 150°C.

4 Again change the receiving tube and heat the oil to 250°C. Collect fractions between 150°C and 250°C.

5 Pour a sample of each fraction onto a watch glass.

6 Note the colour and how thick each fraction is.

7 Light each fraction with a lighted splint and compare the ease with which each fraction burns and the amount of smoke produced.

Cracking

After distilling crude oil, oil companies always find they end up with too many larger molecules like heavy fuel oil and not enough smaller, more valuable ones, such as petrol and naptha.

Chemists have found a way to break up larger molecules into smaller ones by heating them over a catalyst. This is known as **catalytic cracking**.

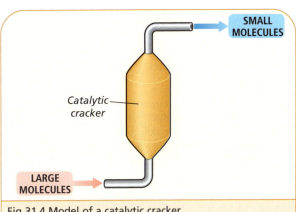

SMALL MOLECULES

Catalytic cracker

LARGE MOLECULES

Fig 31.4 Model of a catalytic cracker

A catalyst is a substance that alters the rate of a chemical reaction and is itself unchanged after the reaction.

Artificial skin is sprayed on an arm. The skin is made of a plastic that can be sprayed on cuts or burns and reduce scarring

CHEMISTRY

Making plastics

The manufacture of plastics from crude oil involves two stages.

1 Simple hydrocarbons are isolated from crude oil. Many of these are produced during cracking. They are known as **monomers** (*mono* = 1).
2 These monomers react together and form long chains called **polymers** (*poly* = many).

All plastics are made up of repeating units linked together to form polymers. This process is known as **polymerisation**.

> **Polymerisation** is the process involving the joining together of many small molecules called monomers to form a large molecule called a polymer.

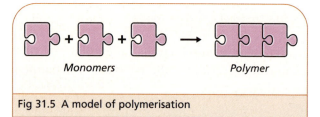

Monomers   Polymer

Fig 31.5 A model of polymerisation

Properties of plastics

Why are so many items made of plastic?

1 They are cheap to make.
2 They can be designed to have a variety of properties which suit their use.

For example:

● Polyethene (or polythene) can be made in high or low density form. Low density polythene is soft and flexible and can be used to make rubbish bags and plastic bottles. High density polythene is harder and rigid and is used in milk bottle crates and dustbins.
● Polystyrene is used in yoghurt pots, for example. Expanded polystyrene is made by simply blowing in some gas during the moulding stage. This plastic is used in fast food packaging, ceiling tiles and insulation. The trapped gas makes it a good insulator.

3 Plastics are easy to maintain as they do not corrode.
4 They can be moulded into almost any shape and size.

5 They are good insulators of heat and electricity. Hence, plastic is used as the insulating material around copper wire and in handles such as saucepans for cooking.
6 They can easily be dyed different colours.
7 They have low densities (are quite light), which makes them easy to work with. So, for example, PVC guttering is much lighter than metal guttering and it doesn't rust or rot.

Table 31.1		
Plastic	**Examples of use**	**Reasons for use**
Polythene	Plastic bags, fertiliser sacks, squeezy bottles, lunchboxes	Strong, cheap and flexible
Rigid PVC	Pipes and gutters, windowframes, floor tiles, curtain rails	Strong, long lasting, weather resistant
Flexible PVC	Insulation for electrical wires, hosepipes, shower curtains	Strong, long lasting, very flexible, electrical insulator
Polystyrene	Food containers, yoghurt pots, electrical components	Strong (but brittle), rigid, electrical insulator
Expanded polystyrene (aeroboard)	Packaging, house wall insulation, disposable coffee cups	Extremely low density, good heat insulator
Nylon	Tights, carpets, ropes, combs, brushes, gear wheels	Can be spun into fibres, hard-wearing, 'slippery'

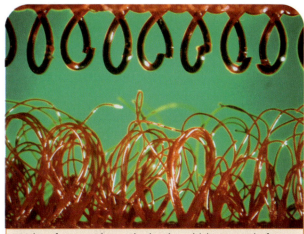
A Velcro fastener shows the hooks, which are made from nylon, and the loops made from cotton

A plastic TV screen, which is only 2 mm deep and can be viewed from any angle

CHEMISTRY

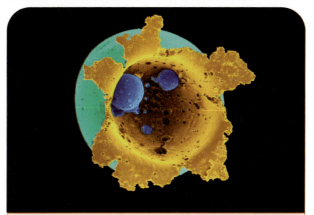
Drug delivery capsules are made from plastic. They encase chemicals and are designed to open in a particular part of the body

A biodegradable pen made of cornstarch, which is derived from maize plants

Environmental impact of plastics

- Plastic does not degrade or break down easily. This is an advantage, but it is also a problem when plastic is being disposed of.

Modern plastics, for example some plastic bags, are made from plants and are biodegradable. This means that over a relatively short period of time the bags will decompose.

In biodegradable plastics, starch is mixed through the plastic. As it absorbs water, the starch expands, causing the plastic to break into small fragments. This action provides a large surface area for bacteria in the soil to digest it.

- Many plastics are a fire risk. When they burn, poisonous fumes are created. PVC (polyvinylchloride) which is used in guttering, pipes and raincoats, releases poisonous hydrogen chloride when it burns. In house fires, many deaths are caused by poisonous fumes and smoke produced as plastic items burn.

Heating plastics

Plastics fall into two categories: thermoplastics, which soften when heated, and thermosetting plastics, which do not soften when heated.

Recycled plastic pellets will be used to make plastic sheeting for the construction industry

CHEMISTRY

Key Points

- Plastics are man-made materials, most being made from crude oil.

- All plastics are made up of units called monomers which join together to form polymers in a process called polymerisation.

- Crude oil is made up of hydrocarbons which can be separated into fractions of different boiling points.

- Plastics can be made quite cheaply, they don't corrode, and they can be moulded to any shape or size.

- Most plastics are non-biodegradable which means they remain intact for very long periods of time.

- Some plastics soften when heated and are called thermoplastics. Others do not soften on heating and are called thermosetting plastics.

The role of chemistry

To most people chemistry is what happens in a laboratory with test tubes, etc; or perhaps a chemist shop springs to mind. One official definition is: 'Chemistry is the study of the composition of substances and of their effect upon one another.'

Everything around us is made of chemicals. Without modern chemistry there would be no medicine, no CDs, no paint. The list goes on and on. Chemistry has had an effect on every area of our lives and will continue to improve our quality of life.

A knowledge of chemistry could lead to a career in medicine, engineering, pharmacy, biotechnology or food production. No wonder it is known as the creative science.

Questions

1 Most plastics are made from _____ _____.

2 Write out and complete the following:
   (a) _____ _____ is a mixture of hydrocarbons. A fractionating _____ is used to separate each _____.
   (b) The larger molecules tend to have _____ boiling points and condense at the _____ of the column, whereas the smaller molecules have _____ boiling points and liquefy near the _____ of the column.
   (c) _____ _____ is the technique whereby large hydrocarbon molecules are broken down into smaller ones which are more useful.

3 (a) What is a catalyst?
   (b) Give one example.
   (c) Would you expect most catalysts to speed up or slow down chemical reactions? Why?

4 Write out and complete the following:
   When a lot of small identical molecules called _____ join together they form a _____. This process is how all _____ are made and is called _____.

5 Geraldine distils a sample of crude oil and collects the first three fractions but gets them mixed up. Can you think of two ways she could find out the correct order of the fractions?

6 What are the advantages of plastic over the other materials in each of the following?
   (a) PVC guttering in place of iron
   (b) Plastic bowls in place of enamelled bowls
   (c) PVC windows in place of wooden windows
   (d) Plastic lens in glasses in place of glass lens
   (e) Plastic bottles instead of glass
   (f) Expanded polystyrene fast-food containers instead of cardboard containers.

7 Give two reasons why fragile items are often packaged in expanded polystyrene.

8 Describe how crude oil was formed. Briefly outline the steps involved in a petrol molecule's journey from crude oil to being used in a car.

Questions

9 Look at the diagram.

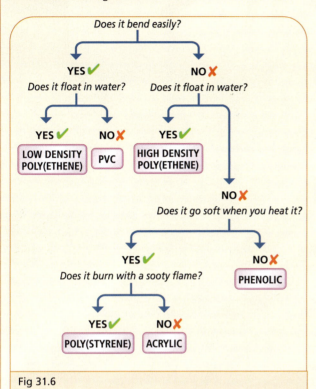

Fig 31.6

(a) How could you tell PVC from low density polythene?

(b) Which plastic will not bend easily or float in water but burns with a sooty flame?

(c) What four properties does the diagram tell you about acrylic?

(d) If a rigid plastic was found floating in a canal, what plastic from the diagram could it be?

(e) One type of polystyrene can float and another type cannot.
 (i) Why is this?
 (ii) What advantages has the type that floats?
 (iii) What could it be used for?

Examination Questions

10 Different plastics have different properties.

The dustpan and brush set shown is made from two different plastics. The bristles are made of type A and the other parts are made of type B plastic.

Give one property of type A and one property of type B plastic that make them suitable for their use in this product.

Fig 31.7

(JC, HL, 2006)

11 The picture shows plastic crates.

(a) Name the raw material used in the making of plastics.

Most plastics are non-biodegradable.

(b) Explain what is meant by the term non-biodegradable.
(Adapted from JC, OL, 2006)

12 Give one negative impact on the environment of the use of non-biodegradable plastics for packaging.

(JC, HL, Sample Paper)

PHYSICS

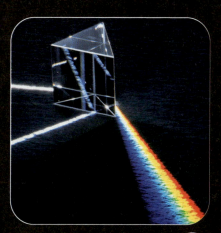

Chapter 32

MEASUREMENT

Introduction

Physics is a type of knowledge. It starts when we ask about what happens in the physical world around us. It grows when we make and discuss theories about what we see, hear and measure. These theories are then tested by experiment.

Physics is sometimes called the science of measurement. This older idea is not the complete picture. But it does make one thing clear: we can have no real description of things in physics if we cannot measure them.

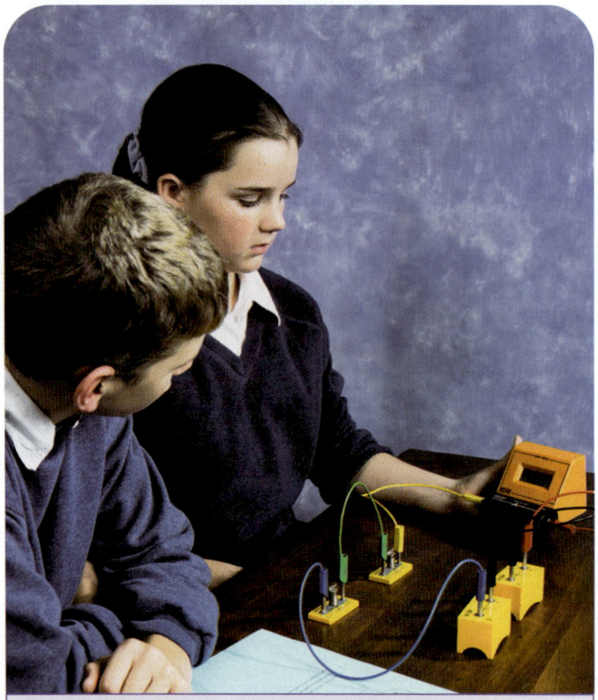

Physics is the science of careful measurement

Two things necessary for measurement

Measurement is important in all of science and in everyday life. It is important that a farmer can measure the size of a field. An engineer must measure the width of a river or the height of a building carefully.

Even a young child will try to measure the amount of liquid in a glass to make sure a brother, sister or friend does not get more! To compare how much liquid is in the glass, she will need to:

(a) measure it **accurately** using a measuring instrument and

(b) be able to describe the amount in units that are understood by others.

Measuring instruments and units of measurement

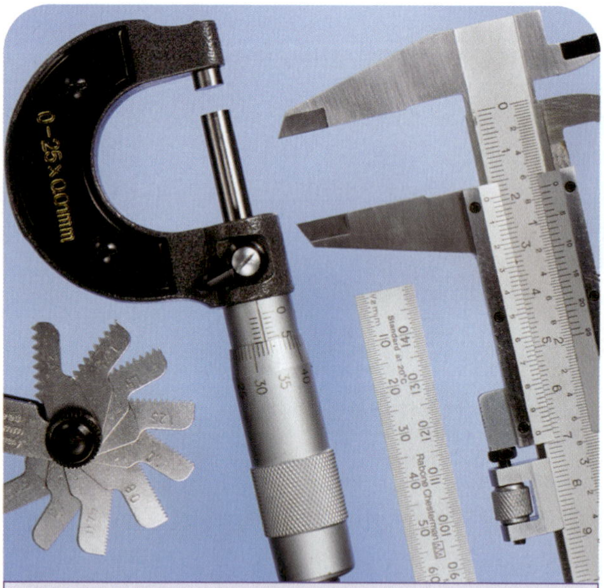

Different measuring devices, including a metric rule and Vernier callipers

There are measuring instruments for all kinds of measurements.

- Simple tapes or rulers measure length.
- Complicated sensors connected to large computers in order to measure the light from distant galaxies.

In every case, we must also be able to talk about our measurement to other people. To do this we use some basic quantities and standard units.

Distances in galaxies are measured by light years, which is the distance it takes light to travel in a year

Length

> **Length** is the straight-line distance between two points.

To measure length, for instance, we use the unit of 1 metre (m) or sometimes the unit of centimetres (cm). Scientists have agreed on the length of a metre. This is the standard unit used in physics.

The distance of 1 metre can be found on a metre stick in the school laboratory. It is quite a long length much longer than a pen, for instance. So when we are measuring smaller lengths we use fractions of 1 metre.

The usual fractions to use are the centimetre ($\frac{1}{100}$ part of a metre) or a millimetre (mm) ($\frac{1}{1000}$ part of a metre). When measuring much longer distances we use the kilometre (km) (1000 metres).

Other units for length

There are, of course, other units of measurement that are used in everyday life. Miles, yards, feet and inches are still used by many people. But in the future probably everybody in Europe will use the metric units.

Length is a basic quantity in physics. Using length we can build new quantities such as area and volume.

Measuring lengths, areas, volumes

Measuring the length of curves

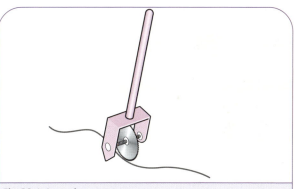

Fig 32.1 An opisometer

As we have said, length is measured in metres and one instrument used is the metre stick. However, the metre stick is useful only for measuring straight distances! How can we measure distances along a curve?

The instrument that is used for measuring small curved distances in the laboratory is called an **opisometer**. This is a wheel on a screw that acts as an axle for the wheel. When the wheel is moved over a curved line it moves up the screw. The distance is found by moving the wheel in a straight line along a metre stick until it is back in its original position.

A larger version of the opisometer is called the trundle wheel. It is used, for example, to measure distances along roads. This also has a wheel on an

PHYSICS

237

axis, but instead of a screw there is a counter which counts every turn of the wheel. The wheel is designed so that each turn of the wheel is 1 m.

ACTIVITY 32.1

To measure the length of a curved line

Method

1 Draw a smooth curved line on a piece of paper.

2 Move the wheel of the opisometer to the end of the screw.

3 'Roll' the opisometer carefully along the curved line from start to finish.

4 Take the opisometer from the paper at the end of the line and place it carefully on the zero mark of a metre stick.

5 'Roll' the opisometer in the opposite direction until the wheel is in its original position.

6 Read off the distance travelled along the metre stick.

Result

This reading gives you the length of the curved line.

Conclusion

The distance travelled along the metre stick is the length of the curved line.

Measuring circular objects

If we want to measure the diameter of a water pipe or a ball-bearing a metre stick is too clumsy. We need another instrument that gives a more accurate reading. Two common instruments used are:

● the Vernier callipers and

● the micrometer screwgauge.

A Vernier callipers looks like a pair of pincers which are put either side of the diameter. The distance between the callipers' ends can be measured from the scale on the side of the callipers.

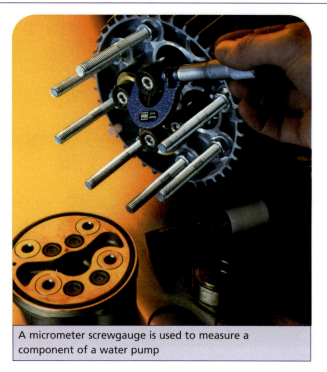

A micrometer screwgauge is used to measure a component of a water pump

Measuring area

Area is a quantity related to lengths.

> **Area** is the amount of surface that covers an object.

To measure the area of a square or rectangular object we multiply the length by the width. For instance, if the length of a matchbox is 5.2 cm and its width is 3.6 cm, then its area is:

$$5.2 \times 3.6 = 18.72 \text{ cm}^2$$

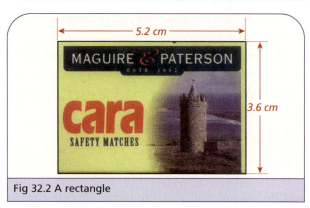

Fig 32.2 A rectangle

The standard unit for area is a 1 metre length by a 1 metre width = 1 square metre. The square metre is quite a large area.

Because a 1 metre length and a 1 metre width each contain 100 cm, the relationship between square centimetres and the square metre is:

$$100 \text{ cm} \times 100 \text{ cm} = 10\,000 \text{ cm}^2 = 1 \text{ m}^2$$

If we wish to find the area of a shape that does not have straight edges we need a different method. One way is to trace the shape onto graph paper which has squares of 1 cm² marked on it. We can then count the number of squares that were covered by the shape.

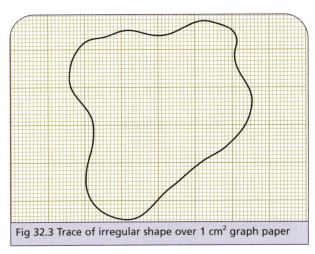

Fig 32.3 Trace of irregular shape over 1 cm² graph paper

Your graph paper is probably divided into even smaller squares of 1 mm². You can get quite an accurate estimate of the area of the shape by counting the number of square millimetres covered by it.

Measuring volume

Volume is a quantity also related to lengths.

> **Volume** is a measure of the space taken up by an object.

Measuring the volume of a cube or a rectangular box is relatively simple. We multiply the length by the width by the height of the object.

If the length of the matchbox is 5.2 cm, the width is 3.6 cm and the height is 1.5 cm, then the volume is

5.2 cm × 3.6 cm × 1.5 cm = 28.08 cm³

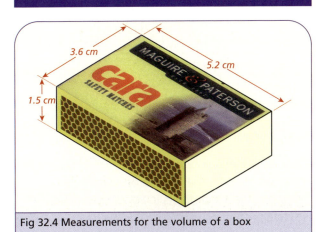

Fig 32.4 Measurements for the volume of a box

The standard unit for volume is the cubic metre (m³). This is a cube that has length 1 m, width 1 m and height 1 m. If you imagine a box of this size you will see that this unit is a large one.

We rarely see 1 m³ of any substance in a laboratory. More often we use cm³ or litres if we are working with liquids. One litre is equal to 1 000 cm³, or the other way around,

1 cm³ = 1 ml = ¹⁄₁₀₀₀ litre

Measuring the volume of an irregularly shaped object

What if an object does not have simple straight edges that make its shape? Then we need other methods to find its volume. This happens with liquids and gases.

If we wish to measure the volume of a liquid we must place it in a container since liquids (and gases) do not have a definite shape themselves.

One way of measuring the volume of a liquid is to use a graduated cylinder. This is a cylindrical container that has a scale in millilitres (ml) along the side.

Care must be taken when reading the volume of a liquid in a graduated cylinder. Most liquids when placed in the cylinder do not have a completely flat top surface.

At the sides of the cylinder some of the liquid sticks to the glass and is pulled upwards or downwards. This means that the surface at the top of the liquid is curved. This curved surface is called the meniscus.

When we read a volume in a graduated cylinder we take the reading from the bottom of the meniscus.

Other basic quantities in physics are mass, time and temperature.

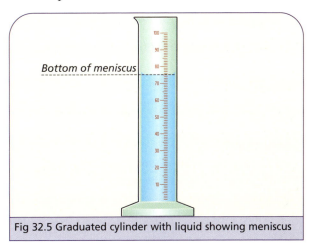

Bottom of meniscus

Fig 32.5 Graduated cylinder with liquid showing meniscus

PHYSICS

Liquids do not have their own shape

►►► ACTIVITY 32.2

To find the volume of an irregularly shaped object that (a) sinks in water (b) floats in water

(a) Method

1 Fill a graduated cylinder about half full of water so that the meniscus is clearly at one of the marks on the scale. Note this reading.

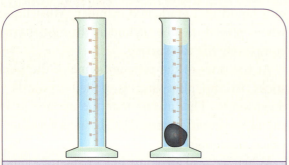

Fig 32.6 Measuring the volume of an irregularly shaped object

2 Lower the object into the graduated cylinder carefully. The object pushes the water up and the meniscus rises. Note the new reading of the meniscus.

3 Subtract the two readings on the graduated cylinder.

Result

The difference between the two readings gives the volume of the object.

Conclusion

We can measure the volume of an irregularly shaped object that sinks in water by using a graduated cylinder.

(b) Method

1 If the object floats in water or if it is too large to fit into a graduated cylinder, it is sometimes easier to use an overflow can to measure it. This is a can with a spout on one side. Fill the overflow can so that a small amount of water comes out of the spout. Wait for all of the water to run off. The water is then up to the edge of the spout.

2 Put a clean, dry graduated cylinder under the spout.

3 Place the object carefully into the overflow can. If the object floats it is necessary to get a pin and press the object below the water level.

4 The water displaced by the object flows out of the spout and into the graduated cylinder.

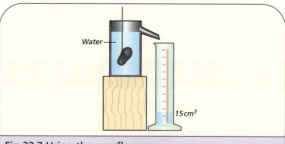

Fig 32.7 Using the overflow can

Result

The volume of water that overflows into the cylinder is equal to the volume of the object in the can.

Conclusion

We can measure the volume of an irregularly shaped object that floats in water by using an overflow can.

PHYSICS

Mass

> Mass is the amount of matter in an object. This amount never changes.

The standard unit for mass is the kilogram (kg), but sometimes for small objects we also use the unit of the gram (g). The mass of a body is linked to its weight. This means that we can compare masses by comparing weights.

One instrument used to measure mass is the mass balance. In modern laboratories these are very sensitive instruments. They give quite an accurate reading to a number of decimal places.

Mass is measured using a very sensitive balance

Mass is not the same as weight: a 1 kilogram mass reads 10 newtons when weighed

Time

> Time is measured by a basic unit called the second (symbol s).

The second is a short interval of time. It is based on the movement of electrons in an atom. In the laboratory we usually measure time using a stop watch or an electronic timer.

Sometimes the unit of a second is too small and readings are taken in minutes, hours, days or even years. We must sometimes change larger units into seconds by multiplying. For instance:

- 3 minutes is $3 \times 60 = 180$ s
- 5 hours is $5 \times 60 \times 60 = 18\,000$ s

Temperature

> Temperature is a measure of how hot or cold an object is.

The usual unit used to measure temperature is the degree Celsius (°C). The unit is based on the freezing point and boiling point of water. The difference between these two temperatures is divided into one hundred units to give the degree Celsius.

Many modern measuring instruments are digital

Measuring temperature

Temperature is usually measured in the laboratory using a thermometer. The most common thermometer in school laboratories has a bulb of coloured liquid, usually alcohol, with a column of liquid above the bulb.

When the bulb becomes more hot or cold, the column of liquid above it rises or falls. The temperature is read off a scale on the side of the column of liquid.

PHYSICS

241

▶▶▶ ACTIVITY 32.3

To measure the temperature of a range of different materials

The materials may include:

● iced water
● boiling water
● the air in the laboratory
● your hand
● a pane of window glass.

In all cases it is important to remember that it is the temperature of the bulb that is being measured. So it is the bulb that must be touching the material.

For instance, if you have your hand on the bulb of the thermometer when you are trying to measure the temperature of the air in the room, you will be measuring the temperature of your hand and not the temperature of the room!

If you are measuring the temperature of boiling water, be careful to put the thermometer into the water at an angle. You could scald your hand if it is in the steam.

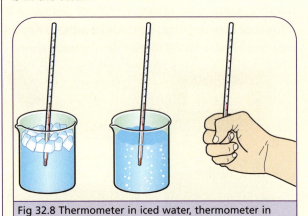

Fig 32.8 Thermometer in iced water, thermometer in boiling water, thermometer in the hand

Data presentation

As in many experiments in physics it is sometimes useful to present your results in a table:

Table 32.1 Temperature readings of different materials	
Material	**Temperature reading on thermometer/°C**
Iced water	
Boiling water	
Human hand	
Pane of window glass	

We can summarise the basic quantities, units and measuring instruments in a table.

Table 32.2 Basic quantities, units and measuring instruments			
Basic physical quantity	**Standard unit**	**Symbol**	**Measuring instruments used in the laboratory**
Length	Metre	m	Metre stick or callipers
Mass	Kilogram	kg	Weighing scales, balance
Time	Second	s	Stop watch, electronic timer
Temperature	Degree Celsius	°C	Thermometer

Nearly every other measurement in physics, as we shall see, is based on these basic quantities.

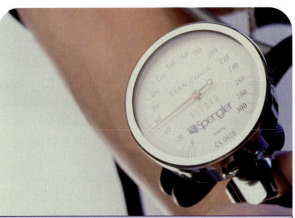

Measurement is also important in health and medicine; measuring blood pressure

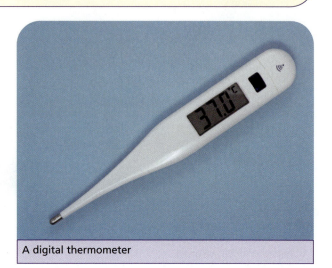

A digital thermometer

PHYSICS

Key Points

- In order to do physics we must be able to (a) measure quantities and (b) communicate these measurements using units.

- The basic quantities in physics are: length, mass, time and temperature.

- The standard units for each of these quantities are the metre, the kilogram, the second and the degree Celsius.

- Area and volume are quantities based on lengths. The standard unit of length is 1 m² and the standard unit of volume is 1 m³.

- Sometimes the standard units are too big or too small. We then use centimetres or millimetres for small lengths and kilometres for large distances. Similarly, we sometimes use cm² or cm³ for area and volumes.

- There are a number of instruments and methods used in the laboratory to measure lengths, areas, volumes, mass, time and temperature depending on the shape and nature of the material.

Questions

1 State four basic measurements in physics.

2 Define the terms mass and temperature.

3 Name the instruments used to measure:
   (a) a curved line
   (b) the diameter of a marble.

4 What are the standard units of:
   (a) area?
   (b) volume?
   (c) temperature?

5 Complete the following table:

Table 32.3		
1 m	=	_____ cm
1 m²	=	_____ cm²
1 litre	=	_____ cm³
1 km	=	_____ m
1 hour	=	_____ s
1 kg	=	_____ g

6 Calculate the following:
   (a) the area of a desk of length 80 cm and width 60 cm
   (b) the area of a window with width 1.5 m and height 1.2 m
   (c) the volume of a cube of side 7 cm
   (d) the volume of water in a right-angled tank of length 1.3 m, width 1.1 m and height 80 cm.
   (**Hint**: Make sure your units are the same!)

7 Describe, using a diagram, how you would measure the volume of a glass marble.

8 Write out and complete the following:
   An _____ is a instrument that is used to measure the length of a curved line.
   A _____ _____ is an instrument used to measure the distance along a road.
   One kilometre is the same as _____ metres and there are _____ cm in a metre.
   The volume of an object is 0.01 m³. This is the same as _____ cm³.
   The measure of how hot or cold an object is called its _____ and has the unit _____. In these units water boils at _____ and freezes at _____.

9 The moon circles the earth in about 28 days. What is 28 days expressed in seconds?

? Questions

10 (a) In Figure 32.9 what is the volume of water in X?

(b) What is the name of the curved surface at the top of the water?

(c) A stone is introduced and the water level rises. This is shown in Y. Use the two water level readings to find the volume of the stone.

(d) What apparatus could you use if the stone was too big to fit into a graduated cylinder?

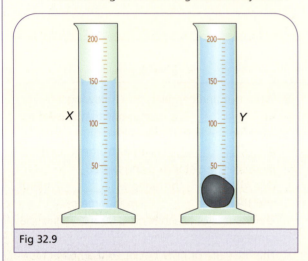

Fig 32.9

11 You are asked to find the volume of a cup by designing an experiment.

(a) Describe two different experiments you might use to get an answer.

(b) What equipment would you use in each experiment?

(c) Which experiment do you think would give a more accurate answer? Why?

(d) Name one thing you might do in both experiments to get a better answer.

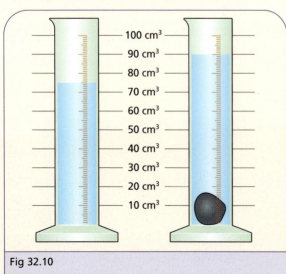

Fig 32.10

Examination Questions

12 (a) The diagram shows a piece of equipment, labelled **A**, containing water. **Name A.**

(b) A stone was then added and a new volume was recorded as shown in B. What was the volume of the stone in cm³? (JC, OL, 2006)

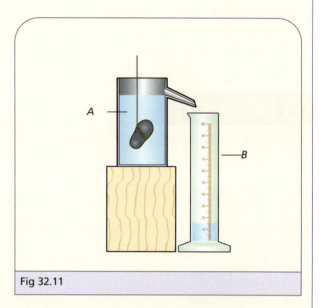

Fig 32.11

13 (a) Find the area of the rectangle drawn below using the measurements given.

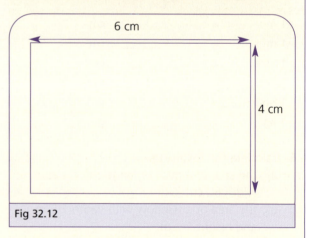

Fig 32.12

(b) In what unit is the area measured? (JC, OL, 2006)

Chapter 33

DENSITY AND FLOTATION

Introduction

In Chapter 32 we learned how the mass and the volume of an object are measured. We can build on this knowledge to find an important connection between mass and volume.

Table 33.1 gives the mass and the volume of two cubes made of different materials.

Table 33.1 Mass and volume of two cubes

Cube of material	Volume (cm³)	Mass (g)
A	30	15
B	30	45

We can see that the volume of both cubes is the same. But the masses are different. Mass measures the amount of material in an object. We see that there is more mass in a volume of B than there is in the same volume of A. This is the idea of density.

Density

> **Density** is the mass of each cubic metre of a material.

The standard unit of density is the number of kilograms (kg) in 1 cubic metre (m) of the material. This is written kg/m³. Mostly we use the unit g/cm³. Table 33.2 shows some substances and their densities.

Table 33.2 Substances and their densities

Substance	Density in kg/m³
Water	1 000
Paraffin oil	800
Ice	900
Brass	8 500
Gold	19 300
Mercury	13 600
Aluminium	2 700

Calculating densities

We can find the density of a material if we know its mass and its volume. If the mass of 3 m³ of water is 3 000 kg, then the mass of one cubic metre is

$$\frac{3\,000 \text{ kg}}{3\text{m}^3} = 1\,000 \text{ kg/m}^3$$

which tells us that the density of water is 1 000 kg/m³.

Given the mass of any volume of a material we can find the density by dividing mass by volume:

$$density = \frac{mass}{volume}$$

This relationship can be written in a shorter form as:

$$D = \frac{M}{V}$$

Using the rules for equations in mathematics, we can multiply both sides of the equation by V to get

$$M = DV$$

which gives us a way of finding the mass of a piece of material if we know its density and volume.

Similarly, we could divide both sides by D to get:

$$V = \frac{M}{D}$$

which gives us a way of finding the volume if we know the density and the mass.

> This triangle can help you to remember how to use the formula for density, mass and volume. Cover the quantity you want and you will be left with its formula, e.g. cover DENSITY and you have MASS/VOLUME. This is a memory aid only. You should also understand the use of the formula!
>

Huge icebergs can float on water

 MANDATORY ACTIVITY 21A

To determine the density of a rectangular block of material

Method

1 Find the mass of the block by placing it on a laboratory balance and note the reading.

2 Measure the length (l), width (w) and height (h) of the block.

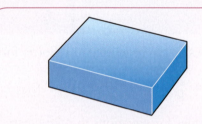

Fig 33.1 Finding the length, width and height of a rectangular block

3 Find the volume of the block using the relationship $V = l \times w \times h$

4 Use the relationship:

$$\text{density} = \frac{\text{mass}}{\text{volume}}$$

Result

This gives you the density of the block.

5 Repeat the experiment on blocks made of different materials.

6 Remember to write a unit of density after your result.

Conclusion

The density of a rectangular object is found by using the relationship:

$$\text{density} = \frac{\text{mass}}{\text{volume}}$$

 MANDATORY ACTIVITY 21B

To find the density of an irregular solid

Method

1 Find the mass of the solid in the usual way using the balance.

2 To find the volume of the irregular solid use the method of the graduated cylinder or the overflow can described in Chapter 32.

3 Use the relationship:

$$\text{density} = \frac{\text{mass}}{\text{volume}}$$

Result

This gives you the density of the irregular solid.

Conclusion

The density of an irregular solid is found by using the relationship:

$$\text{density} = \frac{\text{mass}}{\text{volume}}$$

MANDATORY ACTIVITY 21C

To find the density of a liquid

Method

1 Find the mass of a clean, dry beaker by placing it on a balance (m_1).

2 Transfer a known volume of the liquid to the beaker using a pipette or a burette and write down the mass of the beaker + liquid (m_2).

3 Subtract the two readings to find the mass of the known volume of the liquid ($m_2 - m_1$).

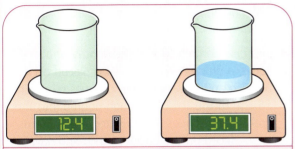

Fig 33.2 Finding the mass of 25 cm³ of a liquid

4 Use the relationship: density $= \dfrac{\text{mass}}{\text{volume}}$

Result

This gives you the density of the liquid.

Conclusion

The density of a liquid is found by using the relationship:

$$\text{density} = \frac{\text{mass}}{\text{volume}}$$

Table of results for activities 21A–21C

It may be useful to put your results into a table with the following headings:

Object	Mass	Volume	Density
Rectangular			
Block			
Irregular			
Solid			
Liquid			

Flotation

We know from experience that some things float while others sink in liquids. We also know that whether a body floats or not does not depend on its size (or volume). Huge ships can float on water while small stones will sink to the bottom. Density explains what is going on.

A substance that floats in one liquid may not float in another. For example, an ice cube will float on water but will sink in paraffin oil. This happens because the ice cube is less dense than the water but more dense than the oil.

> A solid will float in a liquid if the solid is less dense than the liquid. It will sink if it is denser than the liquid.

We can use the connection between flotation and density to say something about the density of liquids. In Figure 33.3 the same round object was placed in two liquids. In the first case the object sank to the bottom, but in the second liquid the object floated.

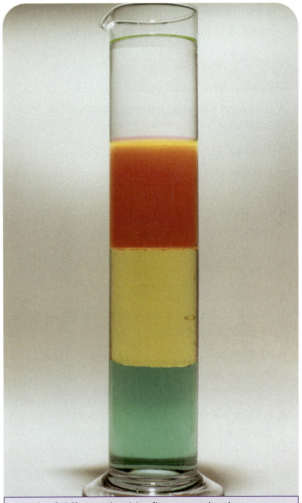

Liquids of different densities float on each other

PHYSICS

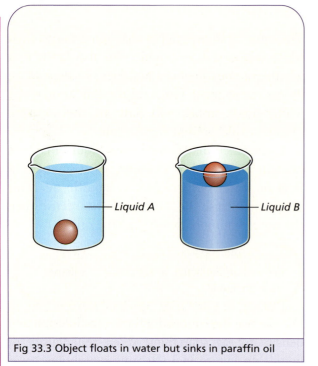

Fig 33.3 Object floats in water but sinks in paraffin oil

- In the first case the object sank because it is more dense than the liquid.
- In the second case the object floats because it is less dense than the liquid. We can say that liquid A is less dense than liquid B.

 Key Points

- Density is the mass per unit volume of a substance.
- The unit of density is kg/m³ or g/cm³.
- Density = $\dfrac{mass}{volume}$ or $D = \dfrac{M}{V}$
- Using the rules of mathematics this relationship can be written as $M = DV$ or $V = \dfrac{M}{D}$
- To find the density of a solid we find the mass from a balance and the volume by calculation or by displacing water.
- To find the density of a liquid we find the mass by subtracting the mass of a beaker from the mass of the beaker and liquid.
- A body will float in a liquid if it is less dense than the liquid; otherwise it will sink.

? Questions

1 (a) What is meant by density?
  (b) Name a unit of density.

2 Find the density of each of the following:
  (a) a rectangular block of wood of length 7 cm, width 5 cm and height 2 cm which has a mass of 35 g
  (b) a piece of polystyrene of length 1.5 m, width 1.2 m, height 0.1 m and a mass of 3.6 g
  (c) 25 cm³ of mercury with a mass of 340 g.

3 Describe an experiment to find the density of an irregularly shaped solid object.

4 In an experiment to find the density of a liquid, a graduated cylinder was placed on a balance and a reading of 150 g was recorded. 25 cm³ of the liquid was then poured into the graduated cylinder. The cylinder and contents were placed on the balance again and gave a reading of 175.5 g. Using this information find the density of the liquid.

5 Write out and complete the following:
  (a) The density of a substance is the _____ of _____ cubic metre of the substance. The standard unit of density is _____ but sometimes we use the units _____.
  (b) To find the density of a substance we use the relationship _____, which means that we must know or measure both the _____ and the _____ of the substance.
  (c) When measuring the density of a liquid we cannot find its _____ directly from a balance since a liquid does not have a definite _____. For this reason we have to place a known _____ of the liquid in a _____ whose mass we have already measured.

6 Use Table 33.2 to find out which has the greatest mass: 45 g of aluminium or 20 g of brass?

7 A stone of density 3.5 g/cm³ has a mass of 105 g. What is its volume?

8 Explain why a brass weight will float in mercury but sink in water.

Questions

9 Figure 33.4 shows three liquids A, B and C which do not mix in a graduated cylinder. If the liquid B is water suggest what the density of A and C might be.

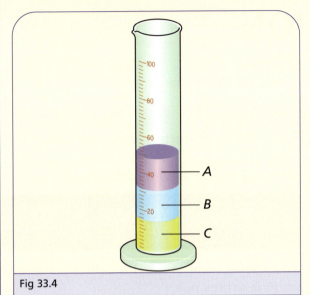

Fig 33.4

10 Two identical round objects are placed in liquids A and B as in Figure 33.5. Which of the liquids has the greater density? Give a reason for your answer.

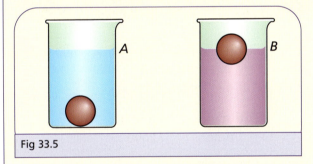

Fig 33.5

11 Figure 33.6 shows three solids floating in water. Which solid has the lowest density? Give a reason for your answer.

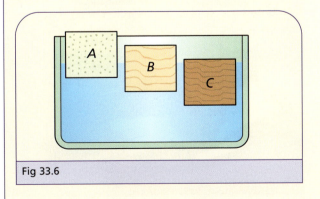

Fig 33.6

12 You are given three liquids – cooking oil, washing-up liquid and vinegar. You are asked to design an experiment to find out which liquid has the greatest density and which has the smallest density.
  (a) Do you need to calculate each density to know which has the greatest and which has the smallest density? Why?
  (b) Describe an experiment to get the result you want without calculations.
  (c) How might you present the results of your experiment?

Examination Question

13 A block of stone rests on a floor as shown in the diagram. The mass of the stone is 180 kg. The dimensions of the block are given.

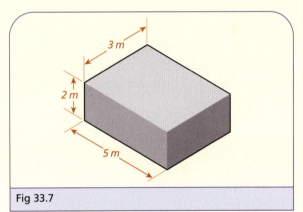

Fig 33.7

  (a) What is the density of the stone in kgm^{-3}?
  (b) Calculate the weight of the stone. In what units is weight measured? (Assume the acceleration due to gravity is 10 ms^{-2}.)
  (JC, OL, Sample Paper)

PHYSICS

Chapter 34

MOTION

Introduction

Motion is one of the clearest signs of life. Something is in motion if it changes its position over a period of time.

There are many possible ways of describing motion. In physics we start with only two of our fundamental quantities from Chapter 32. These quantities are **distance** (length) and **time**. Using these quantities we build new quantities to describe motion.

Some objects when they are moving cover a greater distance than others in a given time. A car, for instance, may travel 60 km in 1 hour, but a satellite circling the Earth may cover a distance of 17 000 km in 1 hour!

Some things move so fast they become blurred

Speed

In order to measure this difference we use the idea of speed.

Speed is the distance travelled by an object in one unit of time.

Example

If a car travels a distance of 400 m in 16 seconds then its speed is:

$$\frac{400}{16} = 25 \text{ metres per second (25 m/s)}$$

We can write this as a formula:

$$speed = \frac{distance}{time}$$

The standard unit of speed is metres per second which is written m/s.

To get the speed of the car in the example above in standard units we change the kilometres to metres and the hours to seconds.

Remember

1 kilometre = 1 000 m, so 400 km is 400 000 m

1 hour = 60 minutes = 60 × 60 = 3 600 seconds

6 hours = 6 × 3 600 = 21 600 s

$$\text{Speed} = \frac{distance}{time} = \frac{400\,000}{21\,600} = 18.5 \text{ m/s}$$

using just one place of decimals.

Racing cars cover many metres per second

PHYSICS

Acceleration

Let's look at a journey between Kerry and Dublin. If a car has an average speed of 18.5 m/s on the journey, what does this mean? It does not mean that it is always moving over 18.5 metres every second of the journey. The car would travel at greater speed on a good clear road or on a motorway, but would travel more slowly going through a town. So, on a particular journey the speed will change. Acceleration is the physical quantity that measures a change in speed.

> **Acceleration** is the change in speed divided by the time taken for the change.

Example

The speed of a car changes from 10 m/s to 20 m/s in 5 seconds. Find its acceleration.

$$\text{Acceleration} = \frac{\text{change in speed}}{\text{time taken for the change}}$$

$$= \frac{(20-10)\ \text{m/s}}{5\ \text{s}}$$

$$= \frac{10\ \text{m/s}}{5\ \text{s}}$$

$$= 2\ \text{m/s/s} = 2\ \text{m/s}^2$$

This means that the speed of the car is increasing by 2 m/s every second. This is usually written 2 m/s² and spoken as 'two metres per second squared'.

Calculations using recorded data

Sometimes we are given data from an experiment carried out on the motion of a body and we are asked to use the data to find speed and acceleration. This job often includes drawing a graph to show the data and then using the graph to make calculations.

Distance–time graphs

The distance that a cyclist travels from a starting point is measured every 4 seconds. The data is recorded in Table 34.1.

Table 34.1 Distance, measured every 4 seconds	
Distance/m	Time/s
0	0
10	4
20	8
30	12
40	16
50	20
60	24
70	28

A cyclist changes speed going up and down mountains

PHYSICS

251

We can illustrate this data on a graph. We put the time on the x-axis and the distance on the y-axis. This means that we plot the points (0, 0), (4, 10), (8, 20) and so on. The result can be seen in Figure 34.1.

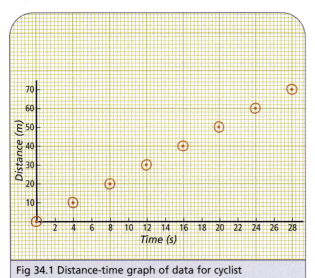

Fig 34.1 Distance-time graph of data for cyclist

We can see that the points on the graph belong to a straight line. This is not always the case, but finding the speed is much easier if the points are in a straight line.

To find the speed we remember the definition:

$$\text{speed} = \frac{\text{distance}}{\text{time}}$$

and take, for example, the distance travelled in the first 8 seconds of the motion. The speed is:

$$\text{speed} = \frac{20}{8} = 2.5 \text{ m/s}$$

If we take the data for the first 20 seconds of the motion we can also find the speed:

$$\text{speed} = \frac{50}{20} = 2.5 \text{ m/s}$$

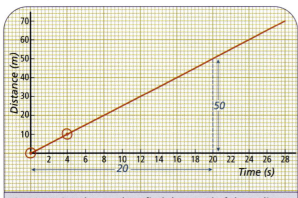

Fig 34.2 Using the graph to find the speed of the cyclist

We see that the speed is the same at every point in the journey. In fact, no matter which data points are taken on the graph the speed is always constant. This happens when the data lies on a straight line. The calculation for the speed will be familiar to students who have done some co-ordinate geometry in mathematics as the slope of the line.

Distance-time curves

If the graph is not a straight line then the speed must be changing. So it does not make sense to use two points to find the speed of the body. The body has a different speed at every point!

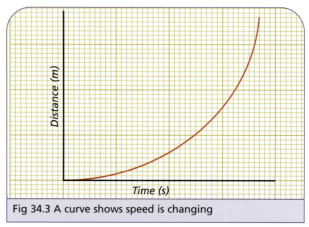

Fig 34.3 A curve shows speed is changing

In any case, the graph can also be used to:

(a) predict the **distance** travelled by the cyclist in a certain time or

(b) the **time** taken to travel a certain distance.

(a) To find out how far the cyclist travelled in 17 seconds, we go along the time axis (x-axis) to 17, go up to the line and then across to the distance axis (y-axis) to read off the distance.

(b) To find out how long it takes the cyclist to cover a distance of 65 m, go up the y-axis (distance axis) to 65, go across to the line and then vertically down to the x-axis to read off the time taken.

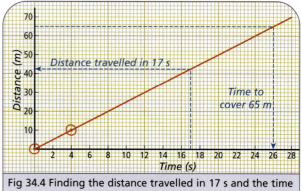

Fig 34.4 Finding the distance travelled in 17 s and the time taken to cover a distance of 65 m

Speed-time graphs

These graphs look very like the distance-time graphs in Figures 34.1 and 34.2, but they have a very different meaning. The two types of graphs should not be confused. On a speed-time graph the x-axis represents the time, but the vertical y-axis shows the speed of the object.

Because of this difference, the information that we can obtain from the graph is not the same as for the distance-time graph.

For example, Table 34.2 gives the data on the speed of a car, starting from rest, at intervals of 2 seconds.

Table 34.2 Speed, measured every 2 seconds	
Time/s	Speed/m/s
0	0
2	4
4	8
6	12
8	16
10	20
12	24

To draw the speed-time graph we plot the points with co-ordinates (0,0), (2,4), (4,8) and so on.

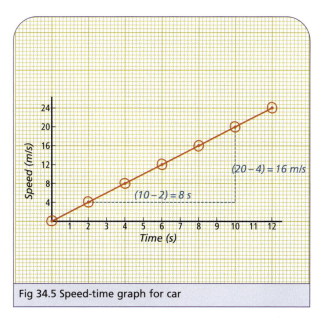

Fig 34.5 Speed-time graph for car

What kind of information can we get from a speed-time graph? The definition of acceleration says it is the change in speed over an interval of time. We can find the acceleration from the graph by taking any two points along our line and finding the change in speed during an interval of time.

Calculating acceleration from the graph

For instance, if we take the points (2, 4) and (10, 20) they represent a speed of 4 m/s^2 after 2 seconds and a speed of 20 m/s after 10 seconds. We can then measure the acceleration as follows:

$$\text{Acceleration} = \frac{\text{change in speed}}{\text{time taken for the change}}$$

$$= \frac{(20-4)\ m/s}{8\ s}$$

$$= \frac{16\ m/s}{8\ s}$$

$$= 2\ m/s^2$$

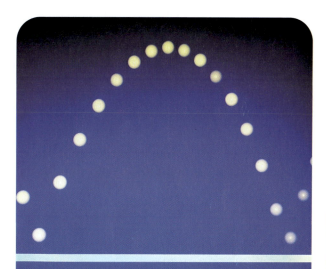

A high-speed camera can show the path of a bouncing ball

If the speed-time graph is a straight line, as in this example, it does not matter which points you take in order to find the acceleration. It will always be the same. In this case you will get an acceleration of 2 m/s^2.

If the graph is not a straight line, then things are a little more complicated and we need more advanced mathematics to find the acceleration. If you continue to study physics, you will learn how to do this.

Speed-time graphs can sometimes look a little strange. Examine the two graphs in Figure 34.6.

Graph A

In graph A, the graph is a horizontal line going through the value 20 m/s on the speed axis. What can we say about the motion from this graph? The horizontal line means that the speed stays at the value 20 m/s no matter what the value of the time is. This means that the object is moving at a constant speed of 20 m/s. We can also say that during this motion the body has no acceleration. Why?

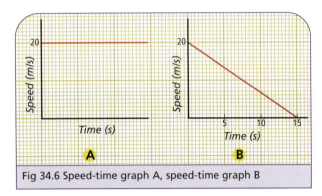

Fig 34.6 Speed-time graph A, speed-time graph B

Graph B

In graph B, we have a straight line, but it starts on the speed axis at the point (0, 20) and ends on the *x*-axis at the point (15, 0). We interpret this information as follows:

At the beginning of the motion (when *t* = 0), the body had a speed of 20 m/s. Over the next 15 seconds the speed of the object dropped to zero. We say the body has a negative acceleration or a deceleration. Since the graph is a straight line, we can measure this in the usual way.

Calculating deceleration from the graph

$$\text{Acceleration} = \frac{\text{change in speed}}{\text{time taken for the change}}$$

$$= \frac{(0-20) \text{ m/s}}{15 \text{ s}}$$

$$= \frac{-20 \text{ m/s}}{15 \text{ s}}$$

$$= -1.3 \text{ m/s}^2$$

Velocity

Velocity is a quantity in physics that needs two pieces of information to describe it.

> **Velocity** is the speed of a body and the direction in which it is moving.

A photo-finish shows distances in very short times

The units of velocity are also m/s in a certain direction.

When you are giving the velocity of a body you need to specify the speed and the direction in which it is going. For example, the velocity of a yacht might be 5 km/hr, east. From this we can see that speed is one part of velocity.

A change in direction is a change in velocity

PHYSICS

Key Points

- Speed is distance travelled in a unit of time. It is calculated by dividing the distance by the time taken. The standard unit of speed is m/s.

- Acceleration is the change in speed in a unit of time. It is calculated by dividing the change in speed by the time taken for the change. The standard unit of acceleration is m/s^2.

- A distance time graph shows time on the x-axis and distance travelled on the y-axis. If the graph is a straight line we can find the speed from the graph by dividing distance by time.

- A speed-time graph has time on the x-axis and velocity on the y-axis. If the graph is a straight line we can find the acceleration from the graph by dividing a change in speed by the time taken for the change.

- Velocity is the speed of a body and the direction in which it is going.

Questions

1 (a) What is meant by speed?
   (b) What are the standard units for speed?

2 (a) What is meant by acceleration?
   (b) What are the standard units of acceleration?

3 Find the speed in m/s in each of the following cases:
   (a) a cyclist who travels 300 m in 25 seconds
   (b) a swimmer who swims 50 m in 90 seconds
   (c) a sprint runner who runs 100 m in 11.2 seconds
   (d) a snail who takes 120 seconds to travel 50 cm.

4 Find the acceleration in each of the following cases:
   (a) a car whose speed changes from 6 m/s to 24 m/s in 6 seconds
   (b) a motorbike whose speed changes from 6 m/s to 40 m/s in 10 seconds
   (c) a parachutist whose velocity changes from zero to 30 m/s in 3 seconds
   (d) a cheetah who goes from 15 m/s to 25 m/s in 2 seconds.

5 A car travels 36 km in half an hour. What is the speed of the car in m/s? The car then increases its velocity to 35 m/s in the next 5 seconds. What is its acceleration?

6 A cyclist is moving in a straight line. The distance travelled by the cyclist each second is given in Table 34.3.

Table 34.3	
Time (s)	Distance (m)
1	3
2	6
3	9
4	12
5	15
6	18

Draw a graph of these data, putting the time on the x-axis.

From the graph find the following:
(a) the distance travelled in 4.5 seconds
(b) the time taken to travel 7 m
(c) the average speed of the cyclist after 3 seconds
(d) the average speed of the cyclist after 6 seconds

State whether you think the cyclist is accelerating. Give a reason for your answer.

PHYSICS

Questions

7 A stone is dropped from the top of a cliff. Table 34.4 gives the velocity of the stone at each second.

Table 34.4	
Time (s)	**Velocity (m/s)**
0	0
1	10
2	20
3	30
4	40
5	50

On a piece of graph paper, draw a graph of velocity against time. Put time on the *x*-axis. Use the graph to find:
(a) the velocity of the stone after 3.5 seconds
(b) the time it takes the stone to reach a velocity of 25 m/s.

8 A car is travelling in a particular direction along a straight road. The graph in Figure 34.7 shows the velocity of the car over the first 10 seconds of the journey. Use the information on the graph to obtain:
(a) the velocity of the car after 10 seconds
(b) the time taken for the car to reach a velocity of 5 m/s
(c) the acceleration of the car.

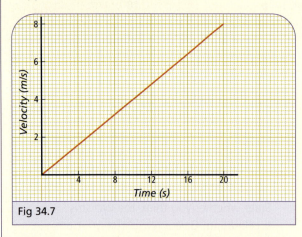

Fig 34.7

9 The graph in Figure 34.8 is a speed-time graph for a train journey. Write out and complete the description of the journey.
The train leaves Heuston Station and accelerates from _____ to a speed of _____ in the first _____ seconds. It then continues the journey at _____ for the next _____ minutes. When it approaches Kildare the train _____ from a speed of _____ to _____ in a time of _____.

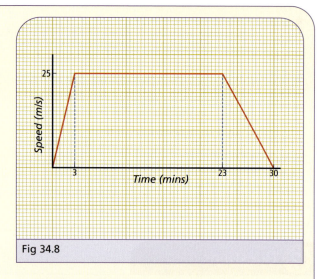

Fig 34.8

10 Dublin is west of Holyhead, and Kerry is south-west of Dublin. Using the points of the compass describe the velocity of each of the following:
(a) a car ferry travelling at 40 km/hr from Dublin to Holyhead
(b) the same car ferry travelling at the same speed from Holyhead to Dublin
(c) a plane travelling at 200 km/hr from Dublin to Farranfore in Kerry
(d) the same plane travelling at 250 km/hr from Farranfore to Dublin.

11 A friend says that she thinks that when a ball is dropped from a height the height of the bounce depends on the height it was dropped from.
(a) Describe an experiment to test your friend's idea.
(b) What measurements would you need to make?
(c) What would you have to do to make sure that the experiment was fair?
(d) What, do you think, would be the best way to present the data you collect from the experiment?

PHYSICS

Chapter 35 •••

FORCE AND MOTION

Introduction

The whole universe is in motion. All kinds of things move on the surface of the Earth. Moons travel around planets, planets travel around stars, and galaxies of stars move through space.

If a body at rest begins to move, something is making it change. Also, if a body is already moving in a particular direction at a certain speed and then changes its speed or direction, there must be a cause for this change.

Orion's Belt, an example of an object at a great distance from the Earth

Force is the idea used to describe the cause of motion or a change in the motion of a body.

As we have seen, motion is described in physics by using the idea of velocity. The velocity of a body is its speed and the direction it is going.

So we can say that forces:

• cause a resting body to gain velocity or
• cause a moving body to change its velocity.

Newton's three laws

The scientist Isaac Newton (1642–1727) wrote a famous book about forces. In it he wrote down three laws for the motion of a body. These are still used today to work out the motion of bodies. We study these three laws later in physics.

In honour of the work of Newton,

The unit used to measure force is called the **newton** and has the symbol N.

Newton is the unit of force.

A horseshoe magnet demonstrates magnetic forces

Types of force and their effects

Scientists have identified many kinds of force.

- If a ball is placed at the top of a slope and released it will begin to move. It gets faster as it goes down the slope. This is caused by the **force of gravity**.
- If a magnet is brought close to iron filings at rest on a piece of paper they will move towards the magnet. This is called a **magnetic force**.
- **Electrical forces** cause charged particles to move in a circuit, this is electrical current.
- **Frictional forces** are used in the brakes of cars and bicycles to slow down or stop.
- **Elastic forces** in springs can be used to provide comfort.
- **Tensile forces** are present in nylon ropes that hold the weight of mountain climbers.

Gravity acts equally on an apple and a feather, but this can be seen only in a vacuum

Motion without force

Is it possible to have motion without a force? Consider a ball rolling down a plane as in Figure 35.1. As the ball rolls down the slope its velocity increases due to the force of gravity.

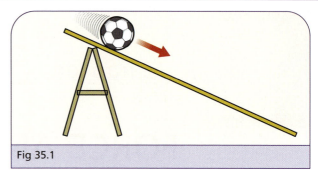

Fig 35.1

If the body then rolls up a ramp as in Figure 35.2, it slows down because of the force of gravity.

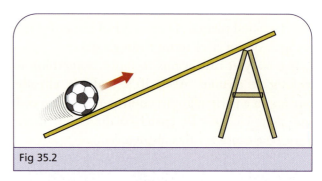

Fig 35.2

If the ball just rolls off the ramp and continues in a straight line, as in Figure 35.3, gravity will not affect it and so it should continue moving at a constant speed even with no force on it.

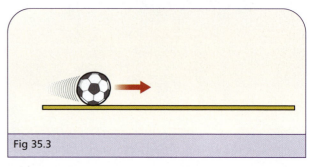

Fig 35.3

If no force acts on a body, then a body in motion will continue at constant speed in a straight line.

The force of friction

In reality, a ball rolling along level ground will stop. This is because there is a force on it that slows it down. We call this force the **force of friction**.

There is a force of friction between all bodies in contact. The force works against one body moving over the other. This is because no matter how smooth the two surfaces may seem they are in fact quite rough when seen under a microscope. The particles in the two surfaces bang and scrape against each other.

PHYSICS

Factors affecting friction

The amount of friction depends on the surfaces that are in contact.

- If we push a puck across ice it takes little effort to get it to go a long distance before it stops.
- On the other hand, if we try to push a large sofa across a carpet, we find that the force of friction that works against the push is quite strong. It may not be possible to move the sofa.

> **Friction** is a force that opposes the sliding motion of one object when it is in contact with another.

Objects move faster without friction

Advantages and disadvantages of friction

Friction can be very helpful.

- We can walk along the ground because the force of friction works against the push we make against the ground.
- Walking on ice is far more difficult as the force of friction is much smaller.
- Cars and bicycles could not move along the ground without the force of friction for the same reason.

The force of friction can also be a nuisance.

- Passing a ball along the ground of a very wet playing surface is a lot more difficult than when the surface is dry.
- The friction due to the gases in the atmosphere can cause the metal on the outside of a space shuttle to melt when it returns to earth.
- Friction between water and a ship will slow a ship down.

Lubrication

Machines, such as car engines, will sometimes grind to a halt if friction between the parts is not reduced. We can reduce friction between surfaces by applying oil or grease to them. This is called **lubrication**. An important example of lubrication is the fluid between our bones that prevents friction.

▶▶▶ ACTIVITY 35.1

To investigate the force of friction

Method

1 Place a wooden block on a table. Attach a spring balance to the side of the block and pull on the spring balance until the block begins to move.

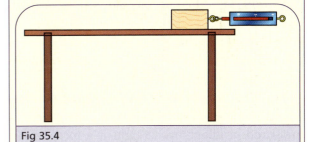

Fig 35.4

2 Note the reading on the spring balance when the block begins to move.

3 Attach a piece of carpet to the bottom of the block and repeat the experiment.

4 Repeat the experiment again, this time with a piece of sandpaper on the surface of the block in contact with the table.

Result

The readings on the spring balance allow you to make comparisons between the different surfaces in contact with the block.

PHYSICS

259

Conclusion

The force of friction that acts against motion depends on the nature of the surfaces that are in contact.

Note

To show the effect of lubrication, place a small amount of oil on the table surface and repeat the experiment with the block. Compare this reading with the first reading taken with the block.

Weight – the force of gravity

One of the forces we are most familiar with is the weight of a body. This force is due to the attraction between two bodies that have **mass**. The weight of an object on the Earth is the attraction between the Earth and the object. Weight is measured in **newtons**.

If the same body were on the moon then its weight would be different. The attraction here is between the moon and the body.

Newton also showed that the force of gravity depends on the **distance** between the centres of the two bodies that are attracted. The weight of a body will be different on the top of a high mountain than at sea level. This is because the distance to the centre of the Earth from the top of the mountain is different to the distance from sea level to the centre of the Earth.

Note that weight is not the same as mass. Mass is the amount of matter in an object and is measured in kilograms. Weight is a force and is measured in newtons.

Calculating force

Forces cause masses to accelerate. We measure force using the equation:

force = mass x acceleration

At the surface of the Earth, we know that the force of gravity gives a body an acceleration of 10 m/s^2.

This means that a body falling under gravity will increase its velocity by 10 m/s every second.

Using the previous equation we can write the equation for weight as:

weight (N) = mass (kg) x 10 (m/s²)

This equation also shows that the unit of the newton (N) is the same as the unit kg m/s^2.

It is important to note that weight is a force, but mass is a measure of the amount of matter in a body. They are different physical quantities with different units of measurement.

Example 1

Find the force of gravity on a bag of flour of mass 1 kg.

Solution

$$F = (1 \text{ kg})(10 \text{ m/s}^2) = 10 \text{ kg m/s}^2 = 10 \text{ N}$$

Example 2

Find the mass of a box that weighs 50 N.

Solution

$$\text{Weight} = \text{mass} \times 10$$
$$\therefore 50 = m(10)$$
$$\Rightarrow \frac{50}{10} = m$$
$$\Rightarrow m = 5 \text{ kg}$$

Elastic forces

When we pull on a spring or on an elastic band we can stretch it. We can also feel the pull as the spring or elastic band tries to return to its original length. This is the restoring force in a spring or elastic band when it is stretched.

Two girls in tandem bungee jump; the elastic obeys Hooke's law

⭐ **MANDATORY ACTIVITY 22**

To investigate the relationship between the extension in a spring and the restoring force

Method

1 Place the metre stick at the base of the metal stand and ensure that it is fixed.

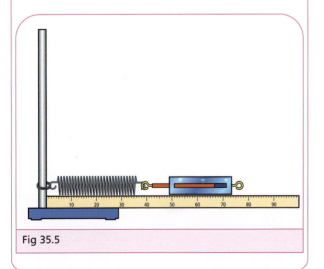

Fig 35.5

2 Slip the keyring with the spring attached onto the stand so that the spring is along the edge of the metre stick.

3 Attach the spring balance (force metre) to the free end of the spring.

4 Measure the length of the spring in its unstretched state.

5 Pull on the spring balance applying a force on the spring and note the extension of the spring as the force is increased.

6 Note your results as in Table 35.1.

Table 35.1	
Extension (m)	**Force (N)**

7 Using a piece of graph paper, plot the values for the extension on the *x*-(horizontal) axis and the values noted for the force on the *y*-(vertical) axis.

Result

You should get the graph of a straight line.

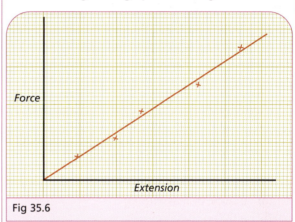

Fig 35.6

Conclusion

The graph of a straight line shows that the extension increases at a constant rate with the applied force. We say that the applied force is directly proportional to the extension. This result is named after the scientist Robert Hooke and is usually stated as follows:

> **Hooke's law:** The extension of an elastic material is directly proportional to the applied force producing the extension.

PHYSICS

Key Points

- A force causes (a) a body at rest to begin moving or (b) a body moving at constant velocity to change this velocity.

- No force is needed to keep a body moving in a straight line with constant velocity.

- Isaac Newton studied forces and stated three laws for motion that are still in use today.

- The unit of force is the newton and its symbol is N.

- Friction is a force that opposes the motion of a body when the body is in contact with another.

- Weight is a force that is due to the attraction between a body and the Earth.

- Elastic forces in materials act to restore a spring or elastic band to its original length.

- Hooke's law states that the extension of an elastic material is directly proportional to the applied force producing the extension.

Questions

1 What is meant by force?

2 Give everyday examples of four different types of force.

3 What is:
   (a) the unit of force?
   (b) the symbol used for this unit?

4 If a ball is rolling with a velocity of 2 m/s and no force acts on it, what will be its velocity after 5 seconds?

5 (a) What is meant by friction?
   (b) Give two advantages and two disadvantages of friction.

6 Write out and complete the following:
Forces cause _____ to _____ and we measure force using the equation
F = ____ x _____. The unit for weight is the _____ and is the same as the unit _____.

7 Calculate the weight of each of the following:
   (a) a book of mass 1.5 kg
   (b) a 500 g carton of milk
   (c) a pen of mass 15 g.

8 Calculate:
   (a) the mass of a table that has weight 300 N
   (b) the force needed to give a 3.5 kg mass an acceleration of 1.5 m/s
   (c) the velocity of a ball after 3 seconds falling under gravity from rest.

9 In Figure 35.7 draw arrows for all the forces you think are at work. Can you name the forces?

Fig 35.7

Examination Questions

10 A pupil measured the weight of an apple of mass 0.2 kg using a spring balance and got a reading of 2 N.

Distinguish between weight and mass.

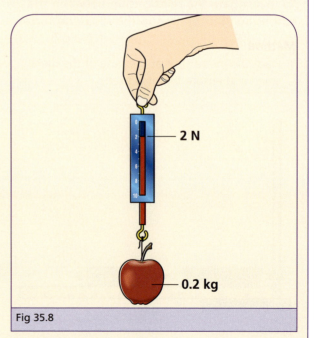

Fig 35.8

(JC, HL, 2006)

? Questions

11 A pupil measured the weight of an apple of mass 0.2 kg.
   (a) What instrument might the pupil use to measure weight?
   (b) What is the difference between mass and weight?
   (c) If the apple were taken to the moon and the weight and mass measured there, which of them would be different? Why?
       (Adapted from JC, HL, 2006)

12 Robert Hooke (1635–1703) made a number of discoveries including the effect of force on elastic bodies now known as Hooke's law. State Hooke's law.
   (JC, HL, 2006)

13 Friction is an example of a force.
   (a) Give another example of a force.
   (b) Give one way to reduce friction.
   (c) After what scientist is the unit of force named?
       (JC, HL, 2006)

14 A student carried out an investigation to examine the relationship between the extension (increase in length) of a spring and the force applied to it.

The diagram shows the apparatus used.

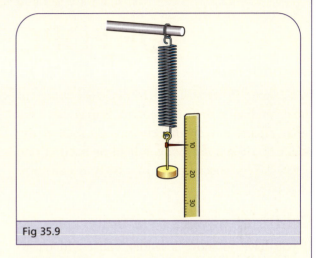

Fig 35.9

The table shows the data collected by the student.

Force (N)	0	2	4	6	8
Extension (cm)	0	4	8	12	16

   (a) Describe how the student could have taken any one of these measurements.
   (b) Draw a graph of the extension (*y*-axis) against the force on a sheet of graph paper.
   (c) What force results in a 6 cm extension of the spring?
       (JC, OL, 2006)

PHYSICS

Chapter 36

TURNING EFFECT
OF FORCES, LEVERS

Introduction

In Chapter 35 we saw that forces cause things to move or change their movement. Forces can have other effects too. In this chapter we look at the effect of a force if a body is held or fixed at one place. A very special place in a body is its centre of gravity.

Centre of gravity of a body

If you have to move a ladder you could try to carry it at one end, at the middle, or at some other point along the ladder. Experience might tell you that if you try to carry a ladder at one end it is very difficult. The best place to grip the ladder is in the middle.

Fig 36.1 Carrying a ladder, holding it at an end

The turning effect of a force

The difference in the carrying is due to gravity acting on the ladder.

1   When you try to carry the ladder at one end you have most of the mass away from you and the force of gravity on all of this mass pushes the ladder down and turns the ladder around the point you are holding. An important effect of a force is the turning effect on a body around a fixed point.

Fig 36.2 Carrying a ladder, holding it in the middle

2   If you hold the ladder in the middle it is much easier to carry since you have an equal amount of mass on either side of you and the turning effect of the force of gravity is balanced on both sides so:

> **The centre of gravity of a body** is a point about which the turning effects due to gravity are balanced. The weight of the body seems to be concentrated at this point.

At the centre of gravity you will be able to balance the body without it turning.

● For example, on a metre stick you should be able to balance the stick at the 50 cm mark without the metre stick turning about this point.

● You can find the centre of gravity of a hammer, for instance, by moving your finger along the shaft of the hammer until you find the point at which the hammer is balanced under gravity.

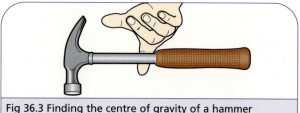

Fig 36.3 Finding the centre of gravity of a hammer

ACTIVITY 36.1

To find the centre of gravity of a thin lamina

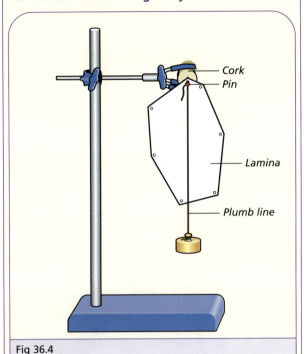

Cork
Pin
Lamina
Plumb line

Fig 36.4

Method

1 Make a hole at each of the corners of the lamina (usually a piece of cardboard).

2 Place the pin in the cork held by the clamp on the retort stand.

3 Suspend the lamina and the plumb line from the pin.

4 Mark two points on the lamina directly behind the string of the plumb line.

5 Repeat this procedure, suspending the piece of cardboard from the other holes at the edges.

6 Draw lines on the cardboard to show the lines of the plumb line while the card was suspended.

The idea is that when the cardboard is hanging from a point, the centre of gravity must be directly below this point, since otherwise there would be a turning effect on the piece of card and it would rotate.

Result

The centre of gravity is the point of intersection of the three (or more lines) that you have drawn.

Conclusion

The centre of gravity of the card is in the centre. You should now be able to suspend the piece of cardboard on the end of a pencil placed at the point that you have found.

Centre of gravity and stability

You can balance a metre stick on your finger if you place your finger at the centre of gravity. If you now push the metre stick from one end, the centre of gravity moves away from your finger. Now there is a turning effect. The metre stick rotates and crashes to the ground. It is clear that the balanced metre stick is not very stable.

The centre of gravity has been moved away from the fixed point provided by your finger. As soon as the centre of gravity is moved away from the fixed point there is a turning effect.

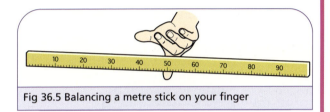

Fig 36.5 Balancing a metre stick on your finger

If you balance the metre stick on a book and then move the centre of gravity across the book, the metre stick will topple over when the centre of gravity mark goes over the edge of the book.

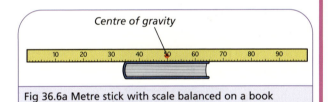

Centre of gravity

Fig 36.6a Metre stick with scale balanced on a book

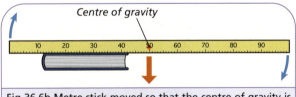

Centre of gravity

Fig 36.6b Metre stick moved so that the centre of gravity is over the edge of the book and toppling

We can conclude that a body will topple when the vertical line through its centre of gravity lies outside the base that is providing its stability.

Stability in design

This effect is important in the design of buildings and vehicles. A structure is more stable if

(a) its centre of gravity is close to the ground and

(b) it has a wide support base.

On the other hand, structures with high centres of gravity and narrow bases are less stable.

PHYSICS

Equilibrium

In physics we often use the term equilibrium when talking about how stable a body is.

● An object is in stable equilibrium if it does not topple easily when a force is applied to it.
● An object is said to be in unstable equilibrium if it will topple over when a small force is applied to it.

One example of this idea is that passengers are not allowed to stand on the top of a bus. The weight of the people upstairs would shift the centre of mass of the bus upwards. When the bus turned a corner it is possible that the vertical line through the centre of mass could lie outside the wheelbase of the bus and it would topple!

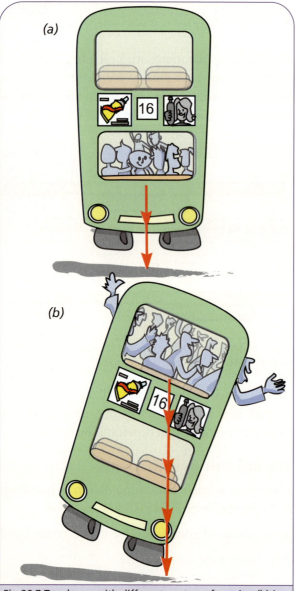

(a)

(b)

Fig 36.7 Two buses with different centres of gravity: (b) is higher than (a) and has a toppling effect if the vertical line is outside the wheelbase

Similarly, when ocean-going yachts are designed they are fitted with a very heavy keel (or blade) under the boat. This keeps the centre of mass very low. In a strong wind the force of the wind will tilt the boat but the boat should not keel over!

Investigating turning effects

If an extended body is kept fixed at a point and a force is applied to another part of the body, then the force has a turning effect on the body.

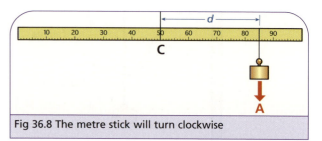

Fig 36.8 The metre stick will turn clockwise

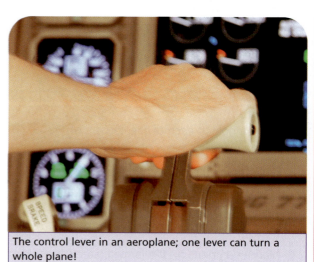

The control lever in an aeroplane; one lever can turn a whole plane!

Levers

The simplest machine possible is made using the idea of the turning effect of forces. This is the **lever**. There are many everyday examples of levers.

> **A lever** can be described as a rigid body that can rotate about a fulcrum (a fixed point).

● The handle of a door has a fixed point where the metal rod goes through the door. When a force is applied to the end of the handle there is a turning effect that opens the door.
● A wheelbarrow has its fixed point on the axle of the wheel. An upward force applied to the handles will turn the wheelbarrow around the fulcrum.
● Revolving doors turn about a central pole when a force is applied to the edge of the door.

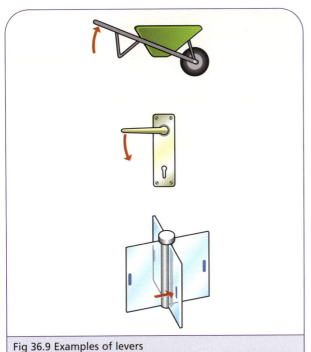

Fig 36.9 Examples of levers

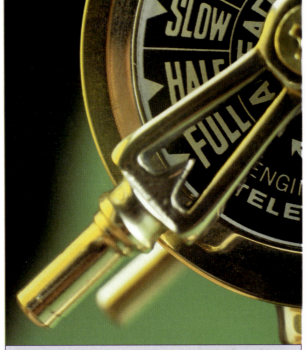

The lever on a ship's telegraph turns quickly through different messages

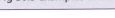 **ACTIVITY 36.2**

To investigate the turning effect of forces on a lever

Method

1 Suspend a metre stick from a stand at its centre of gravity as in Figure 36.10.

2 Hang a 0.5 N weight from a point on one side of the metre stick at a distance of exactly 30 cm from the fulcrum.

3 Now hang a 1 N weight on the other side of the metre stick at a point on the metre stick that keeps the metre stick balanced horizontally.

4 Remove the 1 N weight but leave the 0.5 N weight. Replace the 1 N weight with weights of 0.5 N, 2 N and 3 N, each time placing them so that the metre stick is balanced horizontally.

5 You can also put two weights on one side and balance them with a weight on the other.

Result

When the metre stick is balanced, the turning effect on the right-hand side is equal to the turning effect on the left-hand side.

Conclusion

When the metre stick is balanced, the total of the turning effects on one side is equal to the total of the turning effects on the other side. This is called the law of the lever.

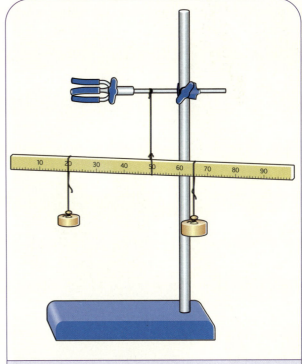

Fig 36.10 Balancing turning effects

PHYSICS

The law of the lever

> **The law of the lever states** that when a lever is balanced, the sum of the clockwise turning effects equals the sum of the anti-clockwise turning effects.

If you remove a weight from the left-hand side of a balanced metre stick, the right-hand side begins to turn clockwise.

If you remove a weight from the right-hand side, the metre stick will begin to turn in an anti-clockwise direction. We can say then that the turning effects are either clockwise or anti-clockwise.

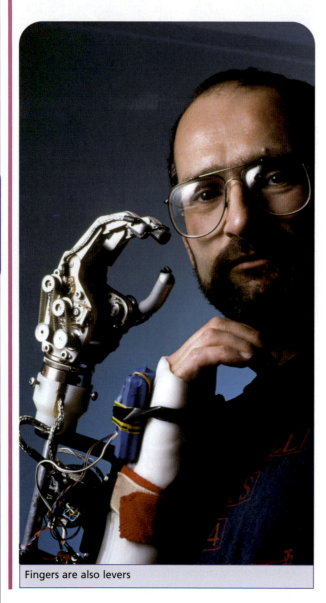

Fingers are also levers

PHYSICS

Key Points

- An important effect of a force is the turning effect on a rigid body around a fixed point.

- The centre of gravity of a body is the point about which the turning effects due to gravity are balanced. The weight of the body seems to be concentrated at this point.

- A body will turn or topple if the vertical line through its centre of gravity lies outside the base giving it stability.

- A body is more stable if (a) its centre of gravity is close to the ground and (b) it has a wide support base.

- A lever is a rigid body that can rotate about a fixed point.

- The law of the lever states that when a lever is balanced the sum of the clockwise turning effects equals the sum of the anti-clockwise turning effects.

? Questions

1. What is meant by:
   (a) the centre of gravity of a body?
   (b) equilibrium?
   (c) a lever?

2. Give four everyday examples of the turning effect of forces.

3. Explain why the centre of gravity of a double-decker bus should be as low as possible.

Fig 36.11

? Questions

4 State the law of the lever.

5 Write out and complete the following:
A lever is a _____ body which is free to rotate about a _____ point called a _____. A coin standing on its side is in a state of _____ equilibrium but a coin lying face down is in a state of _____ equilibrium.

6 Figure 36.12 shows a beam being used to move a rock. Copy the diagram and show (a) the fulcrum of the beam and (b) two of the forces acting on the beam.

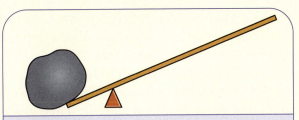

Fig 36.12

7 A spanner is used to turn a bolt. Copy the diagram and indicate the force on the spanner and the forces on the bolt.

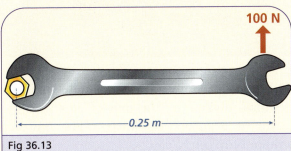

100 N

0.25 m

Fig 36.13

8 Figure 36.14 shows a uniform lever suspended at its centre of gravity with two weights suspended from it. The lever is balanced under these forces.

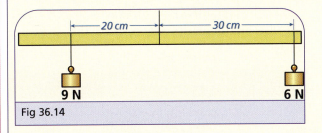

20 cm   30 cm

9 N   6 N

Fig 36.14

Which force gives:
(a) a clockwise turning effect?
(b) an anti-clockwise turning effect?

9 Draw a diagram of a builder's crane lifting a load. Show how the turning effect of the load is balanced so that the crane does not topple.

A crane is based on turning effects

10 Place a ruler on the top of a book. Move the ruler across the book until it begins to topple.
(a) Explain why the ruler begins to topple.
(b) What length of the ruler can extend over the base of the book before it topples?

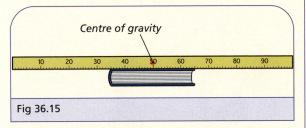

Centre of gravity

10 20 30 40 50 60 70 80 90

Fig 36.15

11 You know that it is easier to push down a door handle if you push as far away as possible from the fixed point. Take a metre stick, tape to attach it to a door handle and a force metre.

Describe an experiment to show the connection between the force needed to push down the handle and the distance from the fixed point to the force.
(a) What are the biggest sources of error in this experiment?
(b) What would be a good way to present the data in this experiment?

PHYSICS

Chapter 37

PRESSURE

Introduction

In the last two chapters we looked at forces and their effects.

(a) A force can cause a body to accelerate.
(b) If a body is fixed at a point, a force can cause it to turn.

We now look at the link between a force and the area on which it acts.

Fig 37.1

It may surprise you, but an elephant standing on your toes would probably do the same damage as an adult cyclist cycling over your toes! This is because the effect of the force depends on the area covered by the force. This is the idea of **pressure**.

Pressure

> **Pressure** is the amount of force acting on a unit of area.

The effect of a force is greater when the area is smaller. This is why when we want to cut something we use a sharp edge. All the force is then concentrated in a small area.

The effect of the force will be less if the area is bigger. This is why mechanical diggers can work on muddy ground without sinking. Their tracks have a large area that spreads out the force on the ground.

We calculate pressure using the equation:

$$pressure = \frac{force}{area}$$

The units of pressure are the units of force, newtons, divided by the unit of area, a square metre or a square centimetre. The units are written N/m^2 or N/cm^2.

If we use the standard unit N/m^2 the numbers we get are sometimes very large. So in many cases we use the smaller unit N/cm^2. The standard unit of N/m^2 is also called a pascal (Pa), after the French scientist Blaise Pascal.

Like the equations that we have seen before, we can use mathematics to rewrite the pressure equation as

$$force = pressure \times area$$

Calculations using the pressure equation

Example 1

A block of mass 5 kg has a base of length 5 cm and width 2 cm. What is the pressure on the base?

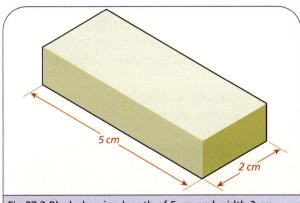

Fig 37.2 Block showing length of 5 cm and width 2 cm

To use the relationship:

$$\text{pressure} = \frac{\text{force}}{\text{area}}$$

we need the force and the area.

The force exerted by the block is due to gravity. As we saw in Chapter 35, we get the force due to gravity by multiplying the mass by g, the acceleration due to gravity which is approximately 10 m/s². This gives

$$F = 5 \times 10 = 50\,\text{N}$$

Now we find the area. Since the distances are given in centimetres we will get the area in cm².

$$\text{Area} = 5\,\text{cm} \times 2\,\text{cm} = 10\,\text{cm}^2$$

Now we can find the pressure.

$$\text{Pressure} = \frac{\text{force}}{\text{area}} = \frac{50\,\text{N}}{10\,\text{cm}^2} = 5\,\text{N/cm}^2$$

Example 2

Work out the pressure exerted by an elephant's foot and the pressure exerted by the wheel of a bicycle given the following information: the elephant has a mass of 3 000 kg and a foot area of 710 cm²; the cyclist has a mass of 70 kg; the area of the wheel in contact with the ground is 16 cm².

For the elephant:

$$\text{Pressure} = \frac{\text{force}}{\text{area}} = \frac{30\,000\,\text{N}}{710\,\text{cm}^2} = 42.3\,\text{N/cm}^2$$

For the cyclist:

$$\text{Pressure} = \frac{\text{force}}{\text{area}} = \frac{700\,\text{N}}{16\,\text{cm}^2} = 43.8\,\text{N/cm}^2$$

So we can see that the effect of the cyclist is greater than the effect of the elephant!

Pressure in liquids and gases

Pressure is a useful quantity to use when talking about the forces exerted by liquids and gases because these always act over an area.

There is one important difference between the pressure exerted by a block on a table and the pressure exerted by a liquid or a gas. In a liquid or gas the pressure is exerted in every direction. We can show this by filling a plastic bag with water and then putting holes in the bag.

We observe two things:

(a) The water comes out of the bag in all directions.
(b) The water comes out of the lower holes with more pressure than from holes near the top, as in Figure 37.3.

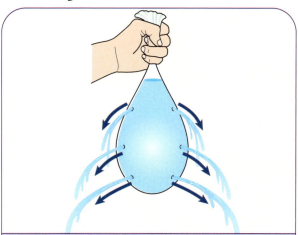

Fig 37.3 Pressure in liquids acts in all directions and the pressure at the bottom of the bag is greater than at the top

Pressure and depth in a liquid

We can explain why the pressure at the bottom of the bag is greater than at the top. At the bottom of the bag there is more liquid above the hole than at the top. This liquid has a weight and the more liquid there is above, the greater the weight and so the greater the force.

Scuba divers and deep-sea divers know this fact well. When a diver goes down, the deeper she goes the more water there is above her. She feels the weight of this load of water above her. As she goes further down the size of this load of water gets bigger, the weight is greater and so the pressure on her increases.

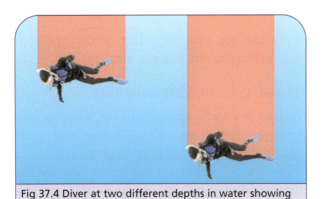

Fig 37.4 Diver at two different depths in water showing columns of water resting on her

The pressure in a liquid increases with depth.

A further example of this principle is used in the building of dams. Because the pressure of water increases with depth, the walls of a dam have to be much thicker at the bottom than at the top.

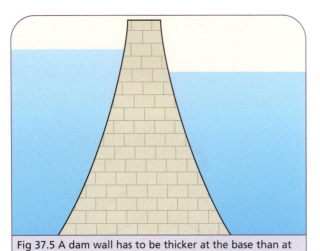

Fig 37.5 A dam wall has to be thicker at the base than at the top

Water supply and pressure

Water will always flow so that it is all at the same level. This fact is used in supplying water to houses and factories.

The reservoir where water is held is usually at a greater height than the houses it supplies. When the water is released from the reservoir it tries to find the same level as the water in the reservoir. This gives it a pressure that will push the water up through the house to the tank under the roof.

This works very well as long as your house is below the level of the reservoir!

Water pressure is much greater at the bottom of a dam than at the top

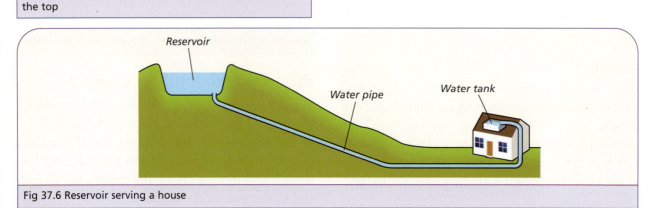

Reservoir

Water pipe

Water tank

Fig 37.6 Reservoir serving a house

PHYSICS

Everyday applications of pressure

The relationship between pressure and force provides a means of applying large forces using liquids. For instance, the pressure in liquids transferred to pistons is used in:

● braking systems on articulated trucks
● lifting gear on fork-lift vehicles
● lifting the bucket on a JCB
● moving the arms of industrial robots.

Look at the piston system in Figure 37.7. If 100 N of force is applied to a piston of area 10 cm², then the pressure is

$$\frac{100\,\text{N}}{10\,\text{cm}^2} = 10\,\text{N/cm}^2$$

This pressure is transferred to a larger piston of area 500 cm². The force on the piston is given by the formula we had above:

force = pressure × area = 10×500 = $5\,000\,\text{N}$

The force has increased 50 times. This is a large force capable of moving a big load.

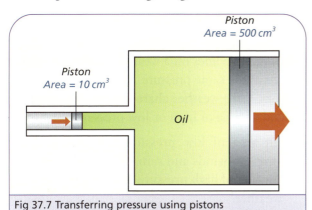

Fig 37.7 Transferring pressure using pistons

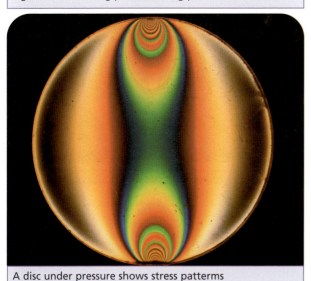

A disc under pressure shows stress patterns

The atmosphere and its pressure

Our planet is covered by a layer of gases. Gas is a kind of matter and all matter has mass. The gases in our atmosphere therefore have mass and weight.

Like all liquids and gases these gases exert a pressure. We call this atmospheric pressure. It is due to the weight of the gases that are above our heads.

It is difficult to estimate the amount of gas that surrounds the Earth, but it is estimated that the total mass of the gases is 500 million million tonnes!

The effect of this huge column of gas above our heads is that there is a pressure of about 1 tonne on the area of our shoulders! We do not feel this pressure because our bodies exert an equal pressure to balance it.

Through some simple experiments, we can show that the atmosphere has a pressure.

Measuring atmospheric pressure

Atmospheric pressure is not constant. It changes according to temperature and moisture in the atmosphere. The instrument used to measure atmospheric pressure is called the barometer.

A barometer measures atmospheric pressure

273

PHYSICS

The mercury barometer

The mercury barometer is based on the height of a column of mercury that can be held up by atmospheric pressure. In Figure 37.8 the column of mercury in the closed tube would drop if it were not held in place by the pressure of the atmosphere pushing on the mercury in the basin.

On average, atmospheric pressure will support a column of mercury 76 cm high. This pressure is equivalent in standard units to $101\,300\ \text{N/m}^2$ or 1013 hectopascals (hPa). If you listen to the sea-area forecast on the radio you will hear the announcer give the pressure in hectopascals.

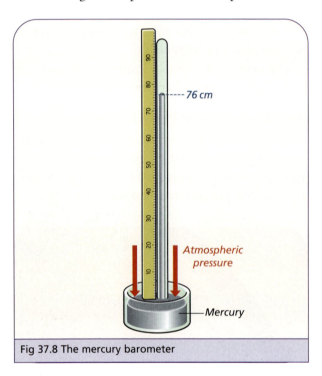

Fig 37.8 The mercury barometer

The altimeter

One use of the barometer is in an aircraft. As an aeroplane climbs up through the atmosphere the pressure drops. This drop in pressure can be measured and translated into a measurement of height above sea level. The instrument that does this is called an altimeter.

Atmospheric pressure and the weather

Atmospheric pressure is a major influence on the weather. Warm, moist air exerts less pressure than cold, dry air. So when the atmospheric pressure drops, warm, moist air is on the way and we can expect rain.

On the other hand, if the barometer shows that the atmospheric pressure is increasing we can expect cool, dry air and better weather.

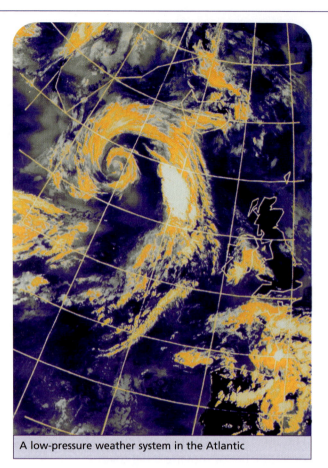

A low-pressure weather system in the Atlantic

Atmospheric pressure and weather charts

Weather charts are pictures of atmospheric pressure. The lines on a weather chart show places that have the same pressure. These lines are called **isobars**. On a weather chart these isobars often go around an area of high or low pressure.

● An area of low pressure will usually bring stronger winds and rain.
● An area of high pressure will bring settled, dry weather.

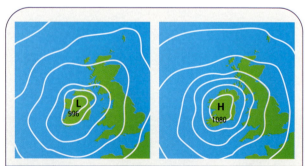

Fig 37.9 Two weather charts showing areas of high and low pressure over Ireland

ACTIVITY 37.1

To demonstrate atmospheric pressure

(a) Using a glass of water and a piece of cardboard

Method

1 Fill a glass to the brim with water.

2 Cover the top of the glass completely with a piece of cardboard.

3 Holding the cardboard in place turn the glass upside down.

4 Remove your hand.

Result

The water will remain in the glass!

Conclusion

Atmospheric pressure is pressing upwards on the cardboard and keeping the water in place.

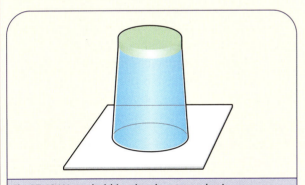

Fig 37.10 Water held in glass by atmospheric pressure

(b) By driving the air out of a can

Method

1 Place a small amount of water in the bottom of a metal can.

2 Boil the water in the can until the steam drives all the air out of the can.

3 Taking great care to avoid burning, seal the can tightly.

4 Allow time for the can to cool.

Result

The can collapses.

Conclusion

When the can cools there is a partial vacuum in the can and the atmospheric pressure outside the can crushes it.

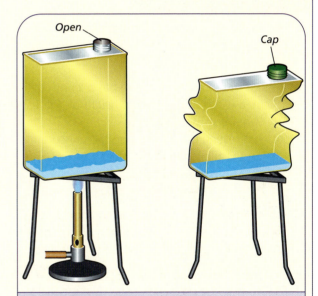

Fig 37.11 The can is crushed by atmospheric pressure when the air is expelled

PHYSICS

Key Points

- Pressure is the amount of force acting on a unit of area. It is calculated using the formula:

$$\text{pressure} = \frac{\text{force}}{\text{area}}$$

- This formula can also be rewritten as:

$$\text{force} = \text{pressure} \times \text{area}$$

- The standard unit of pressure is newtons per square metre, N/m^2 or N/cm^2. The unit N/m^2 is also called a pascal (Pa).

- The pressure in a liquid or gas is exerted in all directions.

- The pressure in a liquid increases with depth.

- Pressure in liquids and pistons can be used to exert great force.

- The atmosphere is made of gases that have mass and exert pressure.

- Atmospheric pressure is measured using a barometer. Normally atmospheric pressure can support a column of mercury 76 cm high.

- An altimeter measures atmospheric pressure at different heights. It is used to show altitude.

- Atmospheric pressure is a major influence on the weather. Low pressure systems usually mean warm, moist air and high pressure systems indicate cool, dry air.

- The lines on a weather chart join places of equal pressure and are called isobars.

Questions

1 Define pressure. What is the standard unit of pressure? What unit is also used for pressure and why?

2 A block of wood of mass 3 kg has a length of 15 cm, a width of 6 cm and a height of 4 cm.
   (a) What is its weight?
   (b) It is resting on a table. Find the pressure on the table due to the block
      (i) if it is lying flat, and
      (ii) if it is standing on its edge.

3 A car of mass 1 000 kg is standing on four tyres. Each of the tyres is in contact with 65 cm² of the ground. What pressure is exerted by each tyre on the ground?

4 Explain why the pressure in a car tyre is usually much less than the pressure in a bicycle tyre.

5 Explain the following in terms of pressure:
   (a) A deep-sea diver needs a protective suit when diving at great depths.
   (b) A dam wall must be thicker at the bottom than at the top.
   (c) A force applied to a small piston can be used to lift a large load.

6 Name three everyday uses of pressure in liquids.

7 Write out and complete the following: The Earth's atmosphere is made of _____. Atmospheric pressure is due to the _____ of the column of _____ above our heads. We do not feel this pressure because our bodies exert an _____ _____ to balance it. Atmospheric pressure is measured using a _____. On average, atmospheric pressure can support a column of _____ measuring _____ cm. This is equivalent to _____ hectopascals. Weather charts contain lines called _____. They join places that have the same _____.

8 Describe two experiments to show that the atmosphere exerts a pressure.

PHYSICS

? Questions

9 (a) When the tap is opened on the burette in Figure 37.12 the liquid flows into the vessel underneath. Draw the level of the liquid in both sides of the vessel when it has settled.

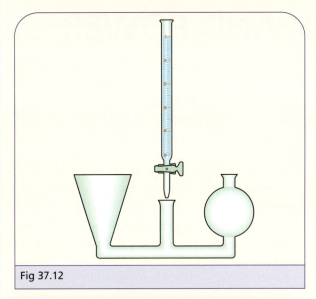

Fig 37.12

(b) What is the weather likely to be on a summer's day in Ireland when
(i) the atmospheric pressure is high?
(ii) the atmospheric pressure is low?

10 A barometer reads 1020 hPa at sea level. What change would you observe in this reading if the barometer were transported to the top of Mount Everest and why?

11 Atmospheric pressure varies with height above sea level. Name one everyday use of this variation.

12 Figure 37.13 shows the weather charts for two different days over Ireland. What type of weather would you expect on (a) day 1 and (b) day 2?

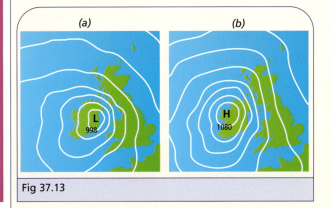

Fig 37.13

13 The pressure that your body exerts on the floor depends on the area in contact with the floor.

Make approximate calculations of the area of:
(a) your feet
(b) your hands
(c) your toes.

Using these measurements find the pressure you exert on the floor when you are standing on your feet, your hands and your tiptoes.
(a) How might you work out an approximate area for your feet, hands and tiptoes?
(b) Which of the pressures is greatest, which is the smallest?

Examination Question

14 (a) Complete the equation in the box below using the words on the right.

| pressure = ———. | Area |
| | Force |

(b) Name the piece of equipment used to measure pressure.
(JC, OL, 2006)

PHYSICS

Chapter 38 •••

WORK, ENERGY AND POWER

Introduction

In physics work is done when a force moves an object. This definition is slightly different to the everyday one. Most of us think that we are doing work if we get tired trying to push a car that will not move, for example. However, in physics work happens only when the object actually moves.

Work

To calculate the work done by a force we multiply the force by the distance travelled in the direction of the force.

> **Work done =**
> **Force × distance moved in the direction of the force**
> **or $W = F \times d$**

The unit of work is newtons × metres or Nm. This unit is also called **the joule** (J), after the English scientist James Joule.

> **Joule:** the unit of work or energy = Newtons × metres.

Example 1

In Figure 38.1 the man is pushing a supermarket trolley with a force of 30 N. He pushes the trolley for 200 m.

Fig 38.1 Pushing a supermarket trolley

Work done = force × distance in the direction
            of the force
         = 30 × 200
         = 6000 J

Example 2

When we lift something we are doing work since we are applying a force to move an object against gravity.

A cook lifts a bag of flour of mass 2 kg from a table to a shelf 50 cm above the table. What work is done by the cook?

Fig 38.2 Lifting flour onto a shelf

The force required is to work against the weight and the weight is 2 × 10 = 20 N

Work = force × distance
     = 20 N × 0.5 m
     = 10 Nm
     = 10 J

Energy

A body or a machine has energy if it can move something. Energy can be stored in many ways.

We call these 'forms of energy'. Here is a list of nine different forms of energy.

Identifying different forms of energy

1 Potential energy

Potential energy is stored in a wound-up spring or a stretched elastic band. An object that has been raised above its resting place under gravity also has potential energy.

- We know that energy is stored in springs and elastic bands because if they are released they can move things and so do work.
- Water that is held up against gravity by a dam also has energy because when it is released it can move the turbine fans of a generator.

2 Kinetic energy

Kinetic energy is stored in a body that is moving.

- The wind has kinetic energy. The moving air molecules can push the sail of a yacht or the sails of a windmill and move them.
- Cars, trucks and aeroplanes have kinetic energy because they can move people and goods.

3 Heat energy

Heat can also move things.

- In a thermometer when the bulb is placed near a source of heat the column of mercury or alcohol in the thermometer tube moves.
- Heat energy in car engines is used to move pistons and finally the wheels.

4 Chemical energy

Chemical energy is stored inside materials in the chemical bonds holding them together. In many chemical reactions some of this energy is released. This is a very important form of energy for human beings.

- Our ability to move and to do work comes from the chemical energy that is stored in the food we eat.

- Chemical energy is also released by organic materials when they are burned. Coal, oil, petrol, gas, turf and wood all contain chemical energy. This form of energy has been a major source for all kinds of machines. It is still very important in industry.

A battery is also a source of chemical energy.

The cooling towers of a power station

5 Sound energy

Sound also stores energy. It causes our eardrums to vibrate so that we can hear.

- Explosions of sound are used in mountain regions to cause controlled avalanches and thus keep ski slopes safe.

Sound energy can cause an avalanche

PHYSICS

PHYSICS

6 Electrical energy

Electrical energy is due to moving charges.

- Many appliances in the home such as hairdryers, mixers, washing machines and televisions use this form of energy to move things. Sometimes we are not able to see the movement directly as in the television. The electrical energy is indeed doing work by moving a beam across the screen to create the pictures we see.

7 Magnetic energy

Magnets have the ability to move things. A simple bar magnet will cause paperclips to move towards it.

- Large magnets on the end of cranes are used to lift heavy pieces of metal such as cars.

8 Solar energy

The light of the sun is our most important source of energy. In a way, all our energy comes from the sun or has come from the sun at some earlier time.

- It is the energy of the sun that makes plants and crops grow. Plants and crops are sources of food for animals and humans.
- Indirectly, the sun is also responsible for many forms of chemical energy. Wood, turf, coal and oil are all the result of plants growing by the energy of the sun, then dying and decaying into the ground.

- Solar energy can also be used directly to store energy in solar cells which can be a source of heat and electricity.

9 Nuclear energy

The tiny particles that are in the core or the nucleus of an atom are held together by extremely strong bonds. If these bonds are made or broken, large amounts of energy are released.

- Splitting the nucleus of an atom is called fission. In nuclear reactors atoms are split to release energy, to make steam, which moves large turbines that make electricity.
- Nuclear energy is also released when atomic particles are pressed together to make a new nucleus. This is called fusion and is the process that happens in the sun, creating solar energy.

The principle of the conservation of energy

Energy is involved in every physical change. When we examine these changes we find that energy is constantly changing its form.

The nuclear energy in the sun is changed into solar energy. This in turn is changed into chemical energy in plants, which can be changed in our bodies into kinetic energy and heat.

It would seem that energy never goes away. It just changes its form. This idea is summed up in the principle of the conservation of energy which says:

> **Energy is neither created nor destroyed but can be converted from one form to another.**

Solar power cells

Sources of energy

From where do we get the energy to do useful work for us? There are many sources of energy in the universe. The most important is, of course, the sun. Other sources of energy also provide us with the means to do work. We classify sources of energy into two groups: non-renewable sources and renewable sources.

Non-renewable sources of energy

The energy from the sun that came to Earth over hundreds of millions of years was taken up by plants that then died and decayed. The chemical energy in these plants has been stored in the decayed matter. This decayed matter has been changed over long periods of time into oil, coal, gas and turf. These sources are called fossil fuels.

Fossil fuels have a high concentration of energy. A disadvantage is that the burning of these fossil fuels causes gases to be sent into the atmosphere causing pollution. Those gases may also be adding to climate change. We have been using up this energy in the past hundreds of years and these sources of energy are being used up very quickly. They are now running out.

Such sources are known as non-renewable sources. In the future we will have to look to other sources.

Renewable sources of energy

There are a number of other sources that can provide energy that does not run out. Among these renewable sources of energy are: solar, hydro-electrical, nuclear, wind, tidal, biomass and geothermal.

Solar energy

> **Solar energy** is energy from the sun.

Solar energy can be stored directly in cells. These can be used to heat houses and provide electricity. It has the advantage of not causing pollution.

At present, it is not possible to generate the same amount of electricity as a generating station without covering many square kilometres with solar cells. This form of energy is also expensive at the moment. However, work is being done to improve the efficiency of the cells and to bring costs down.

Hydro-electricity

> **Hydro-electricity** is energy from the potential and kinetic energy of water.

Hydro-electricity is an excellent source of energy as it uses the natural cycle of water by damming rivers and using the potential energy in the water to create electricity. It has the advantage of not causing pollution and is cost efficient to produce. There are, however, also disadvantages. The building of large dams means the flooding of valleys and the destruction of habitats for plants and animals.

A hydro-electric power station

Nuclear energy

> **Nuclear energy** is energy from making or breaking nuclear bonds.

Nuclear energy is a renewable source and can provide large amounts of energy relatively cheaply.

PHYSICS

But this source is surrounded by controversy. The problem is that radioactive waste is produced in the process. This is harmful or even deadly.

A nuclear power station

The dangerous radiation from this waste can last for hundreds of thousands of years and great care must be taken to dispose of it safely. There are many who think that it is not possible to dispose of the waste in a way that is sure not to cause great damage to future generations.

Wind energy

> **Wind energy** is energy from the kinetic energy of the wind.

Wind energy has been used for hundreds of years in windmills and to drive pumps. It does not cause pollution of the air. In recent years 'wind farms' have been built in a number of countries. One has been proposed on the Kish Bank off the east coast of Ireland. It is estimated that it could provide enough electricity for 200 000 homes.

These 'farms' consist of vertical masts with large propellers that turn in the wind and generate electricity. However, more than a thousand are needed to produce the same amount of electricity as a standard fossil-fuel powered generator.

People sometimes object to wind farms because they change the look of the countryside and can be noisy. Also, they do not work in calm conditions.

Wind power

PHYSICS

Tidal and wave energy

> **Tidal and wave energy** is generated by the movement of the tides and waves.

When the tide goes in and out, huge volumes of water move. This movement could be used to power a turbine to produce electricity.

Similarly, wave motion causes water to move up and down. Once again this kinetic energy could be turned into electrical energy.

Biomass energy

> **Biomass** can generate energy from biological sources.

In Brazil sugar cane is grown for its oil. The oil can be used to make alcohol. Alcohol can then be used as a fuel like petrol.

Farm waste can be treated to make methane gas. This gas can be burned to provide energy.

Geothermal energy

> **Geothermal energy** comes from the heat in the earth beneath the planet's surface.

In some countries, such as Iceland, there are hot water springs fairly close to the surface. The heat from the earth can be used to make steam by pumping cold water down into the earth. It comes back up as steam. The steam can be used to drive turbines.

The hills around Tuscany in Italy are dotted with geothermal power stations. They are a useful energy source for the area but have environmental disadvantages.

Despite all of these renewable sources, a lot of research and development still has to be done on alternative forms of energy before they are capable of satisfying the demand that modern societies make on energy.

PHYSICS

A geothermal plant in Iceland

National energy needs

We need energy for light and heat, for transportation and for industrial and agricultural production as well as for appliances in the home. The way we live today in Ireland demands more and more energy.

It is estimated that the demand for energy in Ireland in the years from 1990 to 2000 grew by over 60 per cent. Demand is still increasing at 3 per cent every year (2007). Our need for electricity alone is hundreds of trillions of joules each year.

Electricity supply in Ireland

Electricity in Ireland is provided for the most part by the Electricity Supply Board (ESB). The ESB runs 23 thermal generating stations burning fossil fuels and gas and three renewable energy stations (hydro-electric and wind).

Providing all of this energy is expensive. We import nearly all the fuel we need to make electricity. It will be necessary in Ireland to find other renewable energy sources in the future such as those listed above.

Energy conservation

At present we do not have enough renewable energy to meet our needs. The energy that is available is expensive and is not distributed evenly around the world. Many poorer countries cannot afford the energy that they need in order to develop their economies.

It is important then that we try to cut down on the energy that we use. This is called energy conservation.

Energy conservation in the home and school

In our homes, much of the energy that we buy is wasted. For instance, a lot of energy is used to heat houses. Much of this heat escapes through walls, roofs, windows and doors. This loss can be prevented by insulation. It involves putting materials that do not let heat through easily into our walls and roofs. We can also prevent heat loss by stopping draughts.

Another way to conserve energy is to switch off lights in rooms that are not being used. We can also switch off the stand-by function on electrical appliances such as televisions and stereo systems at night. Not only is this good for the environment and our economy, it can reduce our own energy bills by a lot.

PHYSICS

⭐ **MANDATORY ACTIVITY 23A**

Converting electrical energy to magnetic energy to kinetic energy

Method

1 Attach a long length of bell wire to one of the terminals of a 6 V battery.

2 Wind the bell wire around a large nail and then to the other terminal of the battery as in Figure 38.3.

3 Place the nail close to a number of metal paperclips on the table.

Result

The paperclips move to the nail and attach themselves.

Conclusion

The chemical energy in the battery has been converted into electrical energy in the wire. This in turn has been converted into magnetic energy in the nail which then causes kinetic energy in the paperclips.

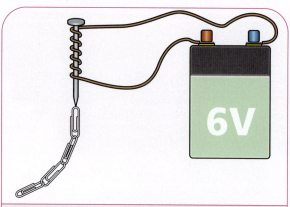

Fig 38.3 Conversion of electrical to magnetic energy to kinetic energy

Energy conversions

As we have seen, energy is converted from one form to another. Electrical energy that comes into houses, offices and factories is converted into heat, light, motion and so on. We are all familiar with many energy conversions in everyday life. For instance, chemical energy in a battery is changed into electrical energy that can produce sound energy in an MP3 player. A hairdryer converts electrical energy into kinetic and heat energy and so on. We can demonstrate some of these conversions in experiments in the laboratory.

MANDATORY ACTIVITY 23B

Converting light energy to electrical energy to kinetic energy

Method

1 Attach one end of a length of wire to the terminal of a solar panel and then to the terminal of a motor to which a fan has been attached.

2 Connect a length of wire from the other terminal of the motor to a switch and then back to the free terminal of the solar panel to make a circuit.

3 Close the switch and observe the result.

4 Cover the solar panel with a cloth so that no light falls on the panel. Observe the result.

Result

The fan on the end of the motor spins when light falls on the solar panel.

Conclusion

The light energy from the solar panel is converted into electrical energy in the wire which is then converted by the motor into kinetic energy.

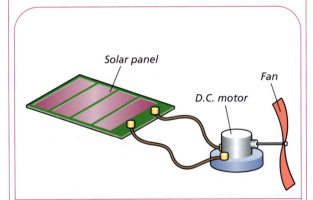

Fig 38.4 Conversion of light to electrical energy to kinetic energy

 MANDATORY ACTIVITY 23C

Converting chemical energy to electrical energy to heat energy

Method

1 Attach a length of wire to one of the terminals of a battery and the other end to one of the terminals of a bulb holder.

2 Attach the other terminal of the bulb holder to a switch and then back to the battery to complete the circuit.

3 Close the switch and place the end of a thermometer bulb on or close to the light bulb.

Result

The thermometer shows an increase in temperature in the light bulb.

Conclusion

Some of the chemical energy in the battery has been converted into electrical energy in the wires and some of this has been converted into heat energy which was detected by the thermometer.

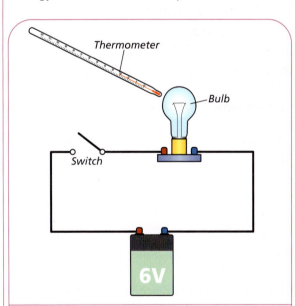

Fig 38.5 Conversion of chemical to electrical energy to heat energy

PHYSICS

285

Power and its connection to work

When work is done it is often important to know how fast it is done. We expect machines to do work for us and we don't usually want them to take all day to do the job! In physics the rate at which work is done is called **power**.

> **Power** is the amount of work done in a unit of time. It is measured in watts.

It is also the rate at which energy is converted since energy is the ability to do work.

The equation for power

The formula used to calculate power is

$$\text{Power} = \frac{Work}{time} \text{ or } P = \frac{W}{t}$$

The unit of power is the unit of work divided by the unit of time, so it is J/s. This unit is also named after a scientist and is called the **watt**.

A 40 watt bulb changes 40 joules of electrical energy to light and heat energy every second.

> **Watt:** the unit of power = J/s.

Calculations

Example 1

The man in Figure 38.1 pushes his supermarket trolley with a force of 30 N for a distance of 200 m. If he covers the distance in 120 seconds what is his average power?

First we find the work done. As above:

$$\begin{aligned} \text{Work} &= \text{Force} \times \text{distance} \\ &= 30\,\text{N} \times 200\,\text{m} \\ &= 6000\,\text{J} \end{aligned}$$

$$\text{Power} = \frac{Work}{time} = \frac{6000\,\text{J}}{120\,\text{s}} = 50 \text{ watts}$$

Example 2

A girl of mass 40 kg climbs a set of stairs of vertical height 3 m in 6 seconds. What is her average power?

Firstly, we calculate the work she has done. She has done work moving her own weight through a distance of 3 m.

Her weight is found using:

$$W = mg = 40\,\text{kg} \times 10\,\text{m/s}^2 = 400\,\text{N}$$

$$\begin{aligned} \text{The work she does} &= \text{Force} \times \text{distance} \\ &= 400\,\text{N} \times 3\,\text{m} \\ &= 1200\,\text{J} \end{aligned}$$

Finally, the power is the work done every second.

$$\text{Power} = \frac{Work}{time} = \frac{1200\,\text{J}}{6\,\text{s}} = 200\,\text{J/s} = 200 \text{ watts}$$

Key Points

- Work is done when a force moves an object.
- Work is calculated as force × distance moved in the direction of the force.
- The standard unit of work is the Nm, also known as the joule (J).
- Energy is the ability to do work.
- There are many forms of energy such as: potential, kinetic, heat, chemical, sound, electrical, magnetic, solar and nuclear.
- The principle of the conservation of energy says that energy is neither created nor destroyed but can be converted from one form to another.
- Non-renewable sources of energy such as oil, gas, coal and turf are limited and are being used up rapidly.
- Renewable sources of energy are unlimited in supply. These include solar, hydro-electric, nuclear, wind, tidal, biomass and geothermal sources.
- The national energy need is very large and is increasing. Many sources have to be imported at a cost to the economy.
- We can save energy by conserving it. Insulation is a way to prevent energy loss in buildings.
- Power is the amount of work done in a unit of time. The unit of power is the watt. We calculate power as

$$\text{Power} = \frac{Work}{time} \text{ or } P = \frac{W}{t}$$

? Questions

1 (a) What is meant by work?
   (b) What is the unit of work?

2 Calculate the amount of work done by a person pushing a pram with a force of 40 N over a distance of 150 m.

3 Calculate the work done in lifting a book of mass 1.5 kg from a table to a shelf 60 cm above the table.

4 (a) What is energy?
   (b) What units are used to measure energy?

5 (a) Name three non-renewable sources of energy.
   (b) Why should we try to conserve these sources of energy?

6 (a) Name four renewable sources of energy.
   (b) Write down one advantage and one disadvantage for each of these sources.

7 Water held behind a dam is allowed to flow downhill and to turn the fins of a turbine generator. The generator produces electricity. Describe the energy conversions in this process.

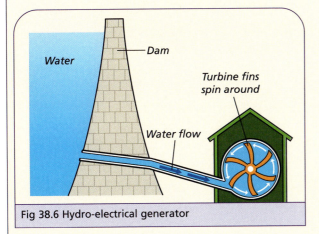

Fig 38.6 Hydro-electrical generator

8 (a) What is the principle of the conservation of energy?
   (b) A car is travelling at night with its lights on. The chemical energy in the car's fuel is the source of energy. Name three energy conversions that take place in the car.

9 Write down six everyday examples of energy doing work or being converted from one form to another.

10 Why is the sun considered the primary source of energy on Earth?

11 Write out and complete the following:
   (a) Energy is defined as the ability to do

   _____.
   The principle of the conservation of energy says that energy can neither be _____ nor _____ but can be _____ from one _____ to _____.
   The unit of energy is the _____.

   (b) Three examples of non-renewable energy sources are _____, _____ and _____. _____ energy is a renewable source of energy which is released when atoms are split.
   The process of splitting atoms is called _____. One major disadvantage of this kind of energy is the danger of _____ to humans and other living things. Alternative forms of energy include _____, _____ and _____ energy.

12 (a) What is meant by power?
   (b) What is the unit of power?

13 A weight-lifter lifts a 20 kg mass from the floor to a height of 2 m above the floor in 3 seconds. What is the average power of the weight-lifter?

14 In an investigation toy cars were allowed to roll down a board raised at one end to make a slope and along the ground. The angle of the slope was changed and the distance that the cars went along the ground when they left the board was measured. At the beginning, the raised end of the slope was 4 cm above the ground and the distance the car travelled was measured. Then the end was raised to 8 cm.
   (a) What do you think was the change in the distance travelled by the car?
   (b) What kind of energy change is happening in this experiment?
   (c) What force brings the cars to a stop?

PHYSICS

? Questions

Examination Questions

15 In Ireland 90% of electricity is generated by burning fossil fuels compared to other European countries which have an average of 50% use of fossil fuels and a 30% use of fossil fuels in the USA.

(a) List two disadvantages, excluding acid rain, of this heavy reliance on fossil fuels for the production of electricity.

(b) Suggest two alternative sources of energy for the generation of electricity in Ireland.
(JC, OL, 2006)

16 A girl of mass 60 kg (weight 600 N) climbed a 6 m high stairs in 15 seconds.

Calculate the work she did and the average power she developed while climbing the stairs.
(JC, OL, 2006)

17 Energy cannot be created or destroyed but it can be changed from one form to another. e.g. chemical energy can be converted into heat energy.

(a) Describe an experiment you could carry out to show the conversion of chemical energy to heat energy.

Draw a labelled diagram of any equipment used.

(b) Give an example from everyday life where electrical energy is converted to kinetic energy.
(JC, HL, 2006)

Chapter 39

HEAT AND HEAT TRANSFER

Introduction

One of the simplest comforts in life is being warm. Warmth is provided by heat. Heat is a form of energy that is transferred from warmer bodies to colder ones. It can therefore also be transformed into other forms of energy. In this chapter we will look at the effects of heat on materials. We will show that it can be converted into other forms of energy.

The expansion of materials when heated

Expansion of solids when heated

When train tracks are being laid a gap is left between the lengths of track. The gap is necessary because the metal in the tracks expands in warm weather. If the gaps were not there the tracks would buckle.

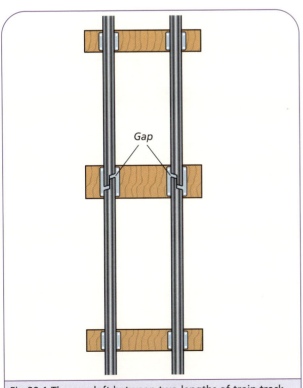

Fig 39.1 The gap left between two lengths of train track

⭐ **MANDATORY ACTIVITY 24A**

To investigate the expansion and contraction of solids when heated and cooled

Method

1 Take a ball and ring apparatus and check that the ball passes through the ring.

2 Using tongs heat the ball over the flame of a Bunsen burner for a few minutes.

3 Using the tongs again try to pass the ball through the ring.

Result

The ball will not pass through the ring because it has expanded on being heated.

4 Allow the ball to cool.

5 Try again to pass the ball through the ring.

Result

The ball once again passes through the ring.

Conclusion

The metal in the ball expands on being heated and contracts when it cools.

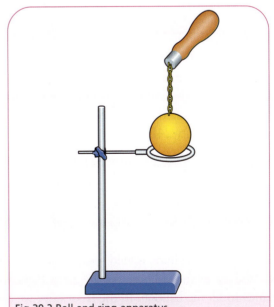

Fig 39.2 Ball and ring apparatus

PHYSICS

Expansion in different metals

Different metals expand and contract at different rates. Copper for instance expands to a greater extent than iron. This fact operates in a thermostat.

A thermostat is used to control the temperature in a room. If the room becomes too cold the thermostat trips a switch to turn on the heating. When the temperature in the room has reached a certain value then the thermostat trips the switch again to turn off the heating.

One thermostat type has a strip made of two metals such as copper and iron riveted together. When the strip is heated the copper expands more than the iron. The strip bends and closes a gap in a circuit. Current can flow around the circuit. The heating is switched on or off. See Figure 39.3.

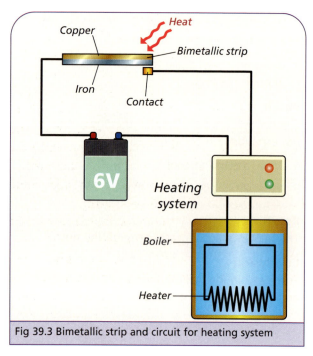

Fig 39.3 Bimetallic strip and circuit for heating system

Expansion of liquids when heated

Liquids also expand when heated and contract when cooled. This can be seen in a thermometer and can demonstrated in a simple experiment. See Mandatory Activity 24B.

Expansion of gases when heated

Gases too expand and contract under the effect of heating and cooling. This is demonstrated very impressively when hot air balloons are prepared for flight.

The collapsed balloon is held open while a burner heats the air in the balloon. As the air is heated it expands to fill the balloon. In the laboratory we can also show this effect. See Mandatory Activity 24C.

★ **MANDATORY ACTIVITY 24B**

To investigate the expansion and contraction of liquids when they are heated and cooled

Method

1 Fill a round-bottomed flask up to the brim with water that contains a little food dye.

2 Insert a cork with a glass tube into the top of the flask and attach to a retort stand, as in Figure 39.4.

3 Heat the flask with a Bunsen burner for a few minutes and observe the glass tube.

Result

The water expands and moves up the glass tube.

4 Remove the heat source and observe the water in the tube.

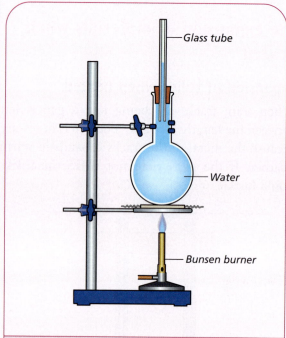

Fig 39.4 Heating of water

Result

The level of the water in the glass tube drops as the water cools.

Conclusion

Water expands on heating and contracts on cooling.

MANDATORY ACTIVITY 24C

To investigate the expansion and contraction of gases when they are heated and cooled

Method

1 Place a cork with a glass tube through it into a round-bottomed flask.

2 Fix the flask to a retort stand in such a way that the glass tubing is submerged in water, as in Figure 39.5.

3 Carefully heat the flask which, of course, contains air and observe the end of the glass tubing in the water.

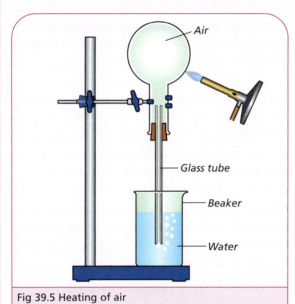

Fig 39.5 Heating of air

Result

Bubbles of air are seen in the water basin.

Conclusion

The air in the flask expands on heating and is forced out of the glass tube into the water.

4 Allow the flask to cool and observe the glass tubing once again.

Result

Water enters the glass tubing and rises along the tube.

Conclusion

When the air in the flask cools it contracts and pulls water into the glass tubing.

We have seen now that solids, liquids and gases expand when heated and contract when cooled. There is one very famous and important exception to this general rule.

Evaporation caused by heat

Expansion of water on freezing

When water is cooled you would expect it to contract. It does this but only when cooled down to 4°C. If water is cooled below this temperature it begins to expand! When 100 cm³ of water freezes at 0°C it expands to make 109 cm³ of ice. This means that ice is less dense than water and will float.

This property of water and ice is not just curious, it is also important. It means that in winter the tops of lakes freeze over but the water below does not. As a result, plants and animals can survive under the ice during the winter. Some scientists think that if water did not have this curious property life could not have developed as it did on Earth!

We can demonstrate that water expands when it freezes. Put a glass bottle full to the top with water into a bag and place it in a freezer overnight. When you examine the bottle on the following day you will see that the bottle has shattered. The water expanded on freezing and burst out of the glass bottle.

The difference between temperature and heat

Heat is a form of energy. Temperature is not an energy form but is related to heat. Temperature is a way of indicating how hot or cold a body is.

> **Temperature** is a measure of the hotness or coldness of a body.

We can demonstrate the difference between heat and temperature in a simple experiment. In the experiment we put the same amount of heat into two amounts of water and see what the result is.

PHYSICS

▶▶▶ ACTIVITY 39.1

To demonstrate the difference between heat and temperature

Method

1 Place 100 ml of water in a beaker, put the beaker on a tripod with a Bunsen burner underneath.

2 Place a thermometer in the middle of the water.

3 Heat the water for exactly 1 minute and take the temperature after 1 minute.

4 Repeat steps 1–3 but this time with 400 ml of water in the beaker.

Result

Although you put the same amount of heat in each time, the temperature was different in each case.

Conclusion

Heat is not the same as temperature.

Measuring the temperature of solids and liquids

You can measure the temperature of objects in the laboratory using a thermometer. To get an idea of the range of temperatures that different objects might have at a particular time you could, for example, measure the temperature of your hand, a pane of glass in a window, lukewarm water, water that has been cooled in a fridge and so on.

Heat rising from candles can actually be seen in this picture

▶▶▶ ACTIVITY 39.2

To determine the melting point of ice

Method

1 Attach a funnel to a retort stand with a beaker beneath.

2 Place some melting ice in the funnel.

3 Place a thermometer into the funnel.

4 Note the constant temperature as the ice melts.

▶▶▶ ACTIVITY 39.3

To determine the boiling point of water

Method

1 Attach a flask containing water to a retort stand.

2 Put a thermometer and a piece of glass tubing into a two-holed stopper and seal the flask carefully with the stopper.

3 Make sure that the thermometer bulb is above the water line.

4 Heat the water with a Bunsen burner and note the constant temperature of the steam as the water boils.

The effect of pressure on the boiling point of water

The boiling point of water is 100°C at normal atmospheric pressure, but this is not always the case. If the pressure on a liquid changes then so does its boiling point. A good example of the use of this fact is the domestic pressure cooker.

In a pressure cooker the steam above the liquid cannot escape and the pressure builds up on the liquid below it. This extra pressure causes the boiling point to be raised and the water boils in the region of 120°C instead of 100°C. Since the food is being cooked at a higher temperature, it cooks more quickly.

> **Increase in pressure raises the boiling point of water.**

We can show the opposite effect in the laboratory. In this case we reduce the pressure on water. The water then boils at a lower temperature.

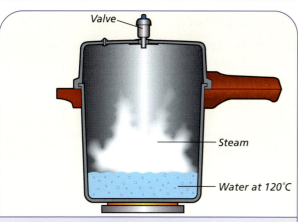

Fig 39.6 A pressure cooker

Changes of state

Normally we experience matter in three states: as a gas, a liquid or a solid. When we heat or cool materials we can get them to change state. Energy is needed in order to change the state of a material. This is demonstrated in the following experiments.

▶▶▶ **ACTIVITY 39.5**

To show the change of state from solid to liquid

Method

1 Place a number of ice cubes in a beaker with a thermometer.

2 Heat the beaker and observe the temperature.

Result

The temperature of the ice cubes rises to 0°C and then remains constant until all the ice cubes have melted. Only then does the temperature rise again.

Conclusion

The heat energy raises the temperature of the ice to 0°C, but then the energy that is being supplied is being used to change the state of the ice.

 ACTIVITY 39.4

To investigate the effect of reducing the pressure on the boiling point of water

Method

1 Boil some water in a round-bottomed flask that has a cork fitted with a thermometer and tubing, as in Figure 39.7.

2 Allow steam and air to be displaced through the tube.

3 Close the clip on the tubing and allow the flask to cool.

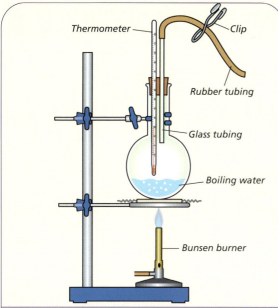

Fig 39.7 The effect of reduced pressure on the boiling point of water

Result

The water will boil on cooling and the thermometer will show a temperature below 100°C.

Conclusion

At reduced pressure the water boils at a lower temperature.

 ACTIVITY 39.6

To show the change of state from liquid to gas

Method

1 Bring water to the boil in a flask that contains a thermometer.

2 Record the temperature of the water every 2 minutes.

Result

The temperature of the water increases as you add heat until it starts to boil at 100°C. Then the temperature does not change, even though heat is being added until all the water has turned to steam.

Conclusion

When the water reaches boiling point, the temperature remains the same because the heat being added is used to change the state of the liquid to a gas.

> When material changes its state energy is taken in (absorbed) or given out (released).

PHYSICS

If steam is cooled from a temperature higher than 100°C, the temperature falls to 100°C and then stays constant for the time it takes for all the steam to change its state to liquid form. Heat is released.

Latent heat

The heat that is needed to change the state of a substance is called the **latent heat**. For example, it is known from experiments that 2257 J of heat are needed to change 1 g of boiling water to steam. This energy is called the latent heat of steam. If 1 g of steam is changed back to 1 g of water, 2257 J of energy are released.

Plotting and explaining a cooling curve

▶▶▶ ACTIVITY 39.7

To plot a cooling curve for stearic acid

Method

1 Set up the apparatus as in Figure 39.8.

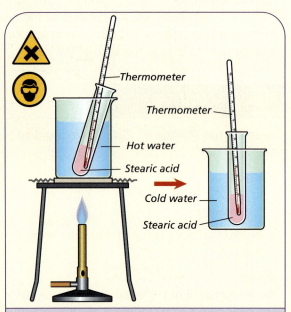

Fig 39.8 Set-up for the cooling curve experiment

2 Heat a piece of stearic acid in a test tube until it turns to liquid by placing it in a water bath and heating the water close to boiling.

3 Remove the test tube from the water bath and place in a beaker of cold water.

4 Record the temperature of the stearic acid every 20 seconds.

5 On a sheet of graph paper record the time passed on the x-axis and the temperature of the stearic acid on the y-axis.

Heat transfer

We know from everyday experience that heat moves. When we stand in front of an open fire we can feel the heat moving from the hotter fire to the colder room.

> **Heat** is a form of energy that moves from hotter objects to colder ones.

The transfer of heat energy happens in three ways: conduction, convection and radiation.

Result

The graph you get should look similar to the graph in Figure 39.9. At the beginning the curve falls as the liquid cools, but then the graph flattens at a constant temperature even though the liquid should be still cooling. This shows that energy is being used to change the liquid into a solid. When the solid is formed it cools again and the temperature drops as shown by the falling curve.

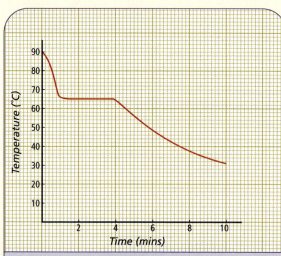

Fig 39.9 Temperature of stearic acid, recorded every 20 seconds

Conclusion

The flat part of the cooling curve shows that the stearic acid is releasing heat when it is changing state. The temperature does not drop during the change from liquid to solid.

Conduction

> **Conduction** is the transfer of heat through a substance without any overall movement of the substance itself.

Heat travels through metals and other solids by conduction. When you heat a spoon by placing it in boiling water the heat moves up the spoon to your hand by conduction.

 MANDATORY ACTIVITY 25A

To show the transfer of heat energy by conduction and that different metals conduct heat at different rates

Method

1 Set up the apparatus as shown, with a drawing pin attached to the end of each metal rod by a little wax or petroleum jelly.

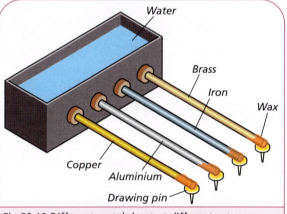

Fig 39.10 Different metals heat at different rates

2 Pour boiling water into the container covering the ends of the rods.

3 Observe the drawing pins.

Result

- As the heat moves along the metal rods the wax melts and the drawing pins drop off. They do not drop off together, showing that some metals are better conductors of heat than others.
- Copper is a better conductor than aluminium, which is a better conductor than iron.

Conclusion

The heat contained in the water has travelled down the rods by conduction.

Convection

> **Convection** is the transfer of heat through liquids and gases by the mass movement of particles.

In convection one area of a liquid or gas becomes hotter than the rest because it is closer to the source of heat. Then all of the molecules in this hot area begin to move upwards so that the heat spreads through the liquid or gas.

 MANDATORY ACTIVITY 25B

To demonstrate the transfer of heat by convection in water

Method

1 Place a crystal of potassium permanganate in the corner of a beaker containing water, as in Figure 39.11. (It helps if you use a straw when placing the crystal in the beaker.)

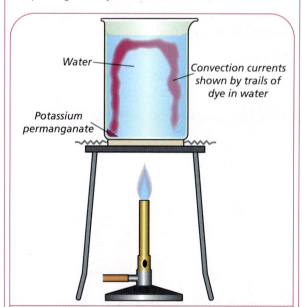

Fig 39.11 Water is heated and begins to move upwards

2 Heat the water and observe.

Result

The potassium permanganate dissolves in the water and acts as a dye. The water at the bottom of the beaker heats up first and begins to move upwards away from the source of heat. This movement is indicated by the dye in the water.

Conclusion

The heat concentrated near the source at the bottom of the beaker spreads through the liquid by convection.

PHYSICS

295

Radiation

> **Radiation** is the transfer of heat by means of waves that can travel through a vacuum.

We know that we receive heat from the sun. But we also know that there is no material between the Earth and the sun that can be used for conduction. There is no liquid or gas in the universe that could carry convection currents. The heat from the sun must be coming to us in a different way. This way is known as radiation. It may seem strange that something can pass through nothing but this is what is happening!

All hot bodies radiate heat. The hotter the body, the greater the heat radiated. The amount of heat that is radiated also depends on the surface of the body.

⭐ MANDATORY ACTIVITY 25C

To show the transfer of heat by radiation of heat from two different surfaces

Method

1 Take a shiny metal can and a similar metal can that has been painted black.

2 Fill both cans with the same amount of boiling water.

Fig 39.12 Heat is radiated more quickly from the dark can

3 Using two thermometers, measure the temperature of the water in the two cans every 4 minutes for 20 minutes.

Result

The temperature of the water falls quicker in the black can.

Conclusion

Heat is radiated by the cans and dark surfaces are better radiators than bright surfaces.

Conduction and convection of heat in water

We showed in Mandatory Activity 25B how we can investigate the transfer of heat by convection. Water is quite good at transferring heat in this way. Is water also a good conductor of heat? To answer this we conduct a further experiment.

⭐ MANDATORY ACTIVITY 25D

To investigate the transfer of heat in water by conduction

Method

1 Place a piece of ice at the bottom of a test tube with water in it.

2 Cover the piece of ice with a piece of wire to keep it in place.

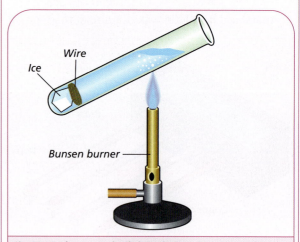

Fig 39.13 The water boils but the ice does not melt

3 Heat the top of the test tube as in Figure 39.13.

Result

The water at the top of the test tube can boil and still the ice at the bottom does not melt.

Conclusion

Water is a poor conductor of heat.

Good and bad conductors of heat

From what we have seen we can say something about the ability of materials to transfer heat.

> **Metals are good conductors of heat.**

Some metals like copper are better conductors than others such as iron. Water is a poor conductor of heat, so is air, wood, fibreglass and polystyrene. These materials are called insulators.

Insulators do not allow heat to travel through them easily.

Insulators are therefore useful for keeping in heat. You can compare the conductivity of materials using an experiment such as Mandatory Activity 25A. Instead of all metal rods, try the experiment with plastic, glass and wooden rods as well as some metal ones.

Heat shield on space shuttle *Discovery*

Heat insulation in the home

Commonly used insulators are:

- air that is trapped between two panes of glass in double-glazed windows
- aeroboard that is placed in the cavity of walls between two layers of bricks or blocks
- loose fibreglass which is used in attics to prevent heat loss through ceilings and roofs
- feathers and duck down, used in bedding and clothing.

You can compare the insulating ability of these materials. Take strips of equal thickness of different materials such as wool, fibreglass and polystyrene and place them around beakers with the same amount of hot water in each. Measure the rate at which the temperature drops with each of the wrappings. The slower the drop in temperature, the better the insulator.

🔑 ⬦ **Key Points**

- Solids expand on heating and contract on cooling.
- Different metals expand at different rates.
- Liquids expand on heating and contract on cooling.
- Gases expand on heating and contract on cooling.
- Water contracts on cooling, like other liquids, down to 4°C, but then it expands. Ice is therefore less dense than water.
- Temperature is different from heat.
- Temperature is the measure of how hot or cold a body is.
- Pressure affects the boiling point of a liquid. An increase in pressure will raise the boiling point. A decrease in pressure will lower the boiling point.
- Matter exists normally in three states: solid, liquid and gas.
- When a substance changes state, energy is released or absorbed.
- A cooling curve has a flat section showing that energy is released when a substance changes state from a liquid to a solid.
- Heat is transferred in three ways: conduction, convection and radiation.
- Conduction is the transfer of heat through a substance without any overall movement of the substance itself.
- Convection is the transfer of heat through liquids and gases by the mass movement of particles.
- Radiation is the transfer of heat by means of waves that can travel through a vacuum.
- Materials that are good at transferring heat are called conductors. Materials that are poor conductors of heat are called insulators.
- To keep heat within a building it is important to use insulators to fill walls and windows and cover attic floors.

PHYSICS

PHYSICS

? Questions

1 What is the effect of heat on solids, liquids and gases? Give one example of the use of this property in the home.

2 Describe an experiment to show that some metals expand more than others when heated.

3 (a) Describe what happens when water is cooled from room temperature to below 4°C.
   (b) Why is this property of water important for fish in lakes?

4 (a) What is temperature?
   (b) How does temperature differ from heat?

5 (a) Describe an experiment to show that water does not always boil at 100°C.
   (b) What physical property causes this change in the boiling point of water?

6 What is a change of state? Describe a change of state that you might see everyday at home.

7 (a) What is the latent heat of a substance?
   (b) If the latent heat of ice is 330 J/g, what does this mean?

8 Name the three ways in which heat can be transferred. By which of these means do we get heat from the sun?

9 Write out and complete the following:
When ice changes to water, energy must be _____. This heat is called the _____ of ice and is measured using the units _____. In reverse, when water freezes, heat is _____.

When you put ice cubes in a drink the ice takes _____ from the drink in order to _____ its state. The drink loses _____ and so it cools down. Perspiration on your skin also takes _____ from the skin to change its state from _____ to _____. This helps to keep your body _____.

10 (a) Name two good conductors of heat and two poor conductors of heat.
   (b) What is an insulator?

11 (a) Name three ways of insulating a house from heat loss.
   (b) Why is it important that houses are well insulated?

12 Draw a sketch of a cooling curve for water when it changes from steam to liquid. Comment on its shape.

13 You are asked to investigate the insulating effect of paper, plastic, cotton wool and cardboard.
   (a) List the apparatus that you would need to carry out the investigation
   (b) Name two things that you would have to do to make sure that the experiment was fair.
   (c) What measurements would you make and how often would you make them?
   (d) Name one way in which you might present your measurements.

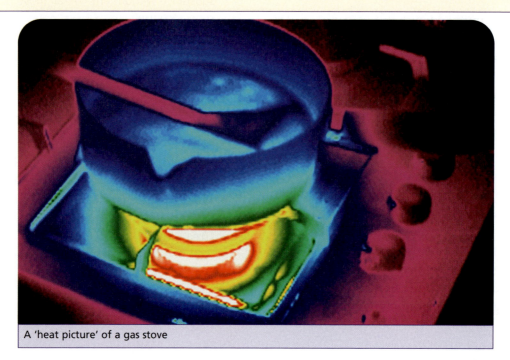

A 'heat picture' of a gas stove

Questions

Examination Questions

14 Define temperature and give a unit used to express temperature measurements.
(JC, HL, 2006)

15 Explain, clearly, the safety role of fuses in household electrical circuits.
(JC, HL, 2006)

16 (a) Name the mode of heat transfer from the hot liquid, through the spoon, to the hand.

Fig 39.14

(b) Heat moves in liquids by convection. Give one difference between convection and the way heat moves along the spoon.
(JC, HL, 2006)

17 The graph is a cooling curve. The substance used in this experiment was naphthalene. Naphthalene has a melting point of 80°C.

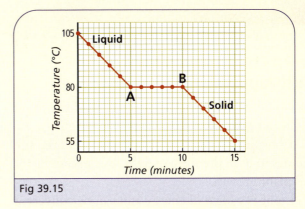

Fig 39.15

The rate of heat loss was constant throughout the experiment.
(a) What is happening to the naphthalene between points A and B on the graph?
(b) What is the heat loss, between points A and B, on the curve called?
(JC, HL, 2006)

18 Describe an experiment to show the expansion of water when it freezes.

You may include a labelled diagram if you wish.
(JC, HL, 2006)

PHYSICS

Chapter 40 ...

LIGHT

Introduction

It would be hard to imagine our world without light. In fact, there would not be much life on Earth without light. Most of the energy of the sun comes to us in the form of light. It is necessary for all plant growth.

Light is a form of energy and can be converted into other energy forms.

- In the case of the solar cell, we know that light energy can be converted into electricity.
- In plants, light energy is converted by photosynthesis into chemical energy.
- When intense sunlight falls on an object it also heats up.

⭐ MANDATORY ACTIVITY 26

To show that light travels in straight lines

Method

1 Put a hole in the middle of three pieces of cardboard of the same shape.

2 Place the three pieces of cardboard in front of a light bulb and view the light through the holes in the cardboard.

3 Move one piece of cardboard and the bulb's light cannot be seen.

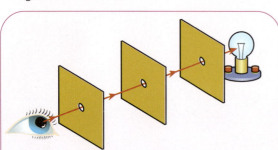

Fig 40.1 Showing that light travels in straight lines

Result

You will see from the light provided that the three holes are in a straight line.

Conclusion

Light travels in a straight line.

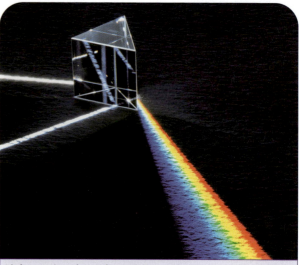

Light passing through a prism

Light travels in a straight line

You may have heard that light travels in waves. Therefore, you might imagine light travelling as a wave travels across a pond.

Light does behave like a wave sometimes. But the wave is so small that for our eyes light seems to travel in straight lines. We call these rays. The beam of a torch and the beam of light that comes from a projector in the cinema both show that light travels on a straight path.

Shadows

Shadows also show that light travels in a straight line. When you stand in sunlight, you can see your shadow on the ground. The shadow is formed because the light cannot pass through you and so the light is blocked from reaching the ground. Your shadow gives the outline of your body.

The eclipse

A very famous example of a shadow is the eclipse. This happens when the moon gets in the way of the light of the sun and then casts a shadow on the Earth or when the Earth casts a shadow of itself on the moon.

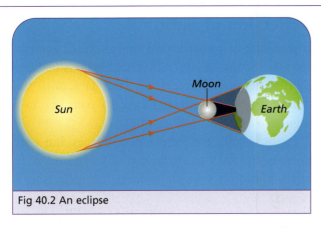

Fig 40.2 An eclipse

Luminous and non-luminous bodies

Some bodies give out light because of reactions happening inside them.

> A **luminous body** creates its own light.

- The sun is a luminous body because it creates its own light by nuclear reactions and emits it.
- A candle is also a luminous body because it generates its own light by the chemical reaction of burning.
- A light bulb is luminous because electrical energy or heat is being converted into light in the bulb itself.

Many other bodies also seem to give out light. The walls of a room can be seen because of light coming from them. However, this light is not being created in the wall. The wall is simply reflecting light from another source. Objects that reflect light but are not the source of light are called non-luminous bodies.

> A **non-luminous body** reflects light.

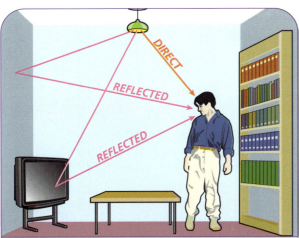

Fig 40.3 Light from bulb and reflected light from walls and furniture

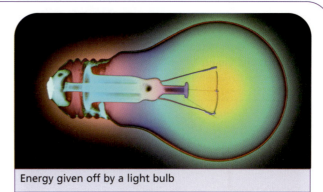

Energy given off by a light bulb

Light and all its colours

If you hold a CD in light you will see all the colours of the rainbow. The CD is a non-luminous body. It is reflecting white light from the sun. Where do the colours come from?

It would seem that white light contains colours that we can see under certain conditions. We can show in a simple experiment that white light is made up of different coloured light.

▶▶▶ **ACTIVITY 40.1**

To show that white light is made up of different colours

Method

1 Allow a beam of light from a light box to fall on a prism.

2 Place a white screen behind the prism, as in Figure 40.4.

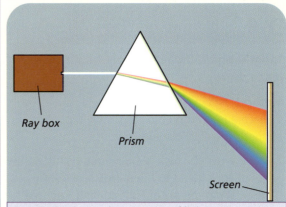

Fig 40.4 Producing the spectrum of light using a prism

Result

You will see all the colours in white light on the screen.

Conclusion

White light is made up of different colours called the spectrum of light.

PHYSICS

The spectrum of light is made up of seven major colours: red, orange, yellow, green, blue, indigo and violet.

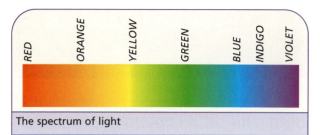

The spectrum of light

Dispersion is the breaking up of white light into its spectrum colours.

A rainbow is formed when sunlight is dispersed by raindrops in the air.

The Canadian Rocky Mountains reflected in a car mirror; this is an example of reflection in a plane surface

Reflection of light

If you look in a mirror or a shiny surface you can see a reflection of yourself. This happens because light is being reflected from your face onto the mirror and then reflected back to your eye in a regular way.

Light is reflected from most surfaces, but only mirrors and smooth surfaces give us an image. In mirrors and on smooth surfaces the reflection is regular.

We can show the path that light follows when it hits a mirror by performing Mandatory Activity 27A.

Applications of reflection in mirrors

Mirrors are useful not only for seeing ourselves in the morning! They have a number of important everyday applications. Here are some of them:

- Mirrors are used in vehicles in order to get a 'rear view' of what is happening on the road behind.
- Dentists use mirrors to look behind teeth to check for cavities.
- The headlamps of a car have a mirror behind the bulb to reflect the light forwards onto the road.

We can even use mirrors to see around corners and over the top of obstacles. A mirror arranged to do this is called a periscope.

 MANDATORY ACTIVITY 27A

To investigate the reflection of light by a plane mirror

Method

1 Place a ray box on a piece of white paper opposite a flat (plane) mirror.

2 Using a pencil, mark two or three points on the ray of light going to the mirror, and do the same along the ray of light coming from the mirror.

3 Change the angle at which the mirror stands to the ray box and repeat step 2.

Result

You will see that the light is reflected from the mirror in straight lines and that there is a regular pattern in the rays.

Conclusion

Light is reflected at the surface of a plane mirror in a regular pattern.

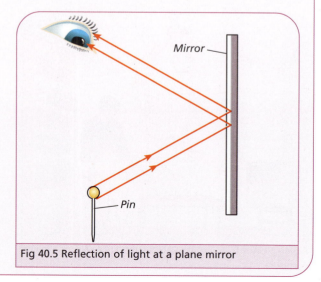

Fig 40.5 Reflection of light at a plane mirror

MANDATORY ACTIVITY 27B

To demonstrate the operation of a simple periscope

Method

1 Arrange two mirrors on two retort stands as in Figure 40.6.

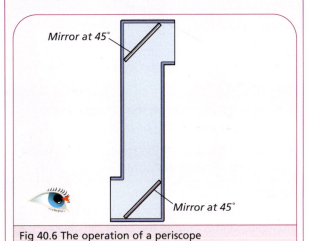

Mirror at 45°

Mirror at 45°

Fig 40.6 The operation of a periscope

2 The upper mirror must be directly above the lower one and each of the mirrors must be placed at an angle of 45 degrees approximately.

3 Place an object, such as a large book, behind the lower mirror.

Result

You will be able to see over the obstacle.

Conclusion

Light from a distant object is reflected in the upper mirror. This reflected light falls on the lower mirror and is reflected again into the eye.

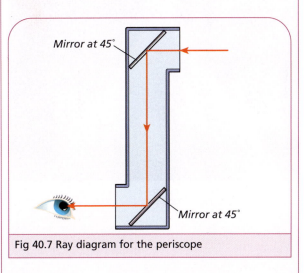

Mirror at 45°

Mirror at 45°

Fig 40.7 Ray diagram for the periscope

Refraction of light

- If you place a pencil in a glass of water it looks as if it is bent where it enters the water.
- Or if you place a coin in the bottom of a cup and move your head back until you can no longer see the coin and hold your head fixed it will 'reappear' if someone pours water into the cup.

What is causing these effects? The explanation is that the light from these objects is bent at the surface of the water. This is what is known as refraction.

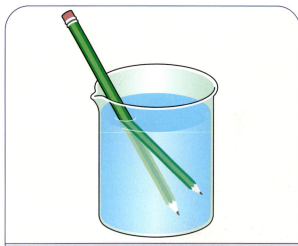

Fig 40.8 Pencil partially submerged in water

Refraction is the bending of light as it passes from one transparent material to another.

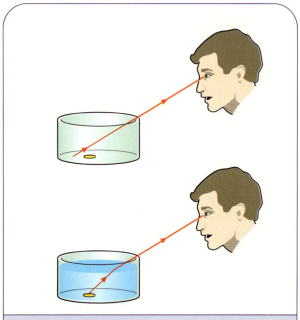

Fig 40.9 Coin in a water basin

PHYSICS

▶▶▶ **ACTIVITY 40.2**

(a) To show the refraction of light as it passes from air to glass and glass to air

Method

1 Pass a ray of light from a ray box into a block of glass as shown in Figure 40.10.

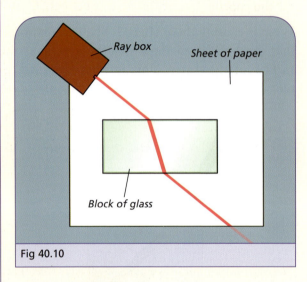

Fig 40.10

2 Note the direction of the light rays outside and inside the glass.

3 Repeat the experiment, this time putting the ray box against the block of glass.

4 Note the direction of the light ray in the glass and after it leaves the glass.

Result

The ray of light is bent one way when it enters the glass, and it is bent in the other direction when it leaves the glass.

Conclusion

Light is refracted when it enters another transparent material.

(b) To show the refraction of light as it passes from air to water and water to air

Method

1 Pass a ray of light from a ray box into a bowl of water as shown in Figure 40.10 (for this activity replace the block of glass with a bowl of water).

2 Note the direction of the light rays outside and inside the bowl.

3 Repeat the experiment, this time putting the ray box up against the bowl of water.

4 Note the direction of the light ray in the water and after it leaves the water.

Result

The ray of light is bent one way when it enters the water and is bent in the other direction when it leaves the water.

Conclusion

Light is refracted when it enters another transparent material.

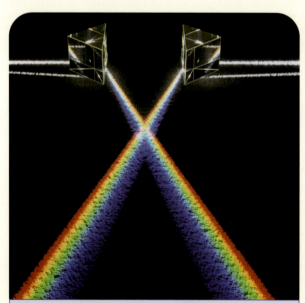

Refraction of light in a prism

PHYSICS

Applications of refraction

Lenses

An important use of refraction is in lenses. A lens is a piece of transparent material that has at least one curved surface.

The convex lens

The convex lens is thicker in the middle than at the edges. When light falls on a convex lens it is refracted and comes together at a point called the focal point.

If you look at something through a convex lens, the lens will make it look bigger. The convex lens can be used as a magnifying glass and in spectacles and contact lenses. Microscopes contain a number of lenses that are lined up to magnify an object.

Optical fibres carrying light

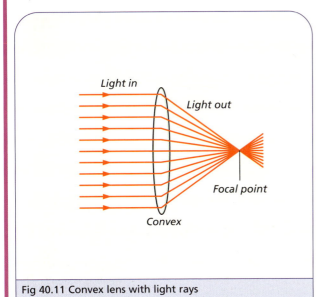

Fig 40.11 Convex lens with light rays

Prisms

A prism uses refraction to form a spectrum of coloured light.

Look at the picure in Activity 40.1. It shows the refraction of the different colours in light.

Mirages

Mirages are images of distant objects that you can see when the ground is very hot compared to the air above it. The light from the object is bent due to refraction giving unusual images. Light from the sky looks as if it is coming from the ground so the ground looks blue like water.

Key Points

- Light is a form of energy and can be converted into different forms.
- Light travels in straight lines.
- A luminous body is a source of light; a non-luminous body can be seen because it reflects light.
- Light can be broken up into colours. This is called dispersion.
- The colours in white light are red, orange, yellow, green, blue, indigo and violet.
- Light is reflected in a regular way from mirrors and shiny surfaces.
- Mirrors have everyday uses such as rear-view mirrors in cars, in headlamps and in dentistry.
- A periscope is an instrument made with two mirrors. It is used to look above or around obstacles.
- Refraction is the bending of light as it passes from one transparent material to another.
- A most important application of refraction is in lenses.

PHYSICS

? Questions

1 Name two forms of energy that can be produced using light.

2 Explain what happens during an eclipse.

3 What is the difference between a luminous body and a non-luminous body?

4 Describe, with the aid of a diagram, an experiment to show that white light contains light of many colours.

5 Name the colours of the spectrum of white light.

6 Name four everyday applications of the reflection of light in mirrors.

7 Write out and complete the following: Light is a form of _____. It travels in _____ lines but it can change direction if it is _____ in a mirror. Light also consists of many _____. The process of breaking up light is called _____. The colours that are then seen are red, _____, _____, _____, _____, _____, _____. A rainbow is formed when sunlight is _____ by _____.

8 What is probably the most important application of refraction?

9 Copy and complete the following ray diagrams showing how the rays continue.

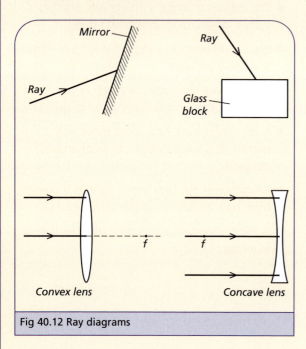

Fig 40.12 Ray diagrams

10 Can you explain why a bear would grab at the water behind where it sees a fish in order to catch it?

Examination Questions

11 When you look in a car mirror at a sign the writing is backwards.
   (a) Is this the same in a periscope?
   (b) Explain how an image is made in a periscope with two mirrors.
   (c) What would the image be like in a periscope with three mirrors?
   (Adapted from JC, HL, 2006)

12 A pupil made a simple periscope using two plane (flat) mirrors. The mirrors were arranged as shown in the diagram. The pupil looked through the periscope at the word 'Science' written on a card pinned to the laboratory wall.

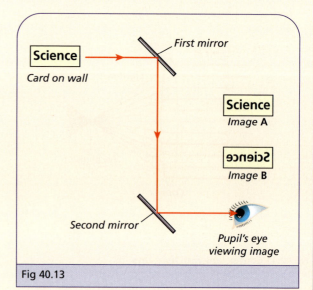

Fig 40.13

Did the pupil see image A or image B when she looked through the periscope? Give a reason for your answer.
(JC, HL, 2006)

Questions

13 The equipment shown in the diagram was set up and used in an experiment on light.

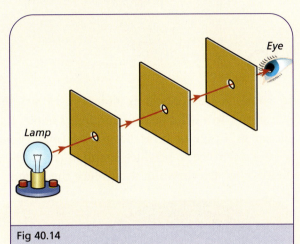

Fig 40.14

(a) What would the eye on the right see if the middle card was moved slightly?

(b) What does the experiment tell us about light?
(JC, OL, 2006)

14 You have been told that red light is refracted less than blue light when it passes from air to glass.

Describe, with the aid of a labelled diagram, how you could investigate this in the laboratory.
(JC, HL, Sample Paper)

15 (a) What is refraction of light?

(b) Give an everyday example of an effect caused by refraction.
(JC, HL, 2006)

16 The properties of light include reflection and refraction.

(a) Figure 40.15 shows three parallel rays of light striking a plane mirror. Complete the diagram showing the effect of the mirror on each of the rays of light.

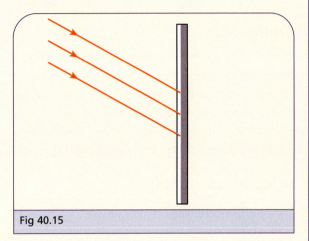

Fig 40.15

(b) Figure 40.16 shows three parallel rays of light entering a glass lens. Complete the diagram showing the effect of the lens on the rays of light.

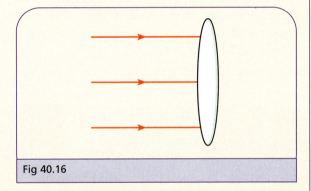

Fig 40.16

(c) Describe, with the aid of diagrams, how you could investigate the refraction of light as it passes from air into a rectangular block of glass and exits the other side.
(JC, HL, Sample Paper)

PHYSICS

Chapter 41

SOUND

Introduction

Sound is another form of energy. We experience it every day provided we are fortunate enough to have our hearing. For most of us, life would be very much poorer if it were not for music and many other daily sounds. Where does sound come from? Can we explain how it is produced?

The creation of sound

> **Sound** is produced when objects vibrate.

One good way of seeing what happens when sound is produced is to pluck a stringed instrument like a guitar. This makes the connection clear because you can see the string vibrate if you look closely.

The vibration of the guitar string causes the air molecules around it to vibrate. This vibrational energy is passed on from one molecule to the next. The vibration gets to your ear and the energy causes a vibration of the eardrum.

In some instruments it is not easy to see the vibrations. In a wind instrument, for example, it is the air in a pipe or tube that is vibrating. When we use our voice to produce sound, it is our vocal chords that are vibrating to produce the sound.

A vibrating guitar string shows the wave nature of sound

The big sound of an orchestra is made up of many sound waves overlapping

PHYSICS

Sound and energy

Sound is a form of energy. Why? Because it can do work and because it can be converted into other energy forms. You can show, for example, that sound energy can be converted into kinetic energy in the following experiment.

 ACTIVITY 41.1

To show that sound energy can be converted into kinetic energy

Method

1 Place a table-tennis ball on the end of a piece of thread attached to a retort stand.

2 Place a speaker which is attached to a sound generator beside the ball.

3 Use the sound generator to create a sound and observe the table-tennis ball.

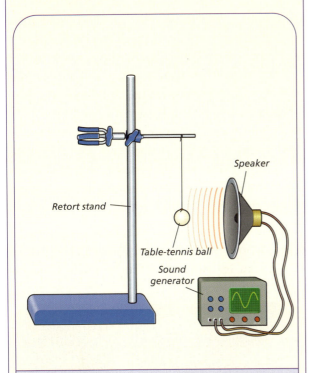

Speaker

Retort stand

Table-tennis ball

Sound generator

Fig 41.1 A table-tennis ball gains kinetic energy from the sound energy emitted by the speaker

Result

When sound comes out of the speaker the table-tennis ball moves.

Conclusion

The sound energy from the speaker has been converted into kinetic energy in the ball.

Sound energy is converted into electrical energy when we use a telephone. The energy in the sound of your voice is converted into vibrations in a small plate. These vibrations are converted into electrical signals that can be sent down a telephone line or transmitted through the air to a telephone mast.

The same conversion from sound energy to electrical energy takes place in a microphone.

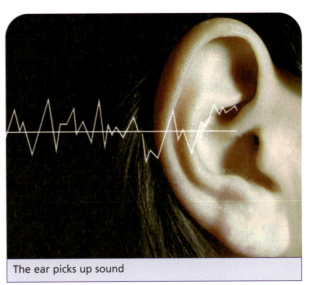

The ear picks up sound

The transmission of sound

Sound is produced by the vibration of particles. This means that sound cannot be transmitted from one place to another unless there is some material in between.

Sound needs a medium in order to be transmitted. This means sound should not be able to travel in a vacuum. This can also be demonstrated using an experiment.

The speed of sound in air

The speed of sound in air at average temperature and atmospheric pressure is about 340 m/s. We say 'about' because this value changes depending on the temperature and pressure at the time.

The speed of sound in other materials

Sound does not travel through all materials at the same speed. It travels at a much greater speed through water. This is why you can hear sounds much more clearly at sea. The speed of sound in water is approximately 1 500 m/s. In a metal such as iron it can be up to 6 000 m/s.

PHYSICS

PHYSICS

▶▶▶ **ACTIVITY 41.2**

To show that sound needs a medium in which to travel

Method

1 Place a ringing or buzzing alarm clock inside a bell jar that is attached to a vacuum pump.

2 Switch on the vacuum pump.

3 Observe what happens to the sound.

Result

As the vacuum pump takes the air out of the bell jar, the sound from the alarm clock becomes more and more faint. Eventually, you cannot hear it at all.

Conclusion

Sound requires a material in order to be transmitted.

to vacuum pump

Fig 41.2 Alarm clock in an evacuated bell jar

The speed of sound and the speed of light

The speed of sound, no matter what material the sound travels in, is much smaller than the speed of light. The speed of light has been measured at 300 000 000 m/s or 300 000 km every second!

The enormous difference in the speeds of light and sound means that if you see an event a long way off that emits a sound, you will see what happened before you will hear it. There are many examples of this.

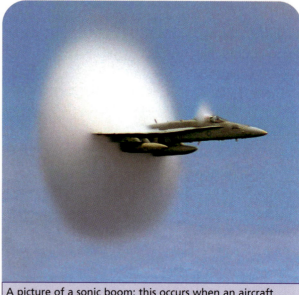

A picture of a sonic boom; this occurs when an aircraft moves at a higher speed than sound

- At a fireworks display you can see the firework separate into hundreds of flying particles before you hear the bang of the explosion.
- If you are standing beside the sea, but still a long way from the beach, you can see the waves crashing on the beach before you hear the sound.
- Finally, a famous example of the difference is the time between when you see a bolt of lightning and hear the thunder that belongs to it. You see the bolt of lightning in the sky first and then you hear the sound of the thunder.

If you count the seconds between seeing the lightning and hearing its thunder, you can get an idea of how far away the storm is! Each second is a distance of about 350 m.

Lightning: you see it first and hear it later

Sound energy can cause pain and damage your ears

Hearing and noise protection

We hear when sound vibrations arrive at our ear-drums. They begin to vibrate with the sound. These vibrations are then changed into electrical energy and are transmitted to the brain. The brain sorts them out and even gives the sounds meaning in some cases.

The ear is a wonderful thing. It can hear very faint sounds and can manage quite loud sounds.

However, sound vibrations can be so strong that they can damage the ear. There are nerves in one part of the ear, which, if they are damaged, can never be repaired. It is very important to protect your ears where there are very loud sounds.

Table 41.1 gives an idea of the scale of sounds that the ear can pick up and shows the threshold of pain. The units that measure sound level are called decibels.

Decibel: a unit of sound level.

Ear protection is generally used by people working in jobs where the sound level is greater than 70 decibels for long periods of time. There is good evidence that people damage their hearing at clubs where the sound level can be greater than 120 decibels.

Table 41.1	
Source of sound or noise	**Sound level (in decibels)**
Jet taking off at a distance of 30 m	140
Threshold of pain	120
Disco loudspeaker 2 m away	120
Pneumatic drill at road works 10 m away	100
Busy street traffic	70
Ordinary conversation	60
Quiet television or radio at home	40
Average whisper	20
Rustling of leaves	10
Threshold of hearing	0

Reflection of sound

Sound, just like light, can be reflected off surfaces.

Echo: the reflection of sound is familiar to us as the echo.

If you stand at a distance from a cliff and shout, you will hear the echo with a slight delay. This is because of the time taken for the sound to get to the wall of the cliff and then back again at a speed of approximately 340 m/s.

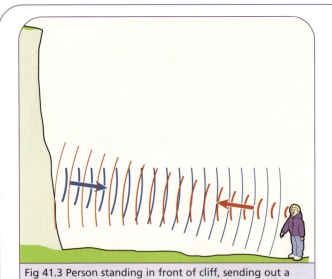

Fig 41.3 Person standing in front of cliff, sending out a sound wave and receiving the reflected wave

Ultrasound and its applications

Ultrasound 'notes' are so high that they cannot be heard by a human being but they are very useful in other ways.

> **Ultrasound** is sound of a very high pitch.

- Bats, for instance, are able to find their way around by emitting ultrasound waves and then waiting for the echo. They can use the echo to know how far away things are.
- Dolphins too use ultrasound waves to find fish and to know where the bottom of the sea is.
- The same idea is used on boats to find shoals of fish, or to calculate how deep the water is under the boat. In this case, an instrument emits an ultrasound wave and then waits for the echo to come back. The time taken for the echo can be used to calculate how far the wave travelled before it was reflected, as in the following example.

Dolphins communicate using sound waves

Example: finding depths using sonar waves

The velocity of sound in water is 1 500 m/s. A ship sends an ultrasonic wave to the seabed and the echo returned in 4 seconds. How deep is the water here?

In 4 seconds the wave travels 4×1500 m $= 6000$ m.

The echo is back after 4 seconds so the wave has travelled down and back. 6 000 m is the distance **down and back** and so the sea is 3 000 m deep.

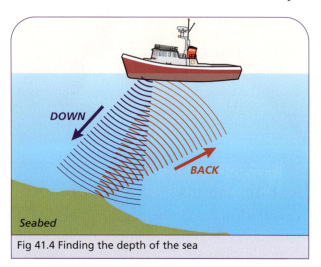

Fig 41.4 Finding the depth of the sea

Sonograms

One of the most important uses of ultrasound technology for human beings is the ultrasound scan or sonogram. This instrument uses ultrasound waves and their reflections to make 'pictures' of internal organs in the body or of a foetus in the womb. Sound waves are better in these situations because there are no harmful effects from them unlike the effects of X-rays.

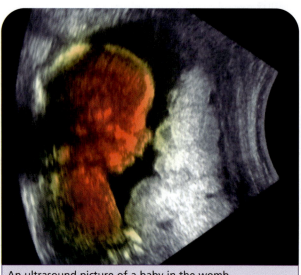

An ultrasound picture of a baby in the womb

PHYSICS

Key Points

- Sound is produced when bodies or particles vibrate.

- Sound is a form of energy. It can be converted into kinetic, electrical and other forms of energy.

- Sound needs to travel in a material in order to transmit its energy.

- The speed of sound in air is approximately 340 m/s but this changes with temperature and pressure.

- The speed of sound differs in different materials.

- The light from an explosion is seen before the sound is heard because light travels very much faster than sound.

- Very loud sounds can cause damage to the ear. Sometimes this damage is permanent. People working or living in very noisy places wear ear protection.

- Sound, like light, can be reflected. An echo is a reflected sound.

- Ultrasound is sound of a very high pitch. Animals such as bats and dolphins use it in order to navigate. It is used by fishing boats and other sea-going vessels to find fish and to measure depths.

- An important use of ultrasound is found in medicine. Sonograms are 'pictures' of internal organs using ultrasound waves.

? Questions

1 How is sound produced?
   (a) in a guitar string
   (b) in a saxophone
   (c) by a drum
   (d) by the voice.

2 Describe an experiment to show that sound can be converted into other forms of energy.

3 Describe an experiment to show that sound cannot travel through a vacuum.

4 Through which of the following materials would sound travel fastest? (a) air, (b) water, (c) iron. Why do you think this is the case?

5 During a storm a person sees a bolt of lightning in the sky and then hears the sound of the lightning 5 seconds later.
   (a) Why does the person see the event first and hear it later?
   (b) If the speed of sound in the air is 330 m/s, how far away was the storm?

6 Why do people who work in noisy places wear ear protection?

7 Write out and complete the following:
   (a) Sound is produced by _____. Sound is a form of _____. It can be converted into _____ or _____ energy.
   (b) Sound cannot pass through a _____ because it needs a _____ to pass on its _____.

   (c) The speed of sound in air is _____ but this can change with _____ and _____. Very loud sounds can damage your _____ and so in noisy workplaces people wear _____ _____.
   (d) An echo is produced when sound is _____. Ultrasound is sound of a very high _____. Two uses of ultrasound waves are for _____ in animals and in _____ in medicine.

8 If you stand on the beach and shout out to sea you will not hear an echo, but if you stand in a valley and shout you will hear an echo. Why is this?

9 What is meant by ultrasound waves?

10 Name three applications of ultrasound.

11 You are asked to design an experiment that will show that the pitch of a plucked elastic band depends on the thickness of the band.
   (a) Describe what equipment you might use in your experiment.
   (b) What variables would you have to keep the same for each elastic band tested?
   (c) What result do you think you would get?

PHYSICS

Chapter 42

MAGNETISM

Introduction

Magnetism was first seen at least 2 000 years ago in rock that was attracted to iron. Albert Einstein, the famous scientist, wrote that he first began thinking about physics when he was given a compass needle as a child. Magnets, and their seemingly mysterious force, still fascinate people.

Apart from natural magnets we can also make magnets from iron, steel, nickel and cobalt metals.

Magnetic forces

If you place the ends of two magnets close to each other you will notice that sometimes the magnets are attracted to each other. One end of one magnet pulls the other towards it. Sometimes the ends of the magnets repel: they push away from each other. This is **magnetic force**.

> **Magnetic force** can be attractive or repulsive.

We talk about these two possibilities by calling one end of a magnet a **north pole** and the other end a **south pole**. Using these names we can say that when two magnets are brought together:

> **Like poles repel each other and unlike poles attract each other.**

So a north pole will attract a south pole, but a north pole will repel another north pole.

When you look at a magnet, how do you know which pole is north and which pole is south? We decide this by suspending a magnet so that it can swing freely horizontally as in Figure 42.2.

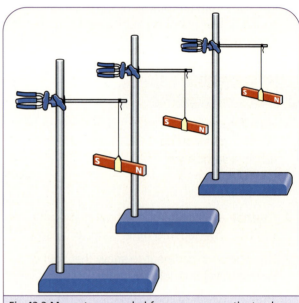

Fig 42.2 Magnets suspended from non-magnetic stands

If you suspend a number of magnets in this way, you find that they all line up in the same direction. In fact, they point in a north-south

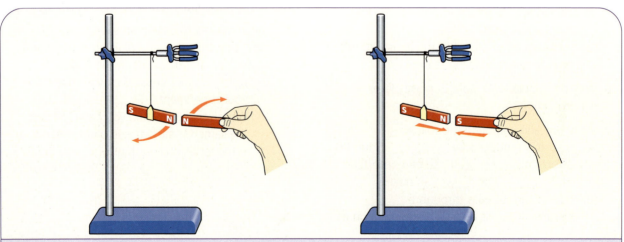

Fig 42.1 Poles of magnets attracting and repelling each other suspended from non-magnetic stands

PHYSICS

direction just like a compass needle. The end of the magnet that points north is called the north pole of the magnet.

Reading a compass

The magnetic compass

A compass is a small magnet or a magnetised needle that is free to move in a protective case. The north pole of the magnet points towards north.

The compass is used in navigation. Aeroplanes, ships or even a walker in the hills can be guided by a magnet.

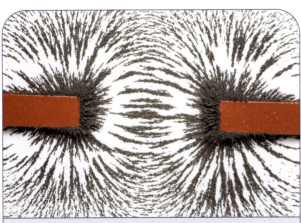

The field around attracting magnets

Effects of magnets on non-magnets

Magnets attract or repel each other but magnets also exert an attractive force on some non-magnets. We can investigate the effect of magnets on materials by performing a simple experiment.

Three metals are attracted by a magnet and can be made into a magnet themselves. They are iron, nickel and cobalt. Alloys which are mixtures of these metals, such as the material in tin cans, are also attracted to magnets.

▶▶▶ ACTIVITY 42.1

To test the effect of a magnet on different materials

Method

1 Collect a number of small samples of wood, paper, plastic, iron, brass and copper or other materials.

2 Bring a magnet close to each of the samples in turn and observe whether the material is attracted to the magnet or not.

3 Make a table of the substances that are attracted to the magnets. These are called magnetic substances.

Table 42.1	
Magnetic substances	**Non-magnetic**

Result

The wood, paper and plastic are not attracted to the magnet. The magnet attracts iron and steel but does not attract copper.

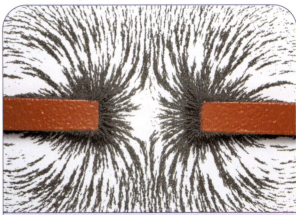

The field around repelling magnets

PHYSICS

Uses of magnets

Magnets are used in electric motors and in generators. This makes them very important. Imagine modern life without motors or generators! You can find magnets in telephones, doorbells, tape recorders and televisions. In Japan and Germany there are even trains that float on a track because of repelling magnets.

Magnets can be used to move scrap metal

Magnetic field

A magnet exerts a force on certain materials that are brought close to it. We can make a picture of this force clearly using a small compass, called a plotting compass, or using iron filings. The diagram we get is an illustration of what is called the field of the magnet.

> **The field of a magnet** is the space around a magnet in which a magnetic force can be detected.

The Earth's magnetic field

How do we explain the fact that a compass, which is a small magnet or a magnetised needle, always points in the same direction? There must be a magnetic field everywhere on Earth.

This is because the Earth itself is a giant magnet! We have said that the north pole of a magnet always points to the north. There must be a magnetic south pole at the Earth's North Pole that is attracting the north pole of every magnet.

★ MANDATORY ACTIVITY 28

To plot the magnetic field of a bar magnet

Method

1 Place a bar magnet in the centre of a large sheet of paper.

2 Draw the outline of the bar magnet on the paper.

3 Take a plotting compass and place it close to the north pole of the magnet.

4 With a pencil, mark the positions of both ends of the needle of the compass.

5 Move the compass, so that the south pole is now at the point where the north pole was in the first position. Mark the positions of both ends of the compass needle again.

6 Repeat this procedure until you have reached the south pole of the magnet. Join the dots to make a smooth curve.

7 Now choose another starting point at the north pole of the magnet and repeat the whole procedure until you have about eight curves around each side of the magnet.

These curves are called the lines of magnetic force. They are the lines that go from the north pole of a magnet to the south pole, showing the direction in which a compass needle points in the magnetic field.

Result

At the end of the experiment you should have an illustration of the magnetic field like the one in Figure 42.3.

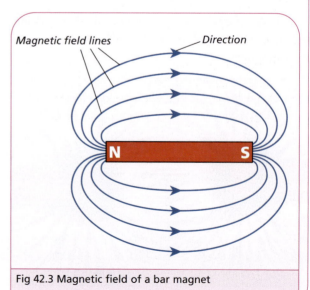

Fig 42.3 Magnetic field of a bar magnet

PHYSICS

Just like a bar magnet, the Earth also has a magnetic field around it. The magnetic field acts on a compass which always points north. We don't really understand why the Earth acts like a giant magnet, even today.

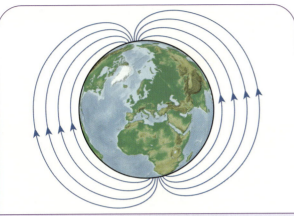

Fig 42.4 The Earth showing magnetic field lines

The magnetic field of the Earth

 Key Points

- Magnets can exert a force on other magnets and on certain kinds of metals.

- The magnetic force of a magnet on another magnet can be attractive or repulsive.

- A magnet is said to have two poles: a north pole and a south pole.

- The north pole of a magnet points in the direction of north.

- Like poles repel each other and unlike poles of a magnet attract each other.

- A compass is a magnet or a magnetised needle that is used in navigation.

- The space around a magnet where a magnetic force can be detected is called a magnetic field.

- A magnetic field can be illustrated by drawing magnetic field lines. These lines go from the north pole to the south pole and show the direction in which a compass would point.

- The Earth acts like a giant magnet and has a magnetic field around it that acts on all compasses to show north.

❓ Questions

1 How would you test whether a piece of metal was a magnet or not?

2 (a) How many poles have a magnet?
   (b) How and why are they named as they are?

3 Name three substances that are attracted to magnets and can be made into magnets themselves.

4 A magnet that is free to move horizontally will always line up in which direction?

5 Define the following terms:
   (a) magnetic field
   (b) magnetic field lines.

6 Describe an experiment to show the magnetic field around a bar magnet. Draw a diagram to show what result you would expect.

7 Write out and complete the following:
   (a) A magnet exerts a force on other magnets and on certain _____. A magnet has two _____, called the _____ _____ and the _____ _____. Like _____ _____ and unlike _____ _____. If a magnet is allowed to swing freely from a string it will line up in the direction of _____.
   (b) The space around a magnet where a force is experienced is called the _____ _____.

   This _____ can be illustrated by drawing _____ _____ lines. These show the direction in which a _____ will point.

PHYSICS

? Questions

8 What is a compass and how does it work?

9 What evidence is there that the Earth acts as a giant magnet?

10 You are given three bars of metal, all painted black. You are told that one is a magnet, one is made of steel and one is made of copper. Using another magnet what experiment would you perform to find out which bar was which?

11 What happens when the north pole of a magnet is brought close to
   (a) the north pole of another magnet?
   (b) the south pole of another magnet?
   (c) a steel bar?
   (d) a carbon rod?
   (e) a plastic pipe?

Examination Question

12 The diagram shows a bar magnet.

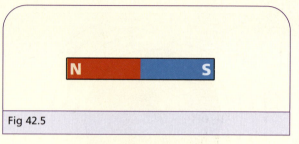

Fig 42.5

   (a) Draw the pattern made if iron filings or plotting compasses were placed around the bar magnet.
   (b) Give one use of a magnet.
      (JC, OL, 2006)

Chapter 43

STATIC AND CURRENT ELECTRICITY

Introduction

As with magnetism, there are some aspects of electricity that have been known since ancient times. The ancient Greeks knew about a material which, when rubbed, could pick up small objects. This is known to us as the yellow stone 'amber'.

Everyday experiences of static electricity

Most of us are familiar with similar experiences. If you rub a balloon against your pullover you will find that small pieces of paper are attracted to the balloon.

The same kind of physics is also involved when you get a shock from a door handle in very dry weather, or when you hear a crackling sound while taking off clothing made from acrylic or polyester.

Static electricity in hair

The force due to charges

You might think that the amber stone is a magnet. Perhaps it is magnetic forces that are attracting objects to it. Some simple experiments show, however, that it is not magnetism.

Amber is not a magnetic substance and a balloon certainly cannot be made into a magnet. There must be another explanation for what is happening. The force at work on the amber and the balloon is caused by **electrical charges**.

The atom and its charged particles

Where do electrical charges come from? This is impossible to answer. It seems that charge was always in the universe. It is a basic property of most of the matter around us.

The smallest particle in any particular substance is the atom. One picture of the atom is of a number of particles at the centre, called protons and neutrons, and of smaller particles, called electrons, somewhere in the space around the centre.

Protons and electrons are charged particles. We say they have opposite charges because they attract each other.

- The proton has a positive charge.
- The electron has a negative charge.

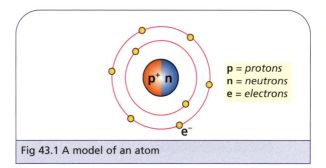

p = protons
n = neutrons
e = electrons

Fig 43.1 A model of an atom

Charging by friction (rubbing)

Charge can explain what is happening with the amber stone and the balloon. When an amber stone is rubbed with a material such as wool, charge is rubbed off one body and transferred to another.

PHYSICS

The charge that moves is the negative charge. This is the charge on the electron. The electron is the smaller and freer particle in the atom.

> **Objects become charged** by the gain or loss of electrons. An object is positively charged if it loses electrons. An object is negatively charged if it gains electrons.

We can show the two-way movement of electrons in a simple experiment.

▶▶▶ ACTIVITY 43.1

To place an opposite charge on two rods by rubbing

Method

1 Charge a polythene rod using a dry cloth and suspend it from a retort stand.

2 Charge another polythene rod in the same way and bring the second rod close to the first. Note that they repel each other.

3 Charge a perspex rod with a dry cloth and bring it close to the polythene rod.

Result

The perspex rod is attracted to the polythene rod.

Conclusion

The polythene rod and the perspex rod have different charges. Oppositely charged bodies attract each other and similarly charged bodies repel each other.

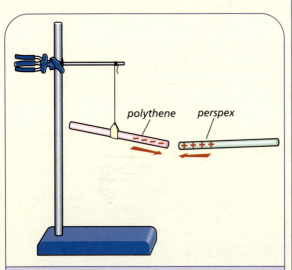

polythene   perspex

Fig 43.2 Polythene rod suspended indicating attraction to perspex rod

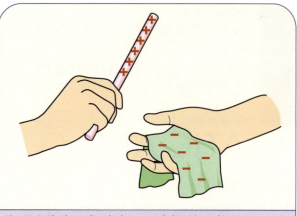

Fig 43.3 Cloth and polythene rod showing charge on each

The experiment shows that certain bodies can be charged positively or negatively by rubbing. They are charged positively when electrons are removed from them. They are charged negatively when they gain electrons. It also shows that there are attractive and repellent forces between charged bodies.

> **Charged bodies** are said to have static electricity.

Insulators and conductors

Objects that can hold a charge, such as the rods in Figure 43.2, are called insulators. Both rods in the experiment are insulators.

The opposite of an insulator is a conductor. In a conductor the charge moves through the body and so it is impossible to build up a concentration of charge.

> **An insulator** holds a charge.

> **A conductor** allows a charge to move.

It can be shown that the polythene rod is negatively charged when it is rubbed. This means that the part of the rod that is rubbed has taken on electrons from the cloth. Because the rod is an insulator, this negative charge, due to all the extra electrons it has obtained, stays on the rod.

Charges and the earth

If you try to charge a conductor by rubbing, any extra electrons it gains move through it and through your hand and your body to the earth. The rod is left uncharged.

Charge cannot be held on a conductor that is connected to the earth. The earth has an interesting property. It accepts any electrons it can get. It also provides electrons to any conductor that has a positive charge and is connected to the earth.

> **The earth** is a source and a sink for charge.

Lightning

One of the most spectacular examples of a charged body is a storm cloud that becomes strongly charged by friction with the wind. This charge can be very great. The attraction between the charge in the cloud and charges in the earth is so big that electrons can jump between the cloud and the earth. This is what we call lightning.

Huge amounts of electrical energy in the lightning damaged this church

Lightning rods

Lightning rods on buildings are strips of a conducting material, usually a metal. A metal conductor allows charge to flow through it more easily than it would flow through the air.

When a charge moves between a cloud and the earth, it is easier for it to pass through the lightning rod. So the charge will move to the lightning rod rather than to the building that the rod is protecting.

Lightning has so much energy that much of it is converted into heat. The heat is so great that it can cause fires.

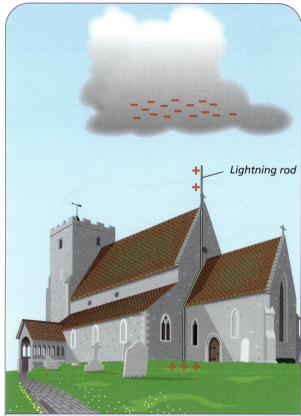

Fig 43.4 Building with a lightning rod and charged cloud above it

Current electricity

> **Electric current** is the movement of charge through a material.

Electrical current has become hugely important in our lives. It is an excellent source of energy that can be converted into many other types of energy. We use these energy conversions every day when we heat and light our houses, cook with electrical cookers, use electrical appliances such as TVs, stereos and disc players.

High-voltage static electricity

PHYSICS

Forces cause electric current

How do we get charges to move? It is force which produces motion, as we saw in previous chapters.

The force that is used here is the force of attraction between positive and negative charges and the repulsive force between negative charges. This is illustrated easily by a battery and a piece of wire.

Fig 43.5 A battery and a piece of wire with battery showing + and – poles

The battery provides the force to move electrons through the wire. In a metal conductor, such as wire, it is electrons that are carrying the charge. In other circumstances, the charge carriers could be other kinds of charged particles.

How a battery works

1 One pole of a battery is **negative** because a chemical reaction at this pole has left a lot of extra electrons there.
2 The other pole of the battery is **positive** because the chemical reaction has stripped this pole of lots of electrons, leaving it with a positive charge.
3 Because of the two charges on the poles, the electrons at the negative pole and those in the wire are attracted to the positive pole and therefore begin to move towards it. We now have electrical current.

The electrons have energy when they move and can do work.

> The number of electrons that pass a point in a second gives a measure of the size of the current.

Electrical conductivity

Some materials allow electrons to move through them easily. These are good conductors. Other materials do not allow the electrons to move and are insulators. We can test the conductivity of various materials in the laboratory.

★ MANDATORY ACTIVITY 29

To test the electrical conduction of various materials

Method

1 Using a battery, a bulb, three pieces of wire and a switch, construct a circuit as in Figure 43.6.
2 Close the switch and check that the bulb lights.
3 Remove the switch and replace with pieces of different materials, for example, a strip of iron, steel, brass, wood, plastic or cardboard.
4 In each case observe whether the bulb lights when each material is inserted into the circuit.

Result

● The bulb lights when a conductor is inserted into the circuit.
● The bulb does not light if an insulator is inserted into the circuit.

Conclusion

Some materials are better electrical conductors than others. Metals are generally good electrical conductors.

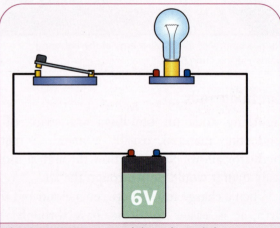

Fig 43.6 Battery connected through a switch to a battery and back to the battery

Electrical circuits, voltage and potential difference

Clearly, if the circuit in the previous experiment is broken, there is no longer any push or pull on the electrons and they stop moving.

> **Electricity flows** through complete circuits only.

The attraction between the poles of the battery does work and gives energy to move the electrons.

> **The pull on the electrons** in a circuit is usually called the voltage or the electromotive force of the battery.

The electromotive force or voltage is measured in volts.

Moving charge has kinetic energy

When the electrons start out on their journey around the circuit they have a certain energy. As they travel along the wire or through light bulbs or appliances they lose some of this energy.

Look, for example, at Figure 43.7 which shows a light bulb in a circuit with a battery. The electrons pass through the very thin wire in the bulb at high speeds. The wire heats up because of friction and glows. So, some of the energy of the electrons is transferred into light and heat.

The bit of energy needed to push the electrons through the bulb alone is called the potential difference across the bulb. This potential difference is also measured in volts.

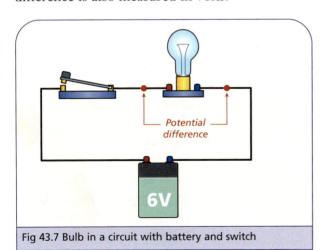

Fig 43.7 Bulb in a circuit with battery and switch

Coloured electric bulbs light when electrical current flows through them

Resistance to current

If an appliance takes a lot of energy we say that it has a high resistance. It is causing the electrons to slow down.

There is then a connection between the potential difference between two points in a circuit, the current flowing and the resistance between the two points.

- Potential difference is measured in volts and is given the symbol V.
- Current is measured in amperes or amp (A) and has the symbol I.
- Resistance is measured in ohms (Ω) and the symbol for resistance is R.

We can find the connection between these quantities by performing an experiment.

PHYSICS

MANDATORY ACTIVITY 30

To establish the relationship between current, potential difference and resistance in part of a circuit (Ohm's law)

Method

1 Set up the electrical circuit shown in Figure 43.8. The **ammeter** is an instrument to measure the current. The **voltmeter** is an instrument for measuring the potential difference (drop in energy) across the resistance. The variable resistor controls the current flowing through the circuit.

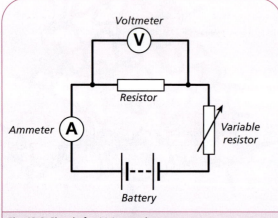

Fig 43.8 Circuit for V, I experiment

2 Move the slider on the variable resistor so that the ammeter shows a small current.

3 Note the current on the ammeter and the voltage on the voltmeter and record these as in Table 43.1.

Table 43.1	
Voltage (V)	**Current (I)**
0	0
4	0.2
10	0.5

4 Draw a graph, using graph paper, with current on the x-axis and voltage on the y-axis. Plot the points corresponding to the pairs of values from the table.

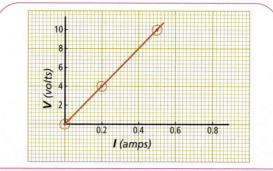

Fig 43.9 Graph of voltage against current with data above showing straight line

Result

The points all stand on a straight line.

This shows that the voltage is directly proportional to the current.

Conclusion

There is a relationship between current voltage and resistance in the circuit. This relationship can be written as

voltage = resistance x current   or   V = RI

This is a memory triangle for the relationship between voltage V, current I and resistance R. To find the formula for any quantity, cover it with your finger and then you see the correct formula, e.g. to get R, cover it up and see it is V/I.

Circuits with bulbs in series and in parallel

Our houses and buildings are all fitted with lights. These lights can be connected to each other in two main ways: in series or in parallel. Let's examine each possibility.

Bulbs in series

Look at the circuit illustrated in Figure 43.10. The two bulbs are connected one after another. The wire out of one bulb goes to the next bulb. We say these bulbs are in series.

When the switch is closed, all of the current (the moving charge) goes through each bulb. The energy from the battery is used, forcing the current through both of the bulbs. The second bulb does not glow as brightly as the first because lots of energy has been used in the first bulb. Finally, if one of the bulbs burns out the circuit is broken and the other bulb does not light.

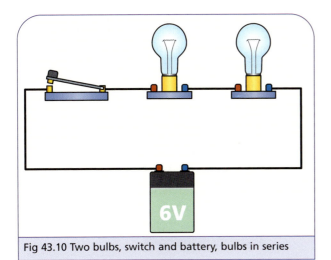

Fig 43.10 Two bulbs, switch and battery, bulbs in series

Bulbs in parallel

Another possibility is to wire the bulbs so that the input to the first bulb is connected to the input of the second bulb, as in Figure 43.11.

We say that these bulbs are connected in parallel. In this case, the potential difference is the same over each of the bulbs. The current through each of the bulbs is different. It splits up at the junction just like water can flow through different channels. If one of the bulbs burns out the other bulb will continue to burn.

The lights in a house are connected in parallel because this saves energy and if one bulb breaks then all of the others still work.

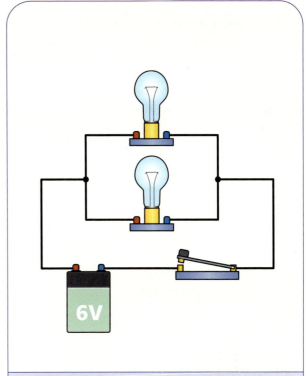

Fig 43.11 Two bulbs, switch and battery, bulbs in parallel

Calculations using the relationship between voltage, current and resistance

Example 1

A machine has a resistance of 60 ohms and uses 4 amps of current. What is the voltage of the supply of electricity?

Solution

We are given R = 60 ohms, I = 4 amps and we are asked to find V. We use the relationship:

$$V = RI$$
$$V = (60)(4) = 240 \text{ V}$$

The relationship $V = RI$ can be written in different ways using some mathematics. If you divide both sides by R you get an equation for the current:

$$I = \frac{V}{R}$$

or if you divide the equation by I on each side you get an equation for the resistance in the circuit:

$$R = \frac{V}{I}$$

PHYSICS

Example 2

A laptop computer takes 3.5 amps of current at 20 volts. What is the resistance of the laptop?

Solution

We are given $I = 3.5$ amps and $V = 20$ V. We need R. We use the equation written with R on the left-hand side:

$$R = \frac{V}{I}$$

$$R = \frac{20}{3.5} = 5.7 \text{ ohms}$$

Example 3

Electricity in Ireland is provided at 230 volts. Calculate the current that flows through a device with a resistance of 50 ohms.

Solution

This time we are given $V = 230$ V, $R = 50$ ohms. We use the equation written with I on the left:

$$I = \frac{V}{R}$$

$$I = \frac{230}{50} = 4.6 \text{ amps}$$

Key Points

- Charge can be positive or negative. Positive charge is a lack of electrons and negative charge is an excess of electrons.

- Unlike charges attract and like charges repel.

- Insulators are bodies that can hold a charge. Conductors are bodies that allow charge to move and so cannot hold a charge.

- All charge tries to go to earth. The earth is a source and a sink for charge.

- Electrical current is the movement of charge through a material.

- Electrical current is a source of energy and this energy can be converted into other energy forms such as heat, light and kinetic energy.

- A battery with its positive and negative poles provides the pull on electrons to get them to move. We say there is a potential difference between the two poles. Potential difference, or voltage, is measured in volts.

- Electricity flows through complete circuits only.

- When current flows through devices there is a resistance to the motion and the charges lose energy.

- The relationship between voltage, current and resistance is $V = RI$.

- This relationship can be written as
  $$I = \frac{V}{R} \quad \text{or} \quad R = \frac{V}{I}$$

- Devices can be connected in series or in parallel.

Questions

1. Name the smallest particle of a given substance and the two charged particles that are part of it.

2. How do bodies become charged? Explain what is happening in terms of transfer of charge.

3. What is meant in electricity by
   (a) an insulator?
   (b) a conductor?

4. Why can a conductor that is in contact with the earth not hold a charge?

5. What is a lightning rod and what is its purpose?

6. (a) What is the difference between static and current electricity?
   (b) Name five uses that are made of current electricity in the home.

7. What is the electromotive force, or voltage, of a battery?

8. What is meant by the resistance of an appliance?

PHYSICS

? Questions

9 Write out and complete the following:

(a) An _____ is a body that can be charged by _____ it. _____ are bodies that cannot hold a charge. The earth is a _____ and a _____ for charge. Lightning is the movement of charge between _____ and the _____. Current electricity is due to the _____ of charge.

(b) Current electricity is a source of _____ that is used in appliances. If an appliance needs a lot of _____ we say it has a high _____. The force needed to move electrons around circuits is provided by _____. _____ is measured in volts.

(c) The connection between voltage, current and resistance in a circuit is: voltage = _____ × _____. Bulbs or appliances in circuits can be connected in _____ or in _____.

10 Calculate the following:

(a) the resistance of a bulb in a torch connected to a 1.5 volt battery and taking 0.3 amps of current

(b) the current flowing through a resistance of 6 ohms connected to a 12 volt battery

(c) the voltage needed to send a current of 2 amps through a resistance of 10 ohms.

11 The measurements made by a student in an experiment to show the relationship between potential difference, current and resistance in a simple circuit are shown in Table 43.2.

Table 43.2						
Voltage (V)	1.0	2.0	3.0	4.0	5.0	6.0
Current (A)	0.1	0.2	0.3	0.4	0.5	0.6

(a) Use the table to draw a graph, on graph paper, of voltage against current. Put voltage on the *y*-axis.

(b) Describe the graph. What does the graph show about the relationship?

(c) Calculate the resistance of the resistor used in this experiment.

12 An experiment is to be designed to see if you can insulate against magnetic force like you can insulate electrical forces. Different materials are to be put in front of the pole of a magnet and the magnetic field is to be tested behind the insulating material.

(a) Name four different materials that could be used as insulators.

(b) Name two ways in which you might measure the magnetic field.

(c) What would you have to do to make sure of a fair experiment for each of the materials?

Examination Questions

13 A student set up the circuit drawn below to investigate different materials to see which were electrical conductors and which were electrical insulators.

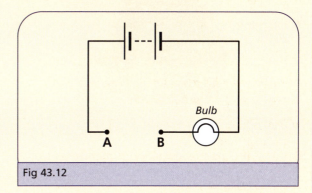

Fig 43.12

(a) What would you expect to observe when an electrical conductor is connected between the contact points A and B? Give a reason for your answer.

(b) What would you expect to observe when an electrical insulator is connected between the contact points A and B? Give a reason for your answer.

(JC, OL, 2006)

PHYSICS

? Questions

14 Components, e.g. bulbs, in electrical circuits can be connected in series or in parallel.
  (a) It is noticed that, when one headlight fails (blows) in a car, the second remains alight. State the way the headlights are connected and give a reason why this mode of connection is used.
  (b) All of the bulbs go out in an old set of Christmas tree lights, when one of the bulbs fails (blows). In what way are the bulbs connected in this set of lights?
  (c) Explain why, when one bulb blows, they all go out.
  (JC, HL, 2006)

Fig 43.13

15 (a) Calculate the resistance of the filament of a car headlamp when 12 V produces a current of 5 A in it.
  (b) In what unit is resistance measured?
  (JC, HL, 2006)

16 Georg Ohm published his law in 1827.

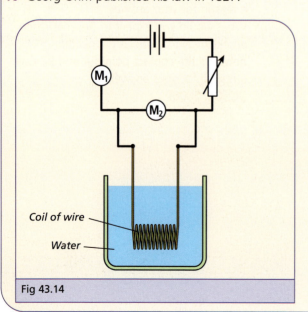

Coil of wire

Water

Fig 43.14

A student set up the circuit in Figure 43.14 to investigate the relationship between the potential difference (voltage) across a metal conductor and the current flowing through it. Two meters, M_1 and M_2, were inserted in the circuit. The data collected in this investigation was used to plot a graph of current against potential difference.

  (a) Identify the meters labelled M_1 and M_2 in the circuit.
  (b) Why is it desirable to have the metal conductor immersed in a liquid such as water?

  The graph of the data produced in this investigation is shown in Figure 43.15.

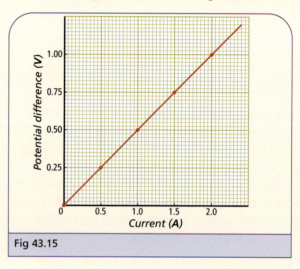

Fig 43.15

  (c) Use the information in the graph to calculate the resistance of the conductor. What are the units of resistance?
  (JC, HL, Sample Paper)

PHYSICS

Chapter 44

USES AND EFFECTS OF ELECTRICITY

Introduction

We now look at how electricity works in the home and in experiments. In that way we learn more about the use of electricity. Electricity plays a very important role in modern life so it is useful to know something about how it works, the dangers that can be involved, and the cost of it.

We will also look at the effects of electricity.

Electricity in use

Direct current (DC)

In a simple circuit with a battery, a switch and a bulb, the potential difference between the poles of the battery makes the charges move around the circuit.

The charges in a metal conductor are electrons. These move towards the positive pole because they are negatively charged. The charges move in one direction only around the circuit. This is called **direct current**.

> **Direct current** is the flow of charge in one direction only around a circuit.

Alternating current (AC)

The electricity that is produced in generating stations is not produced by a battery! In the generating station, turbines are made to move between huge magnets and electricity is produced. Because of the way in which it is produced the direction in which the charges flow changes all the time.

In Ireland, electrical current in the mains supply changes its direction of flow one hundred times every second. The current flows 50 times in one direction and 50 times in the other direction.

> **Alternating current** is current that constantly changes direction.

Plugs and fuses

Large appliances in the home such as cookers, ovens and power-showers are connected directly to the mains supply. Most small electrical appliances in the home have a plug on them. It is sometimes necessary to change the plug on an appliance and everyone who studies physics should be able to do this safely.

An electrical plug in Ireland has three pins.

- One pin connects to the mains supply to pick up the current coming into the house and allows it to flow into the appliance through wire with a brown coating.
- The other pin picks the current coming back out of the appliance and allows it to flow back to the power station through wire with a blue coating.
- The third, and largest, pin is the earth pin. This pin acts as a safety device. If, because of a fault, the current becomes connected to the outside of the appliance, it is necessary to lead this current to the earth through a wire that is striped green and yellow.

If the current cannot go to earth through a wire it will go through a person who touches the appliance. This is what happens when a person gets an electric shock.

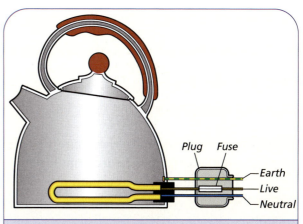

Fig 44.1 The wiring of an electric kettle

PHYSICS

In addition to the three pins, a plug also contains a fuse. It is connected to the pin that takes the incoming current. A fuse is also a safety device. It consists of a piece of wire that will melt if the current becomes too big.

If there is a fault in the circuit and the current is flowing through the earth wire to the earth, great amounts of current flow. In this case, the fuse melts and the circuit is broken so the current stops flowing.

Wiring a plug

- The wire carrying the current to the appliance is called the live wire (brown).
- The wire carrying the current from the appliance is called the neutral wire (blue).
- The earth wire is striped green and yellow and carries current only if there is a fault.

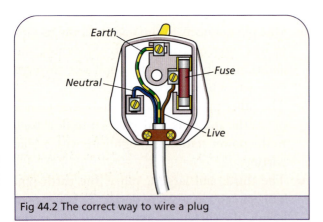

Fig 44.2 The correct way to wire a plug

Figure 44.2 shows how a plug can be wired safely.

1 Remove the cover of the plug and loosen the screws holding the cord grip. Loosen the screws on each of the three pins.

2 Strip the top 4 cm of the insulation on the flex. Cut 1 cm from the top of the blue and brown wires, leaving the striped wire a little longer.

3 Strip back 1.5 cm of insulation from each of the wires. Twist each wire and bend over the top.

4 Connect the brown wire to the live pin with the fuse on it. Connect the blue wire to the pin without the fuse. Connect the green and yellow wire to the large pin at the top. This is the earth wire.

5 Tighten all screws well. Make sure that the flex is well secured in the cord grip. Place the cover on the plug and screw until it is closed tightly.

Circuit breakers

Every three-pin plug is fitted with a fuse that will melt if too much current flows through it. A similar safety device is found in all electrical circuits in a house. In an average house there may be seven or more different circuits that carry electricity.

- One circuit may carry current for a washing machine alone.
- Another circuit might have all of the ground-floor lights on it.
- A third circuit might have the electrical sockets in the kitchen on it.

The starting point of each circuit is at the fuse board. Here there is a circuit breaker connected to the live wire. The circuit breaker has now taken over from the fuse as a safety device in a circuit. It has the advantage that when too much current flows through the circuit it breaks the circuit but can be reset. This is done by pushing the switch back up when the problem has been solved. A fuse, on the other hand, when it has melted cannot be reused and a new one is needed.

Circuit breakers on a mains electricity board

Units of electricity and their cost

Electricity is expensive to produce. The turbines that produce electricity must be moved at high speeds. This requires energy which comes, mostly, from burning fuels, but also from flowing water or the energy of the wind. The person using the electricity, the consumer, must pay for it.

The Electricity Supply Board (ESB) is the main provider of electricity in Ireland. The ESB charges its customers based on unit cost.

The unit that is used is the kilowatt hour, which has the symbol kWh. One watt of power is the same as providing one joule of energy per second. A kilowatt is 1000 watts and the kilowatt hour is 1000 joules of energy per second flowing for 1 hour.

> **A kilowatt hour** is the amount of energy provided when 1 kilowatt of power is used for 1 hour.

The ESB charged 12.73c for 1 kilowatt hour of power in 2006.

We can find the cost of running appliances in the home by looking at the power rating on the appliance and doing a simple calculation.

Example 1

What is the cost of running a 2.5 kW electric heater for 5 hours if a unit of electricity costs 12.73c?

Solution

The unit of electricity is 1 kWh costing 12.73c.

A 2.5 kW heater will cost 2.5 × 12.73c to run for 1 hour.

The heater will then cost:

2.5 × 12.73 × 5 = 159c = €1.59

Here is a table of the power ratings for some common electrical appliances:

Table 44.1	
Appliance	**Power rating**
Electric heater	2.5 kW
Computer	70 W
Dishwasher	1.5 kW
CD player	15 W
Hairdryer	1200 W at full speed; otherwise 600 W

Example 2

Find the cost of running a dishwasher for 1.5 hours. Use the table to find the power rating of a dishwasher.

Solution

The power rating of a dishwasher = 1.5 kW

∴ it costs 1.5 × 12.73c = 19.1c per hour

∴ it costs 19.1 × 1.5 = 28.65c for 1.5 hours

The effects of electricity

We can study the effects of electricity under three headings:

1 The heating effect
2 The chemical effect
3 The magnetic effect.

An electricity substation changes the electricity from the power station so that it can be used in homes

1 The heating effect

Electricity is the flow of charge. In metal conductors the charges are electrons. When the electrons move through the wire they have to pass other particles and so there is friction. This friction causes heating.

The electric bar heater is based on the idea that the friction between the charges and metal will be greatest if the wire is very thin. A bar heater uses thin wire coiled on an insulating material. When electricity flows the wire heats up so that it glows red-hot. This is an example of the heating effect of electricity.

Other examples of the use of the heating effect of electricity are the electric kettle, the electric iron and the immersion heater.

The heating effect is also used in the fuse which was described earlier. If the flow of electricity through a circuit is too great, the heating effect of the current in the fuse causes it to melt and thus break the circuit.

2 The chemical effect

A second effect of electricity is its ability to break chemical compounds into charged parts called ions. These charged particles are then attracted to opposite charges.

PHYSICS

The breaking down of a compound by passing electricity through it is called electrolysis. This property of electricity is used in electroplating and electro-painting. It can also be used to break water into hydrogen and oxygen gas.

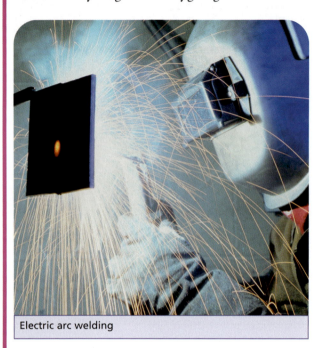
Electric arc welding

Electroplating

In electroplating a coat of one metal is placed on top of another metal. Silver plating puts a coat of silver on top of a less valuable metal.

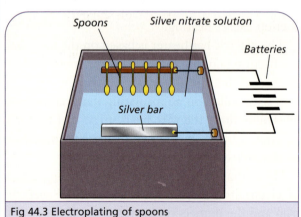
Fig 44.3 Electroplating of spoons

The spoons in Figure 44.3 are attached to the negative pole of a strong battery. The positive pole is attached to a silver plate and both are immersed in solution of silver nitrate.

When the electricity flows the silver nitrate is broken down into silver ions and nitrate ions. The silver ions are positively charged and are attracted to the spoons which are negatively charged. The silver sticks to the spoons.

Electro-painting

Cars are painted in a similar way. The paint is made of ions and the whole body of a car is attached to a very large battery. The paint and the car body are in a large 'bath'. When electricity flows the ionised paint is attracted to the car body and sticks to it. The car body is covered with a coat of paint.

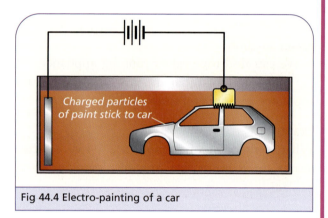
Fig 44.4 Electro-painting of a car

3 The magnetic effect

A third important effect of electricity is the magnetic effect. This can be demonstrated using a battery, wire and a compass as in Figure 44.5.

1 Attach the two ends of a length of wire to the poles of a battery.
2 Hold the wire above a compass so that the direction of the wire is perpendicular to the direction of the compass needle.

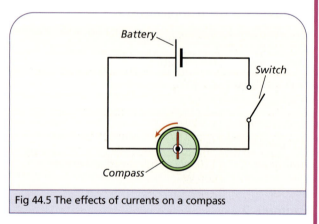
Fig 44.5 The effects of currents on a compass

Result

The compass needle moves as if there were a magnet near it. This shows the magnetic effect of an electric current.

PHYSICS

The electromagnet

One use of the magnetic effect of electricity is the electromagnet. This is a coil of wire that is wound around a large metal ring. When electricity flows through the coil of wire the ring behaves like a magnet. Electromagnets are used, for instance, in scrapyards to lift and separate metals.

An electromagnet attracts metals when current flows

Key Points

- Direct current is the flow of charge in one direction only around a circuit.

- Alternating current is current that constantly changes direction around a circuit.

- An earth pin in a plug is a safety device to allow stray current to flow to earth safely.

- A fuse is also a safety device. The metal in a fuse will melt and break a circuit if the current in the circuit is greater than an allowed amount.

- The wire carrying current to an appliance is called the live wire; the wire carrying it away from an appliance is called the neutral wire.

- When wiring a plug, the brown wire must be connected to the live, the blue wire must be connected to the neutral and the green/yellow wire must be connected to the earth pin of the plug.

- A circuit breaker is a safety feature of a circuit. It will break a circuit if too much current is detected. It has the advantage over a fuse that it can be reused.

- The unit of electricity is the kilowatt hour.

- A kilowatt hour is the amount of energy provided when 1 kilowatt of power is used for 1 hour.

- There are three main effects of electricity:
  – the heating effect
  – the chemical effect
  – the magnetic effect.

- The heating effect of electricity is used in fuses, electric kettles and irons.

- The chemical effect is used in electroplating and in electro-painting.

- The magnetic effect is used in electromagnets.

? Questions

1 What is meant by:
  (a) direct current?
  (b) alternating current?

2 (a) Explain why a plug has an earth pin.
  (b) In the case of an electric kettle, to what part of the kettle would the earth wire be connected?

3 Describe how to wire a three-pin plug correctly.

4 In the diagram below, what is shown by the letters A, B, C, D and E?

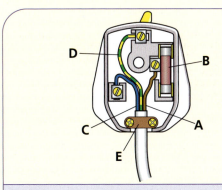

Fig 44.6

5 (a) What is the function of a circuit breaker?
  (b) What is the difference between a circuit breaker and a fuse?

6 What is the unit of electricity that is used by the ESB?

PHYSICS

? Questions

7 Find the cost of operating the following appliances for the lengths of time indicated:
   (a) a 2.5 kW electric heater for 6 hours
   (b) a 70 W computer for 10 hours
   (c) a 1.5 kW dishwasher for 2 hours
   (d) a hairdryer at full speed for half an hour (consult Table 44.1 for the power rating).

8 Write out and complete the following:
   (a) Alternating current is current that changes _____ in a circuit. Electricity flows around _____.
   (b) We can connect appliances to the supply using a _____. This has three pins on it. One pin is connected to the _____ wire which is coloured _____, one is connected to the _____ wire which is coloured _____ and the third is connected to the _____ wire which has the colour _____.
   (c) One of the pins has a _____ attached to it. This is a safety device. If the current in the circuit is too great a strip of _____ in the _____ melts and _____ the circuit.
      In some cases _____ are used instead of _____. They have the advantage that they can be _____.

9 What are the three main effects of electrical current?

10 (a) Describe how a hairdryer works.
     (b) Name two energy changes that occur in a hairdryer.

11 Describe how a piece of cutlery might be electroplated. Why would a cutlery manufacturer wish to do this?

12 (a) Describe how an electromagnet works.
     (b) Why is it important on a boat to keep the compass at a distance from all electrical circuits?

13 You are to design an experiment to see whether the thickness of a wire has an effect on electrical resistance. Describe:
   (a) the apparatus that you would need to conduct the experiment
   (b) what you would need to do to make sure the experiment was fair in each case of different wires
   (c) what you think the outcome would be.

Examination Questions

14 The diagram shows a three-pin plug with the back removed.

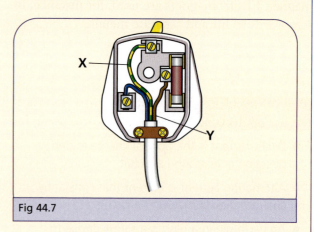

Fig 44.7

   (a) What are the correct names for the cables labelled X and Y?
   (b) Give one reason why the back covering (casing) of a plug is made from plastic.
      (JC, OL, 2006)

15 Appliances vary in the amount of electricity they use depending on their power rating. A tumble-dryer has a high power rating of 2.5 kW.
   (a) Name another appliance found in the home that has a high power rating.
   (b) Name an appliance found in the home that has a low power rating.
      (JC, OL, 2006)

16 The ESB charges for electricity at a rate of 12 cent per kWh.

   A tumble-dryer of power rating 2.5 kW is used for 2 hours each week for 4 weeks.
   (a) How many units of electricity are used?
   (b) What is the cost, in cent, of using the tumble-dryer?
      (JC, OL, 2006)

PHYSICS

Chapter 45

ELECTRONICS

Introduction

One of the obvious signs of a wealthy society is the use of electronic devices. These are found, for example, in

- washing machines
- CD players
- mobile phones
- alarm systems
- calculators.

In all of these devices there are tiny electrical circuits carrying a current of moving electrons. These circuits are built to do a number of different tasks.

Electronics is about sending small amounts of current around tiny circuits in order to control devices. Modern technology allows us to place thousands of tiny circuits onto a small silicon chip. These tiny circuits perform thousands of tasks each second.

In each circuit there are a number of components to help control what the circuit is to do. In this chapter we look at some of these components and their uses.

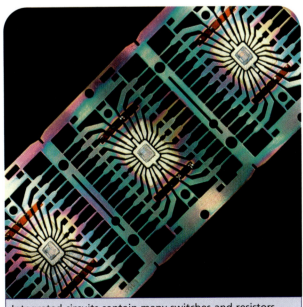

Integrated circuits contain many switches and resistors

Switches

One of the most obvious components in a circuit is the switch. A switch in an electronic circuit works just like a light switch.

- When a switch is closed current will flow.
- When a switch is opened the circuit is broken and no current can flow.

> A **switch** is a device for completing or breaking a circuit.

In a washing machine there are two switches. One switch is the on-off button on the machine. The other switch is connected to the door. These two switches are connected one after the other. They are said to be in series. Both switches must be closed for the circuit to be complete. Current flows only when both switches are closed. Figure 45.1 illustrates this.

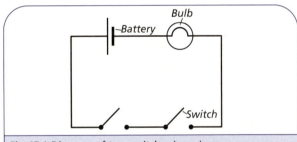

Fig 45.1 Diagram of two switches in series

Resistors

> A **resistor** is a component that is used to control the amount of current flowing in a circuit.

As its name suggests, a resistor contains a material that provides great resistance to the flow of current.

Resistors come in different sizes depending on the task they have to do in a circuit.

There are also variable resistors whose resistance can be changed while they are connected in a circuit to allow different amounts of current to flow.

PHYSICS

Resistors can be used to get smaller voltages from larger ones. This can be important in circuits that have different devices which need different sizes of voltage. For example, a dimmer switch on a light uses a resistance to change the current to the bulb making it brighter or dimmer.

▶▶▶ ACTIVITY 45.1

To demonstrate how to use resistors to get lower voltages from a 6 V battery

Method

1 Connect a battery and two 100 ohm resistors as in the diagram.

2 Connect a voltmeter on either side of one of the resistors and measure the voltage across the resistance (or the potential difference between the points A and B).

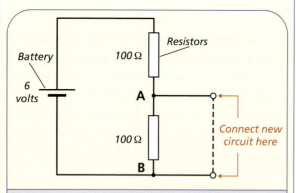

Fig 45.2

Result
The voltmeter measures 3 volts across the resistor.

3 Replace the 100 ohm resistor with different size resistors and measure the voltage across these resistors.

Conclusion
Resistors can be used to divide up the voltage of a battery. If, for example, you needed 3 volts for a particular circuit, you could connect the circuit across the 100 ohm resistor in the diagram.

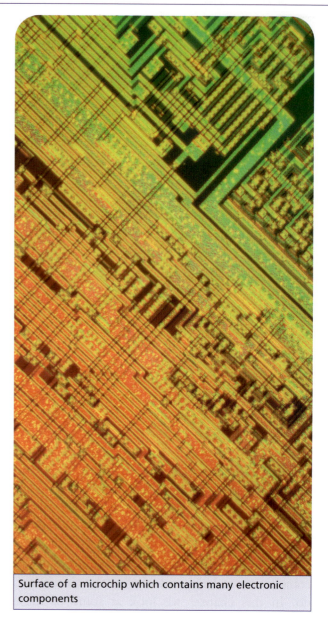

Surface of a microchip which contains many electronic components

The diode

Another device that is used in electronic circuits to control current is the diode.

> A **diode** is a device that allows current to flow in one direction only.

A diode, like many other components in an electronic circuit, is quite small and is made from materials called semi-conductors. The physics of semi-conductors is left for later study. We are interested here in the uses of the diode.

One of the most important uses of the diode is to change alternating current into direct current. As we saw in the last chapter the electricity that we get from the power station is alternating current. It changes direction one hundred times each second.

PHYSICS

There are a number of appliances in the home that cannot run on alternating current and need direct current.

Examples of such devices are:

● computers
● transistor radios
● tape recorders.

We can do an experiment to demonstrate this property of a diode in a circuit.

In order to draw a diagram of the circuit we use certain symbols for the devices in the circuit. Table 45.1 shows the symbols we use for the diode and other components that we will discuss in this chapter.

Table 45.1 Symbols for electronic components	
Component	**Circuit symbol**
Battery	(The long, thin line represents the positive pole and the shorter, thicker line is the negative pole)
Switch	
Bulb	or
Resistance	50 Ω
Diode	
Light-emitting diode	
Light-dependent resistor	
Alternating current supply	

►►► ACTIVITY 45.2

To demonstrate the properties of a diode

Method

1 Connect a battery, bulb, diode and switch as shown.

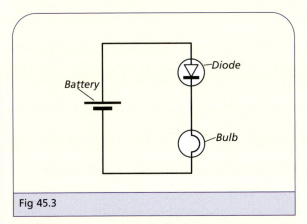

Fig 45.3

2 Close the switch and observe the bulb.

Result

If the negative pole of the diode is connected to the negative of the battery then the bulb lights.

3 Turn the diode around and connect it in the circuit with the poles connected in the opposite way.

Result

The bulb does not light.

Conclusion

● A diode allows current to flow in one direction only. When a diode is connected to allow current to flow, it is said to be **forward biased**. If it is connected in such a way that no current flows, it is said to be **reverse biased**.

● When a diode is built into a circuit it is important to know which end of the diode is the positive terminal and which is the negative. To show this a silver band is placed on the diode at the negative terminal.

The light-emitting diode (LED)

> **A light-emitting diode** is a diode that gives out light when current passes through it.

This is another example of an energy conversion. When current passes through circuit components the electrons lose kinetic energy and this is usually converted into heat.

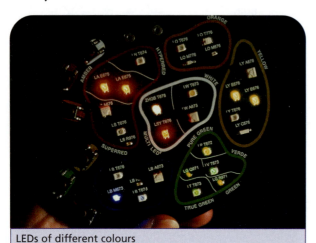

LEDs of different colours

There are some substances, however, in which the energy is converted into light. This is what happens in the LED. The light given out by LEDs is normally red or green. Other less common colours are yellow and blue. When the diode allows current to pass, light is emitted and the plastic case glows.

Like all diodes the LED allows current to flow only when it is connected in forward bias. This property can be put to use to identify which pole of a battery is positive or negative.

- In the circuit in Figure 45.4(a) if the red LED lights up then A is the positive pole of the battery.
- In the circuit in Figure 45.4(b) if the green LED lights up then A is the positive pole of the battery.

You will also see the symbol for a resistor in series with each of the LEDs. This is because an LED needs very little current. If an LED receives too much current then it will be damaged. To limit the amount of current that goes through the LED, a large resistor is always connected in series with it.

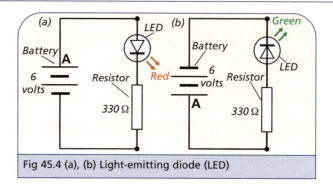

Fig 45.4 (a), (b) Light-emitting diode (LED)

Uses of the light-emitting diode

LEDs are used on most electronic equipment. They are seen most often as power indicators which show that a device has current flowing to it. LEDs need very little current – much less than a bulb. They are cheap to manufacture and are very reliable.

Another common use of the LED is in calculators and in alarm clocks. They are used to light up the numbers on the display. Any number can be illustrated by an array of seven strips of light. Each strip is a light-emitting diode. By lighting up different combinations of the array of strips all the numbers from zero to nine can be represented.

The light-dependent resistor (LDR)

Another component that is used very often in electronic circuits is the light-dependent resistor.

> **A light-dependent resistor (LDR)** is a variable resistor whose resistance changes with the amount of light that falls on it.

If very little light falls on an LDR, its resistance is very high and it will allow very little current to flow. If intense light falls on an LDR, the resistance is very small and it will allow a lot of current to flow through it.

We can do an experiment to show the change in resistance of an LDR depending on the amount of light falling on it.

A light-dependent resistor (LDR)

Uses of the light–dependent resistor

The light-dependent resistor is used in many circuits that should react to light.

- The most common use is in the light meter on a camera. The circuit is designed so that the light falling on an LDR controls the current and thus the reading that is produced for the correct settings.
- Streetlights are also controlled by LDRs. When it begins to get dark an LDR reacts by reducing the current flowing in a control circuit. This triggers the switching on of the lights.

 ACTIVITY 45.3

To measure the resistance of a light-dependent resistor in different degrees of brightness of light

Method

1 Connect a light-dependent resistor to an ohmmeter. An ohmmeter is an instrument for measuring resistance.

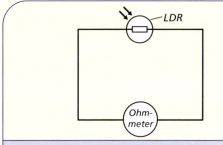

Fig 45.5

2 Note the reading on the ohmmeter for the resistance of the LDR.

3 Change the amount of light that is shining on the LDR by
   (a) covering the LDR with a cloth
   (b) bringing the LDR close to a bright light
   (c) changing the distance of the LDR from the light source.

4 Note the reading on the ohmmeter in (a), (b) and (c).

Conclusion

The more light that falls on the LDR the lower the resistance.

Investigating control circuits

We can combine the components that have been introduced in this chapter in simple series circuits and observe the effects.

In Figure 45.6 a battery, switch, diode, LDR and bulb (or a buzzer) are connected in series, that is, one after the other.

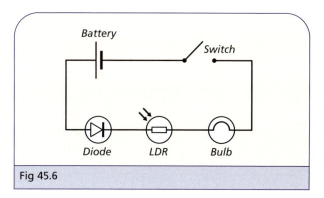

Fig 45.6

When the switch is closed, in normal daylight, the bulb lights up. This is because the diode is connected in forward bias and the LDR has light falling on it.

If the LDR is covered with a sheet of paper or a cloth, the light goes out.

Also, if the diode is reversed and the switch is closed again the bulb does not light. The diode only allows current to travel through it in one direction.

The bulb can be replaced by a light-emitting diode. In this case we place a large resistance in the circuit in series, as the LED can take only a small current.

In each of these cases we are controlling the current in a circuit. This is the central idea in electronics.

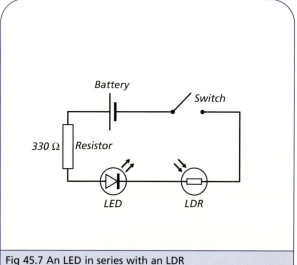

Fig 45.7 An LED in series with an LDR

PHYSICS

Key Points

- Electronics is about controlling devices by directing small amounts of current in circuits.

- A switch is a device for completing or breaking a circuit.

- A resistor is a circuit component that is used to control the amount of current flowing in a circuit.

- A diode is a device that allows current to flow in one direction only.

- When a diode is connected in such a way that it allows current to flow, it is said to be forward biased. If it is connected so that no current flows, it is reverse biased.

- A light-emitting diode (LED) is a diode that gives out light when current passes through it. They are used as signal lights to show that current is flowing.

- A light-emitting diode needs much less current than a bulb but must be protected in a circuit by putting a large resistance in series with it.

- A light-dependent resistor (LDR) is a variable resistor whose resistance changes with the amount of light falling on it.

- LDRs are used in light meters and to control the switching on of streetlights.

Questions

1 (a) What is meant by the term electronics?
  (b) Define the following terms
   (i) switch
   (ii) resistor
   (iii) diode
   (iv) light-emitting diode
   (v) light-dependent resistor.

2 Name four everyday appliances that contain electronic circuits.

3 In the circuit diagram below name the components A, B, C and D.

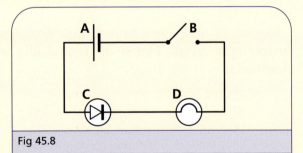

Fig 45.8

4 Draw a circuit diagram with two switches and a bulb so that (a) the bulb will light only if both switches are closed and (b) the bulb will light if either of the switches is closed.

5 Figure 45.9 shows a circuit with a two-way switch.
  (a) Name the devices labelled X and Y.
  (b) Using the letters in the diagram explain how the two-way switch works.
  (c) Give one use for a two-way switch in the home.

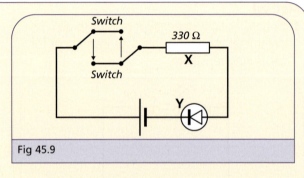

Fig 45.9

6 Describe how you would use two resistances to divide the potential difference of a battery.

7 (a) Name one use of the diode.
  (b) In the circuit diagram below will the bulb light up when the switch is closed? Why?
  (c) What change would you make to the circuit so that the lamp will light when the switch is closed?

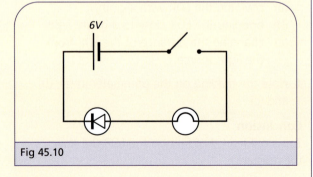

Fig 45.10

❓ Questions

8 (a) Why must a large resistance be placed in series with a light-emitting diode?

   (b) Name two uses of an LED.

9 Using a circuit diagram that has two LEDs in parallel, describe an experiment to determine which pole of a battery is which.

10 Write out and complete the following:

   (a) A switch is a device for _____ or _____ a circuit. A diode is a device that allows _____ to flow in ____ direction only. A diode connected to allow current to flow is _____ biased. If it is connected the other way around it is _____ biased.

   (b) An LED is a _____ _____ diode. It needs much less _____ than a bulb. To protect an LED in a circuit we connect a _____ in _____ with it.

   (c) An LDR is a _____ _____ resistor. Its resistance is very high if _____ light falls on it. Its resistance is very _____ if light falls on it. One use of an LDR is in a light _____ on a camera.

11 Draw a circuit diagram with a battery, switch, LED and LDR connected in such a way that current will flow. How will you know whether current is flowing in the circuit?

Examination Questions

12 A pupil carried out an investigation into the effect of a diode on d.c. and on a.c. circuits using an LED. The following circuits were initially set up.

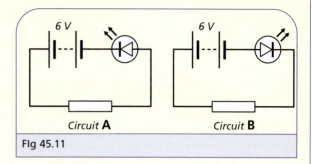

Circuit **A**   Circuit **B**

Fig 45.11

   (a) What is observed in circuit A and in circuit B?

   (b) Why are resistances needed in the circuits?

   (c) When the batteries in circuits A and B were replaced by 6 V a.c. supplies, the LEDs glowed dimly in both circuits. Explain this observation.
   (Adapted from JC, HL, 2006)

13 The diagram shows the symbol of a LED.

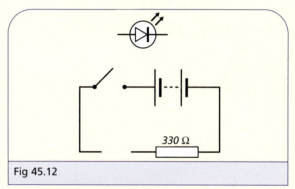

330 Ω

Fig 45.12

   (a) Complete the circuit on the right by drawing in the LED so that the LED will light when the switch is closed.

   (b) Why is there a resistor connected in series with the LED?
   (JC, OL, 2006)

PHYSICS

GLOSSARY

acceleration The change in speed divided by the time taken for the change.

acid A substance which turns blue litmus red and has a pH less than 7. Acids have a sour taste and a sharp feel.

acid rain Rainwater with a pH of less than 5.5.

activity series A list of metals placed in order of how reactive they are.

adaptation A structure or habit that helps an organism to survive in its habitat.

aerobic respiration Oxygen is needed to release energy from food.

alkali A substance which reacts with an acid to form a salt and water. Alkalis are bases that dissolve in water.

alkali metals The elements of Group 1 of the periodic table, e.g. sodium.

alkaline earth metals The elements of Group II of the periodic table, e.g. calcium.

alloy A mixture of metals.

alternating current A current that constantly changes direction.

altimeter An instrument that measures change in pressure and translates it into height above sea level.

amp The unit of current.

anaerobic respiration Oxygen is not used to release energy from food.

area The amount of surface that covers an object.

asexual reproduction New individuals are formed from only one parent.

atom The smallest part of an element that still has the properties of that element.

atomic number The number of protons in an atom of that element.

base A substance that neutralises an acid to form a salt and water. Bases turn red litmus blue.

beating tray A device used to collect animals.

biotechnology The use of living things to produce useful products.

bond Force of attraction that holds atoms together.

canines Teeth that are long and pointed and used to grip and tear food.

carnivore An animal that eats other animals only.

catalysts Substances that alter the rate of a chemical reaction but are not used up themselves.

cellulose The substance that makes cell walls and provides fibre in the human diet.

cell wall Found outside the cell membrane in plant cells.

centre of gravity of a body A point about which the turning effects due to gravity are balanced.

chemical change A change (reaction) in which a new substance is formed.

chemical energy Energy stored inside materials in the chemical bonds holding them together.

chlorination The addition of chlorine to water to kill bacteria.

chloroplasts Contain the green pigment called chlorophyll.

chromatography Separation techniques used to separate a mixture of dissolved substances in a solution.

competition Takes place when two or more organisms require something that is in short supply.

compounds Made up of two or more different types of atom chemically combined.

concentrated solution A solution which contains a large amount of solute in a small amount of solvent.

conductor (electrical) An object that allows a charge to move.

conductor (heat) An object that allows heat to flow through it.

conservation The protection and wise management of natural resources.

consumers Animals that get their food by eating plants or other animals.

contraception The prevention of fertilisation or pregnancy.

convection The transfer of heat through liquids and gases by the mass movement of particles.

corrosion An undesirable process whereby a metal changes to its oxide or some other compound by combining with oxygen from air.

covalent bond A bond that consists of a pair of electrons shared between two non-metal atoms.

crystallisation The formation of crystals by cooling a saturated solution or by evaporating off solvent.

decibel A unit of sound level.

decomposers Organisms that feed on dead plants and animals.

density The mass of each cubic metre of a material.

diffusion The name used to describe the way particles in gases and liquids spread throughout the space in which they are placed.

digestion The breakdown of food.

dilute solution A solution containing a small amount of solute in a large amount of solvent.

diode A device that allows current to flow in one direction only.

direct current The flow of charge in one direction only around a circuit.

dispersal The carrying of the seed (and the surrounding fruit) as far away as possible from the parent plant.

dispersion The breaking up of white light into its spectrum colours.

distillation A process used to separate two miscible liquids with different boiling points, e.g. alcohol and water.

echo The reflection of sound.

ecology The study of the relationships between plants, animals and their environment.

electrical energy Energy caused by moving charges.

electric current The movement of charge through a material.

electrolysis A chemical reaction produced by using electricity.

electromotive force Force from a battery to cause current.

electron A negatively charged particle in an atom.

electronic configuration The arrangement of electrons in an atom.

elements Substances made up of only one type of atom.

energy The ability to do work. It is measured in joules (J).

enzyme A chemical (protein) that speeds up chemical reactions without the enzyme being used up.

equilibrium An object that does not topple easily when a force is applied to it.

evaporation The changing of a liquid to a gas.

excretion The removal from the body of the waste products of chemical reactions in the body.

fertilisation The joining of the male and female gametes to form a zygote.

filtration A method used to separate insoluble solids from liquids.

fluoridation The addition of small amounts of fluorine to water to prevent tooth decay.

food chain A list of organisms in which each organism is eaten by the next one in the chain.

food web Consists of two or more interconnected food chains.

force The idea used to describe the cause of motion or a change in the motion of a body (force = mass × acceleration).

friction A force that opposes the sliding motion of one object when it is in contact with another.

fuel A substance that burns in oxygen and produces heat.

gamete A sex cell.

genetics The study of how traits or characteristics are inherited.

geothermal energy Energy that comes from the heat in the earth beneath the planet's surface.

geotropism The way in which a plant changes its growth in response to gravity.

germination The growth of a seed to form a new plant.

habitat The area where a plant or animal lives.

halogens The elements of Group VII of the periodic table, e.g. chlorine.

hard water Water that does not lather easily with soap.

heat A form of energy that is transferred from warmer bodies to colder ones.

herbivore An animal that eats plants only.

Hooke's law The extension of an elastic material is directly proportional to the applied force producing the extension.

hydrocarbons Compounds made up of hydrogen and carbon.

hydro-electricity Energy from the potential and kinetic energy of water.

implantation The attachment of the embryo to the lining of the uterus.

incisors Teeth that have sharp edges used to cut and nibble food.

indicators Chemicals that show by changing colour whether a substance is acidic, alkaline or neutral.

insulator (electrical) An object that holds a charge.

insulators (heat) Do not allow heat to travel through them easily.

interdependence Living things depend on each other for survival.

invertebrates Animals that do not have a backbone.

ion An atom that has lost or gained electrons and is then charged.

isotopes Atoms that have the same number of protons but different numbers of neutrons.

joule The unit of work or energy = newtons × metres.

kilowatt hour The amount of energy provided when 1 kilowatt of power is used for 1 hour.

kinetic energy Energy stored in a body that is moving.

latent heat Heat that is needed to change the state of a substance.

law of the lever States that when a lever is balanced, the sum of the clockwise turning effects equals the sum of the anti-clockwise turning effects.

length The straight-line distance between two points.

lever A rigid body that can rotate about a fulcrum (a fixed point).

ligaments Connect bone to bone.

light-dependent resistor (LDR) A variable resistor whose resistance changes with the amount of light that falls on it.

light-emitting diode A diode that gives out light when current passes through it.

line transect A rope marked at regular intervals and laid out across a habitat to estimate the number of plants present.

luminous body A body that creates its own light.

magnetic field The space around a magnet in which a magnetic force can be detected.

mass The amount of matter in an object. This amount never changes.

mass number The number of protons *and* neutrons in an atom of that element.

matter Anything that occupies space and has mass.

meniscus The curved surface of a liquid in a vessel.

miscible liquids Liquids that mix, e.g. alcohol and water.

mixture Contains two or more different substances mingled together but not chemically combined.

molars Larger teeth that are used for chewing, crushing and grinding food.

molecule Made up of two or more atoms chemically combined.

motor nerves Nerves that carry messages away from the brain.

neutron An uncharged subatomic particle found in the nucleus of an atom.

newton The unit of force.

noble gases The elements of Group VIII of the periodic table, e.g. helium.

non-luminous body A body that reflects light.

nuclear energy Energy from making and breaking nuclear bonds.

nucleus The part of the cell that controls the cell's activities (biology).

nucleus The central part of an atom that is made up of protons and neutrons (chemistry).

nutrition The way in which an organism gets its food.

ohm The unit of electrical resistance.

omnivore An animal that eats both plants and animals.

ovulation The release of an egg from the ovary.

periodic table A table that is an arrangement of elements in order of increasing atomic number arranged in rows called periods and columns called groups.

phloem The tissue that carries food in plants.

photosynthesis The way in which green plants make food.

phototropism The way in which a plant changes its growth in response to light.

pH scale A scale, which runs from 0 to 14, which indicates the level of acidity or basicity of a solution.

physical change A change in which no new substance is formed.

pitfall trap A device used to collect animals.

plasma The liquid part of blood which transports chemicals and heat.

plastics Man-made materials made from crude oil.

pollination The transfer of pollen from a stamen to a carpel.

pollution The addition of harmful materials to the environment.

polymerisation The process involving the joining together of many small molecules called monomers to form a large molecule called a polymer.

pooter A device used to collect animals.

potential energy Energy stored in a wound-up spring or a stretched elastic band.

power The amount of work done in a unit of time. It is measured in watts.

pregnancy The length of time the baby spends developing in the uterus.

premolars Large rounded teeth used for chewing, crushing and grinding food.

pressure The amount of force acting on a unit of area.

producers Plants that make their own food.

proton A positively charged subatomic particle found in the nucleus of an atom.

quadrat A frame used to estimate the number of plants in a habitat.

radiation The transfer of heat by means of waves that can travel through a vacuum.

red blood cells Cells that transport oxygen.

resistor A component that is used to control the amount of current flowing in a circuit.

respiration The release of energy from food.

rusting The corrosion of iron and steel to form a brown oxide (rust). It occurs in the presence of water and oxygen.

salt A compound formed when the hydrogen of an acid is replaced with a metal.

saturated solution A solution that contains as much dissolved solute as possible at that temperature.

sensory nerves Nerves that carry messages to the brain.

sexual reproduction Two sex cells joining together.

soft water Water that lathers easily with soap.

solar energy Energy from the sun.

solute A substance that dissolves.

solution A mixture of a solute and a solvent.

solvent A liquid in which a solute dissolves.

sound Produced when objects vibrate.

speed The distance travelled by an object in one unit of time.

sweep net A device used to collect animals.

switch A device for completing or breaking a circuit.

temperature A measure of how hot or cold an object is.

tendons Connect muscle to bone.

tidal and wave energy Energy generated by the movement of the tides and waves.

time It is measured by a basic unit called the second (symbol s).

titration A process used to determine the quantity of an acid that is required to exactly neutralise a fixed quantity of base.

transpiration The loss of water vapour from a plant.

transpiration stream A flow of water from the roots to the leaves of a plant.

tropism The change in growth of a plant in response to an outside stimulus.

ultrasound Sound of a very high pitch.

vacuoles Membrane-bounded compartments that support the cell and store substances such as food, wastes and water. Mostly found in plant cells.

valency of an element The number of electrons an atom of that element wants to gain, lose or share in order to have a full outer shell (to be chemically stable).

velocity The speed of a body and the direction in which it is moving.

vertebrates Animals that have a backbone.

volt The unit of potential difference or electromotive force.

voltage Force needed to pull current through a resistance.

volume A measure of the space taken up by an object.

water cycle The natural circulation of water on the planet.

watt The unit of power = J/s.

weight The force of gravity.

wind energy Energy from the kinetic energy of the wind.

xylem The tissue that carries water and dissolved minerals in plants.

INDEX